SOLVENT SUBSTITUTION FOR POLLUTION PREVENTION

SOLVENT SUBSTITUTION FOR POLLUTION PREVENTION

U.S. Department of Energy

Office of Technology Development
Environmental Restoration and Waste Management
Washington, DC

and

U.S. Air Force

Engineering and Services Center
Tyndall Air Base, Florida

NOYES DATA CORPORATION
Park Ridge, New Jersey, U.S.A.

Library of Congress Catalog Card Number: 92-24212
ISBN: 0-8155-1319-4
ISSN: 0090-516X
Printed in the United States

Published in the United States of America by
Noyes Data Corporation
Mill Road, Park Ridge, New Jersey 07656

10 9 8 7 6 5 4 3 2 1

Library of Congress Cataloging-in-Publication Data

Solvent substitution for pollution prevention / U.S. Department of Energy, Office of Technology Development, Environmental Restoration, and Waste Management and U.S. Air Force Engineering and Services Center.

p. cm. -- (Pollution technology review, ISSN 0090-516X ; no. 212)

Includes bibliographical references and index.

ISBN 0-8155-1319-4

1. Solvent wastes--Environmental aspects. 2. Factory and trade waste--Management. 3. Air quality management. I. United States. Dept. of Energy. Office of Technology Development Environmental Restoration and Waste Management. II. Air Force Engineering and Services Center (U.S.) III. Series

TD898.8.S65S64 1992

628.5'2--dc20 92-24212

CIP

FOREWORD

This book provides up-to-date information on current technologies and practices in solvent substitution for those involved in handling, using or managing hazardous and toxic solvents.

The subject matter of the book's 43 papers includes a wide range of solvent substitution activities such as: paint stripping, coating applications, printed circuit boards, metal cleaning, metal finishing, manufacturing, hardware and weapons production and maintenance, compliance monitoring and process control monitoring.

The book is divided into six sections which cover alternative technologies, alternative solvents, solvent recovery and recycling, dealing with low VOCs (volatile organic compounds), treatments for environmentally safe disposal of toxic solvents, and special issues.

The information in the book is from *Solvent Substitution—A Proceedings/Compendium of Papers* (based on The First International Workshop on Solvent Substitution, December 4–7, 1990, Phoenix, Arizona), sponsored by the U.S. Department of Energy, Office of Technology Development, Environmental Restoration and Waste Management and the U.S. Air Force Engineering & Services Center, prepared by the Weapons Complex Monitor Forums for the U.S. Department of Energy. Proceedings papers, as well as solicited papers, are included.

The table of contents is organized in such a way as to serve as a subject index and provides easy access to the information contained in the book.

Advanced composition and production methods developed by Noyes Data Corporation are employed to bring this durably bound book to you in a minimum of time. Special techniques are used to close the gap between "manuscript" and "completed book." In order to keep the price of the book to a reasonable level, it has been partially reproduced by photo-offset directly from the original report and the cost saving passed on to the reader. Due to this method of publishing, certain portions of the book may be less legible than desired.

PREFACE

The DOE Environmental Restoration and Waste Management Office of Technology Development and the Air Force Engineering and Services Center convened the First Annual International Workshop on Solvent Substitution on December 4–7, 1990, at the Executive Conference Center, The Pointe at Tapatio Cliffs, Phoenix, Arizona. The primary objectives of this joint effort were to:

- Share information and ideas among attendees in order to enhance the development and implementation of required new technologies for the elimination of pollutants associated with industrial use of hazardous and toxic solvents.
- Aid in accelerating collaborative efforts and technology transfer between government and industry for solvent substitution.

This highly successful 2½ day event brought together over 300 leading national and international experts from industry, federal and state government agencies, various branches of the Armed Services, research laboratories, universities and public interest groups.

There were workshop sessions focusing on Alternative Technologies, Alternative Solvents, Recovery/Recycling, Low VOC Materials and Treatment for Environmentally Safe Disposal. The 35 invited papers presented covered a wide range of solvent substitution activities including: hardware and weapons production and maintenance, paint stripping, coating applications, printed circuit boards, metal cleaning, metal finishing, manufacturing, compliance monitoring and process control monitoring.

This publication includes the majority of these presentations. In addition, in order to further facilitate information exchange and technology transfer, the U.S. Air Force and DOE solicited additional papers under a general "Call for Papers." These papers, which underwent review and final selection by a peer review committee, are also included in this combined Proceedings/Compendium.

For those involved in handling, using or managing hazardous and toxic solvents, this document should prove to be a valuable resource, providing the most up–to–date information on current technologies and practices in solvent substitution.

NOTICE

The materials in this book were prepared for a workshop, and/or subsequent proceedings, sponsored by the U.S. Department of Energy and the U.S. Air Force. On this basis the Publisher assumes no responsibility nor liability for errors or any consequences arising from the use of the information contained herein.

Mention of trade names or commercial products does not constitute endorsement or recommendation for use by the Agencies or the Publisher. Final determination of the suitability of any information or product for use contemplated by any user, and the manner of that use, is the sole responsibility of the user. The book is intended for information purposes only. The reader is warned that caution must always be exercised with potentially hazardous materials, and expert advice should be obtained before implementation of new materials and processes utilized as substitutes for solvents.

All information pertaining to law and regulations is provided for background only. The reader must contact the appropriate legal sources and regulatory authorities for up-to-date regulatory requirements, and their interpretation and implementation.

The book is sold with the understanding that the Publisher is not engaged in rendering legal, engineering, or other professional service. If advice or other expert assistance is required, the service of a competent professional should be sought.

CONTENTS AND SUBJECT INDEX

SECTION I
ALTERNATIVE TECHNOLOGIES

SECTION III
SOLVENT RECOVERY AND RECYCLING

SECTION IV
DEALING WITH LOW VOCs

SECTION V
TREATMENT FOR ENVIRONMENTALLY SAFE DISPOSAL OF TOXIC SOLVENTS

SECTION VI
ISSUES TO CONSIDER

Section I

ALTERNATIVE TECHNOLOGIES

SURFACE CLEANING BY LASER ABLATION

H. C. Peebles, N. A. Creager, and D. E. Peebles
Sandia National Laboratories
Albuquerque, New Mexico

ABSTRACT

Nd:YAG laser cleaning of metal oxides from 304L stainless steel surfaces has been characterized. Thin chromium oxide films can be completely removed from the surface using a single 10 nsec pulse of laser radiation with an average surface irradiance greater than 120 MW/cm2. Laser etching of thicker iron oxide films exhibit a self-limiting effect that prevents overetching into the stainless steel substrate.

INTRODUCTION

Recent international agreements will virtually eliminate the use of chlorinated hydrocarbons in the cleaning of manufactured parts in the near future. Effective alternative cleaning technologies must be developed to replace cleaning procedures which currently utilize these solvents. At the present time, two approaches are being pursued. The most common approach seeks to identify non-chlorinated substitute solvents which are environmentally acceptable. Possible candidates include aqueous and terpene based cleaners as well as simple alcohols. A less common approach involves total elimination of the solvent, substituting instead an alternative cleaning technology. Alternative technologies include dry processing techniques such as plasma etching and laser ablation. This paper will focus on the application of laser ablation as a cleaning technique for material surfaces. Data will be presented characterizing the removal of metal oxides from 304L stainless steel surfaces.

TECHNIQUES AND PROCEDURES

Laser ablation involves the use of very short pulses of high peak power laser radiation to rapidly heat and vaporize thin layers of material surfaces. Laser cleaning of a surface is accomplished by rastering the laser beam across the surface of the material. When used as a cleaning technique, this process must be performed in a chamber containing an inert gas environment in order to prevent recontamination of the surface by reactive gas-phase species. The ablated surface material forms a dense cloud of hot vapors which, upon cooling, condenses into submicron diameter particles. This particulate waste must be removed from the near surface region of the part by entrainment into a flowing gas stream or recontamination of the surface will occur. One possible method for the removal of the particulate waste is shown schematically in Figure 1. Inert gas from the near surface region of the part is drawn into a nozzle coaxial with the incident laser beam. The gas entering the nozzle exits through a side port and is then passed through a canister filter which removes the particulate material from the waste stream. The filtered inert gas is returned to the chamber.

Because the incident laser radiation can be focused to a small spot size, laser ablation can clean with high spacial selectivity, allowing application on partially or completely assembled parts which could be damaged by other cleaning methods. The waste generated by this process is limited to the volume of material vaporized from the surface. For a typical 1 mm thick iron oxide film with a material density of 5.2 g/cm3 (Fe_2O_3), the waste generation rate will be approximately 480 mg per square foot of surface area cleaned. Thus, as a cleaning process, laser ablation exhibits high potential for waste minimization.

The experimental apparatus used to characterize the etching of iron and chromium oxides from 304L stainless steel is shown schematically in Figure 2. Target substrates were placed in a small stainless steel chamber fitted with an 8 mm thick fused quartz window. The chamber was purged to remove oxygen and other reactive gas contaminants in the ambient atmosphere by evacuation to 1 micron pressure followed by backfilling the chamber to 1 atmosphere with 99.999% pure helium gas. This purge process was repeated a total of three times. All etching experiments described in this paper were performed at 1 atmosphere pressure in helium gas. Laser etch craters were formed on the substrate surface by directing laser radiation onto the surface at a near normal angle of incidence through the quartz window in the chamber. The laser used in these experiments was a Spectra Physics Model DCR-2A Q-switched pulsed Nd:YAG laser emitting radiation at 1.064 mm wavelength. The laser beam has a divergence of 0.5mradmrad and a full width at half height pulse length of 10 nsec. The laser beam intensity profile is gaussian in shape with a diameter (measured at the points where the intensity is 1/e2 of the maximum value) of 5.9 mm. A 2.5 mm diameter aperture was placed in the optical path of the laser beam at the center of the beam to form a pseudo flat-top intensity profile. An approximate relative intensity profile for the DCR-2A output beam indicating the clipping points of the aperture is shown in Figure 3. All laser exposures occurred at a single spot on the substrate. The laser beam was not focused with a lens.

Prior to laser etching, substrate surfaces were prepared by polishing to give a surface flatness of 200 nm and an arithmetic surface roughness less than 5 nm. The straw yellow oxide film was formed on polished 304L stainless steel substrates by baking the substrate in air at 650RC for 6 minutes. This produces a film which is composed predominantly of chromium oxide and is less than 5 nm in thickness. The iron oxide film was produced on the polished stainless steel surfaces by baking the substrate in air at 800R C for 30 minutes. This produced an oxide film that was blue in color with a film thickness of approximately 3 mm. Laser etch craters were profiled with a Dektak 3030 profilometer using a 2.5 mm radius diamond tip with a vertical applied force of 20 mg. The surface profiles have a vertical resolution of 0.1 nm and a horizontal resolution of 5 mm.

RESULTS AND DISCUSSION

Annealed 304L stainless steel substrates with a thin (< 5 nm) overlayer of chromium oxide were irradiated with multiple pulses of Nd:YAG laser radiation. Topographical surface profiles along a line through the center of the irradiated spot are shown in Figure 4 for multiple pulse laser exposures of 1, 8, and 20 laser pulses. The pulse energy for each laser pulse was 100 mJ, corresponding to a time averaged irradiance of 240 MW/cm2 at the center of the laser beam and 170 MW/cm2 at the edge of the beam. Note in the figure that the horizontal scales are all identical, and are roughly a factor of 10,000 larger than the vertical scales. The vertical scale increases for each successive profile. The vertical dashed lines accompanying each profile mark the visible boundaries of the region where the chromium oxide film was totally etched away by the incident laser radiation. These boundaries were determined by microscopic inspection of the light reflection properties at the surface of the laser irradiated spot during acquisition of the surface profile.

A single 100 mJ laser pulse was sufficient to completely remove the chromium oxide film over the area directly irradiated by the incident laser beam. Single pulse experiments performed at lower pulse energies revealed that a laser pulse energy of 50 mJ or greater is required to completely remove the chromium oxide film from the 304L stainless steel surface. This pulse energy corresponds to a central beam irradiance of 120 MW/cm2.

The surface profile in Figure 4 corresponding to a single 100 mJ pulse of laser irradiation shows no significant etch crater. This is because the chromium oxide film is much thinner than the apparent surface roughness of the 304L stainless steel substrates. As a result, the etch crater corresponding to removal of the chromium oxide film cannot be detected in the profile. Note, however, that the apparent roughness of the irradiated surface has increased significantly when compared to adjacent areas which did not receive laser irradiation.

The surface profile corresponding to 8 consecutive laser pulses in Figure 4 shows a definite etch crater resulting from laser etching into the 304L stainless steel substrate. Note that the profile of this etch crater exhibits definite peaks and valleys. The apparent surface roughness in the laser irradiated region has increased dramatically when compared to that observed for a single pulse of laser irradiation. Many peaks are observed in the etch crater profile which extend above the plane of the original substrate surface. The height of the peaks generally increases toward the center of the laser irradiated spot, suggesting that the extent of peak formation increases with increasing surface irradiance.

The depth of the etch crater and the magnitude of the surface roughness in the laser-irradiated region continually increase with increasing laser pulse exposure as seen by the progression in the etch profiles shown in Figure 4. A scanning electron micrograph of the etch crater resulting from a 10 pulse exposure to laser irradiation is shown in Figure 5a. The surface features responsible for the increased surface roughness appearing at the center of the laser irradiated region are readily apparent in the figure. This region exhibits a large number of exposed grain boundaries, each corresponding to the interface between differently oriented single crystal grains. The central region of the laser-irradiated spot is shown at increased magnification in Figure 5b. The linear boundaries bisecting crystal grains in the micrograph reveal the presence of crystal twins. Preferential etching by the laser radiation along grain and twin boundaries is responsible for the exposure of these boundaries in the micrograph. The presence of crystal grains extending above the original plane of the highly polished substrate surface indicates that some vertical growth of the crystal grains occurs during the laser etching process. This growth could be the result of thermally driven diffusion of metal atoms along the substrate surface or the redeposition of laser vaporized material onto a more stable crystal plane.

Chemical reagents are also known to etch preferentially along grain and twin boundaries. An identical 304L stainless steel substrate was polished to a 0.05 micron alumina grit and electrolytically etched in 10% oxalic acid for 60 seconds at 5 volts DC. The resulting surface microstructure is shown in the optical micrograph in Figure 6. The grain size is approximately defined by ASTM GS 7. Comparison of Figure 6 with Figure 5b shows that the chemically etched surface and the laser-etched surface exhibit a similar microstructure. Laser etching does not significantly alter the microstructure in the near surface region of 304L stainless steel.

A plot of laser etch depth in 304L stainless steel as a function of laser exposure for 100 mJ laser pulses is shown in Figure 7. Etch depth is defined in this case as the central crater depth averaged over the apparent surface roughness. The data shown in Figure 7 suggests a linear relationship between etch depth and laser pulse exposure. The least-squares fit of a line to the data gives a laser etch rate of 0.0068 mm per laser pulse for 304L stainless steel.

A plot of laser etch depth in an iron oxide film on a 304L stainless steel substrate versus laser exposure for 100 mJ laser pulses is shown in Figure 8. For laser exposures less than 30 pulses, the data show a linear dependence between etch depth and laser exposure. The least-squares fit of a line to the data in this segment of the figure gives a laser etch rate of 0.10 mm per laser pulse. Note that the etch rate in the iron oxide layer is 15 times greater than that observed in 304L stainless steel. As a result, one would predict that as soon as the laser penetrated the iron oxide layer, etching would effectively terminate just past the metal-metal oxide interface. This behavior is demonstrated in Figure 8. For laser exposures greater than 30 pulses, the etch rate is effectively zero, indicating a 3 mm thick oxide layer. This self-limiting effect is highly advantageous in laser cleaning operations because it provides a simple and ready method for endpoint control. If the opposite were true, i.e. laser etching in the substrate proceeded much more rapidly than laser etching of the contaminant overlayer, it would be very difficult to remove the contaminant layer without seriously overetching into the substrate. Most contamination layers on metal substrates will exhibit this self-limiting effect due to the characteristically high reflectivity of metals at 1.064 mm wavelength.

SUMMARY AND CONCLUSIONS

Nd:YAG laser cleaning of metal oxides from 304L stainless steel surfaces has been characterized. Thin chromium oxide films can be completely removed from stainless steel surfaces using a single 10 nsec laser pulse with an average surface irradiance greater than 120 MW/cm2. Laser etching proceeds preferentially along specific crystallographic planes resulting in increased surface roughness for highly polished surfaces. No laser induced change in the near surface microstructure of the stainless steel is observed. Laser etching of thicker iron oxide films from stainless steel surfaces exhibits a self-limiting effect. The laser etch rate in the iron oxide is 15 times greater than in the 304L stainless steel substrate resulting in effective termination of laser etching just past the metal-metal oxide interface.

ACKNOWLEDGEMENTS

The authors gratefully acknowledge numerous discussions with C. V. Robino which proved useful in the interpretation of the metallographic data presented in this paper. This work was performed at Sandia National Laboratories supported by the U.S. Department of Energy under contract number DE-AC04-76DP00789.

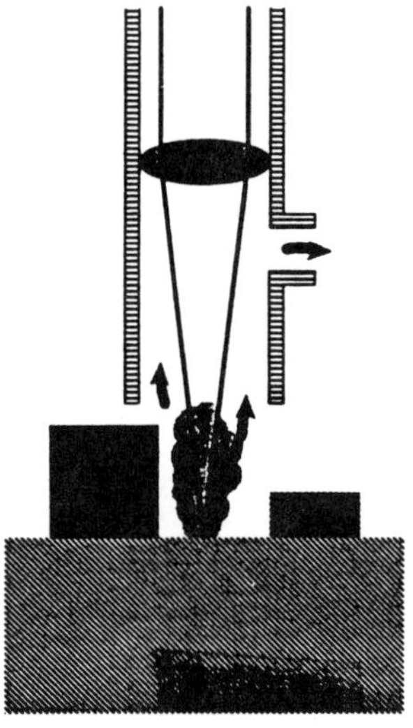

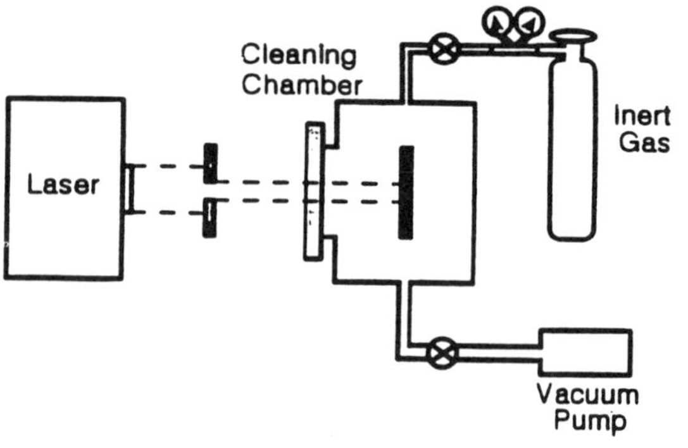

Figure 2. Experimental apparatus used to characterize laser etching of iron and chromium oxides on 304L stainless steel surfaces.

Figure 1. Schematic of particulate waste collection method for laser ablative cleaning of surfaces.

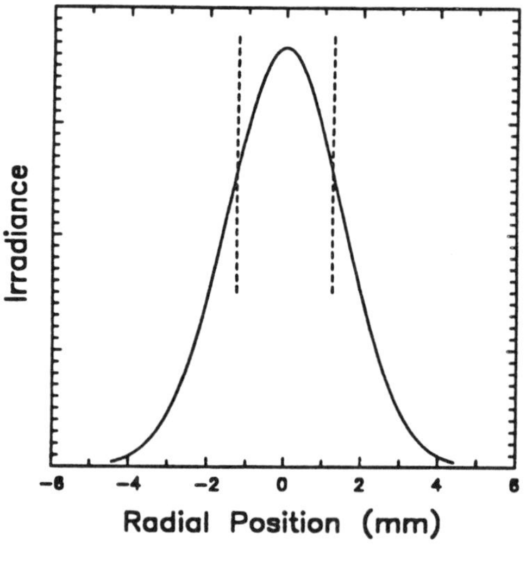

Figure 3. Approximate laser beam relative intensity profile. Dashed lines represent the laser beam profile cut-off points for a 2.5 mm beam aperture.

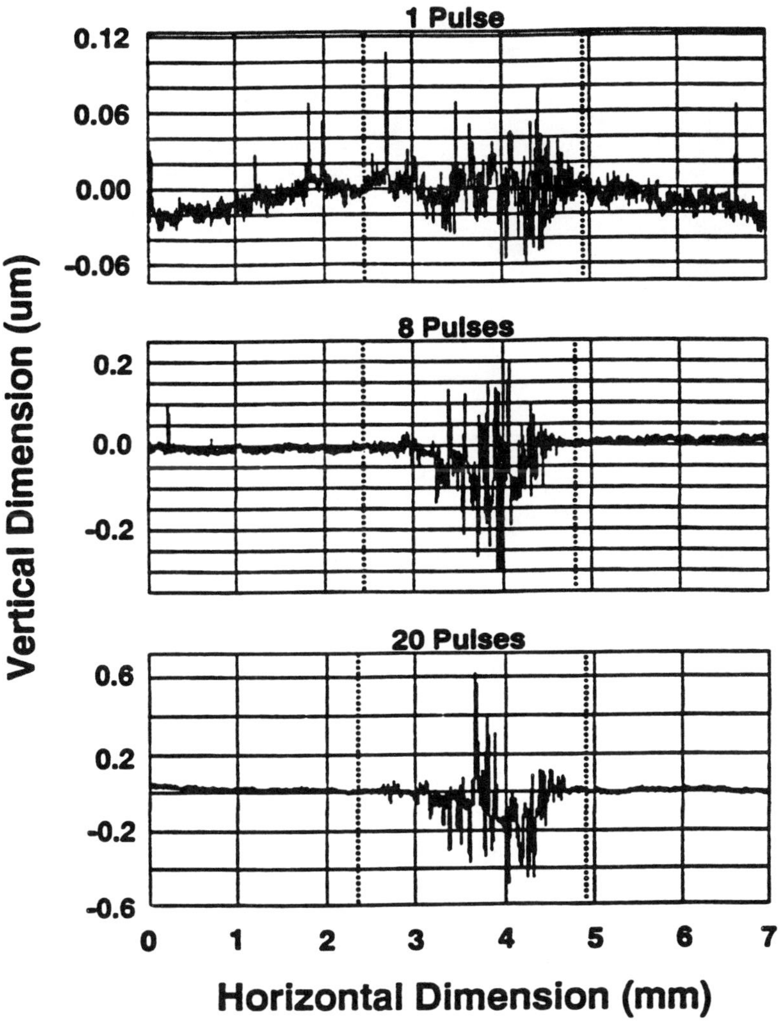

Figure 4. Laser etch crater topographical profiles resulting from 1, 8, and 20 consecutive 100 mJ laser pulses on a chromium oxide covered 304L stainless steel substrate.

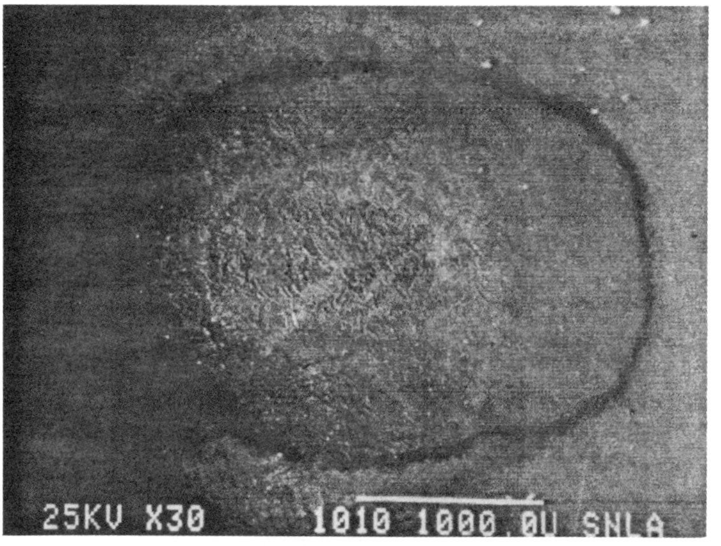

(a)

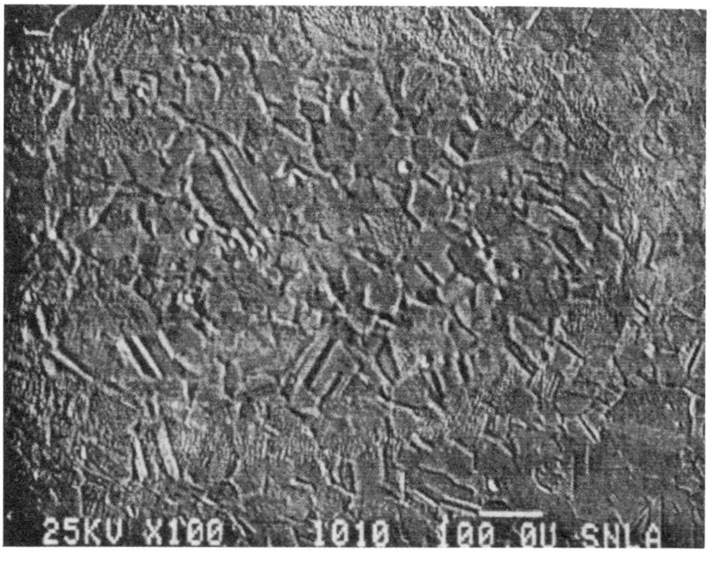

(b)

Figure 5. (a) Scanning electron micrograph of the etch crater resulting from 10 consecutive laser pulses on a chromium oxide covered 304L stainless steel surface. (b) View of the central region of the etch crater at high magnification.

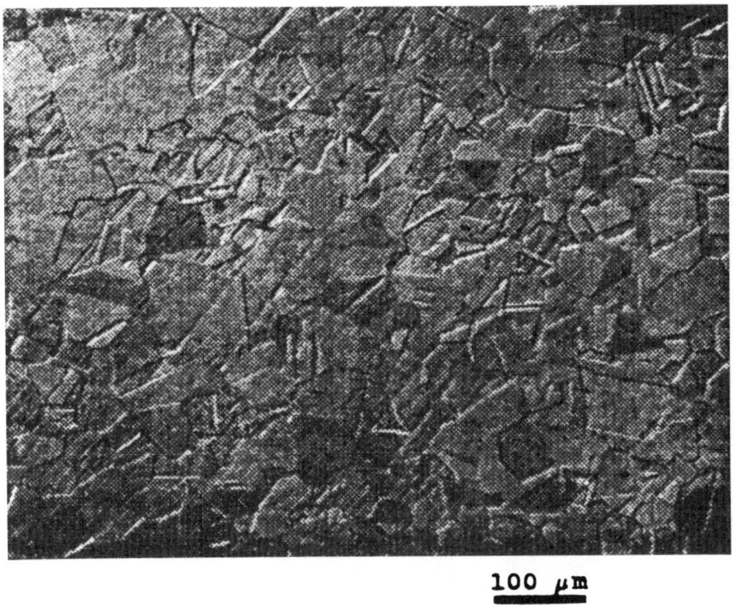

Figure 6. Bulk microstructure of electrochemically etched 304L stainless steel

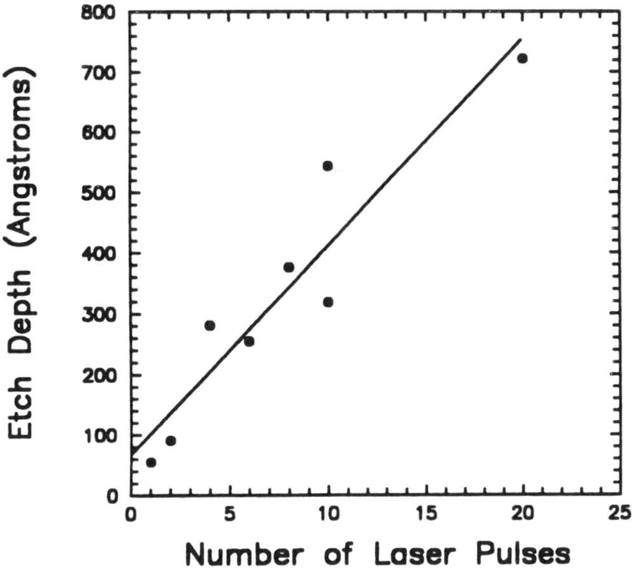

Figure 7. Etch depth versus pulse number for laser etching of chromium oxide covered 304L stainless steel substrates using 100 mJ laser pulses. The slope of the line indicates an etch rate of 6.8 nm per laser pulse in the stainless steel substrate.

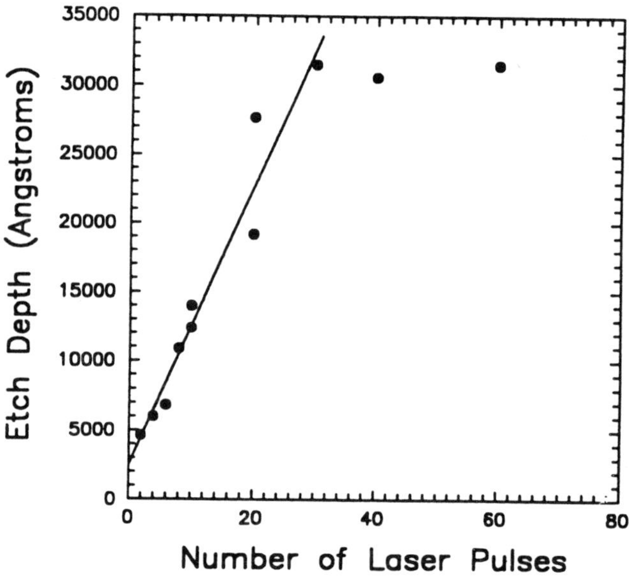

Figure 8. Etch depth versus pulse number for laser etching of 3000 nm thick iron oxide films on 304L stainless steel substrates using 100 mJ laser pulses. The slope of the line indicates an etch rate of 100 nm per laser pulse in the iron oxide film.

CO_2 PELLET BLASTING FOR PAINT STRIPPING/COATINGS REMOVAL

Wayne N. Schmitz
McDonnell Aircraft Company (MCAIR)
St. Louis, Missouri

MCAIR's MOTIVATION

A major element of every MCAIR program is Integrated Logistic Support (ILS) of the fielded weapon system. As more aircraft are manufactured with composite materials to reduce weight while maintaining high strength structures, the requirement to effectively (completely) remove paint, primer and rain erosion coatings to facilitate bonded repairs becomes critical.

While composite structures are not subject to corrosion or fatigue cracking, the remaining metal portions of the airframe must be inspected periodically to preclude catastrophic failures from metal fatigue. Again, surface coatings must be completely removed to facilitate inspection.

Current paint stripping activities employing the use of phenol-based methylene chloride chemicals are not acceptable because:

- Composite materials are susceptible to damage, and

- The use of hazardous materials resulting in maintenance personnel injury and toxic waste generation must be curtailed.

MCAIR, subsequently, began a search for new paint stripping technologies to satisfy ILS requirements.

THE SEARCH

Our search for a new process to strip paint/primer and any variety of surface coatings was conducted under the following constraints:

- The process must not compromise the structural integrity of the aircraft-this requirement is far more stringent than merely not damaging the surface of the substrate as it includes substructure.

- Toxic waste and the use of hazardous materials must be eliminated.

- Disposable materials, i.e., removed paint, etc. plus any worn out, contaminated media must be reduced by 90%.

- The process must reduce maintenance manhours, overall stripping cost, and aircraft cycle time by 50%.

The technologies investigated fell into three categories:

- Dry Media Blasting
  - CO_2 pellets
  - Plastic grit
  - Wheat starch
  - Walnut shells and the like

- Liquid Media Blasting
  - Medium pressure (7,000 Psi) Water jet
  - High pressure (32,000 Psi) Water jet
  - Water ice slurry
  - Sodium bicarbonate slurry

- Pulse Light Energy
  - Lasers
  - Xenon flashlamps

Each of the technologies exhibited one or more of the following problems:

- Potential to damage substrates and/or substructure (as a secondary effect)
- Media intrusion/airframe contamination
- Polysulfide sealant/rubber seals damage
- Corrosion promotion
- Aircraft pre-cleaning/post stripping clean-up requirements
- Hazardous operator environment
- Special facilities; toxic waste capture system; media/removed coatings separation/recycling requirement
- Spent media/toxic waste disposal costs

THE MCAIR CHOICE/RATIONALE

A comparative study of the technologies listed above was conducted to determine which of the processes could effectively strip military specification paint and primer within the specified constraints. Two additional categories were added to those identified problems in order to arrive at the "bottom line" effectiveness of each process: 1) Life cycle cost benefits for a total weapon system, and 2) Aircraft thru-put rate.

MCAIR chose CO_2 pellet blasting as the technology offering the greatest benefits both in terms of maintenance cost reduction and environmental issues compliance. Because the CO_2 pellets are made from liquified CO_2 gas, a natural atmospheric element, and sublimate instantly on contact back to that gas, they represent an operator/environmentally safe process. Since there is no media of which to dispose, only removed paint chips remain and their volume, compared to toxic waste generated by chemical stripping processes, represents a 96 percent reduction. CO_2 pellet blasting is generally benign to most substrates though there are some concerns which will be discussed later. Since there is no solid media, intrusion is not a factor, and the vast percentage of aircraft masking is eliminated as is post-stripping clean-up, media disposal cost, and the requirement for a media separation/recycling system. Elimination of these tasks reduces maintenance manhours by at least 50 percent. Another benefit of CO_2 pellet blasting is its ability to remove a broad range of aircraft surface coatings, sealants and adhesives. Best of all, there is no need to pre-clean the aircraft; the process instantly removes grease, oil, etc. while stripping paint. The economic bottom line to these benefits is an overall stripping cost of \$5/ft^2 compared to \$19 plus/ft^2 for current chemical processes.

CO_2 PELLET BLASTING CONCERNS

Every paint stripping technology we investigated exhibited negative characteristics to a greater or lesser extent and none of the processes has been thoroughly tested with respect to all potential effects on aircraft structures. For all of its excellent benefits, CO_2 pellet blasting still requires further testing on fatigue life degradation, crack growth potential and the possibility of inducing micro-cracking in composite materials.

Some of the effects of CO_2 pellet blasting are visual. At the blast pressures required to effectively remove paint/primer, soft aluminum skins less than 0.032 inch thick show evidence of peening. Thermoset composite materials are easily damaged unless very close attention is paid to dwell time and stand-off distance. One other aspect of CO_2 pellet blasting is a relatively slow stripping rate (0.5 ft^2/min) on alclad-coated aluminum skins and thermoset composites. Clearly, further optimization of the process is required before CO_2 pellet blasting is used to remove paint/primer from the wide range of aircraft substrates.

FUTURE PLANS

Because CO_2 pellet blasting offers outstanding environmental gains, MCAIR will continue research to enhance its performance. Preliminary results of combining CO_2 pellet blasting with other paint stripping technologies look promising, proving once again that there are no simple solutions and no one process is a panacea for all problems.

THE EVALUATION OF ALTERNATIVES TO OZONE-DEPLETING CHLOROFLUOROCARBONS

Robin L. Sellers
Naval Avionics Center
Indianapolis, Indiana

INTRODUCTION

In the spring of 1988, the electronics industry was facing a serious set of problems. The Montreal Protocol, an international agreement signed on 16 September 1987, required a 50% reduction in production of chlorofluorocarbon 1,1,2-trichloro-1,2,2-trifluoro-ethane (CFC-113) by 1998.[1] In 1986, the electronics industry used an estimated 80 million kilograms of CFC-113 to remove flux from printed circuit board assemblies, representing 45 percent of worldwide CFC-113 consumption.[2] As the largest worldwide user of CFC-113, the electronics industry was confronted with a significant challenge.

In addition, the electronics industry was facing the challenge of implementing surface mount technology (SMT). With SMT's smaller stand-off heights and higher density designs, the traditional cleaning processes and tests for measuring cleanliness were being questioned.

Alternative cleaning materials and processes were being used by some commercial electronic manufacturers for both plated-through-hole technology and SMT; however, the U.S. military specifications limited the choices available to the military electronics manufacturer. Further analysis of the situation showed that U.S. military specifications also influenced a substantial portion of the commercial electronics industry. Due to their association with highly reliable products, military specifications had become de facto world standards, driving an estimated 50 percent of the world's use of CFC-113.[2]

Due to the unique cooperation and hard work of representatives from the Environmental Protection Agency, the Department of Defense, and industry, this set of problems was transformed into an opportunity to improve electronics manufacturing processes and military specifications.

AD HOC SOLVENTS WORKING GROUP AND THE CLEANING AND CLEANLINESS TEST PROGRAMS

In the spring of 1988, the Environmental Protection Agency (EPA) was preparing to issue the regulations that would implement the Montreal Protocol provisions within the United States. The pronouncement of proposed reductions in the supply of CFC-113 brought strong comments from the suppliers of CFC-113 products, electronic manufacturers, and the military's technical community. Faced with changing cleaning materials and processes, the U.S. military and electronic manufacturers wanted assurance that the new materials/processes would perform as well as the CFC-113 they had used for years. However, there was no widely accepted measure of "how clean is clean" or "how clean is clean enough."

To answer these questions and expedite the transition from CFC-113 to other materials and processes, Dr. Stephen Andersen, EPA, formed the ad hoc Solvents Working Group. The ad hoc Solvents Working Group is comprised of representatives from materials suppliers, equipment manufacturers, commercial electronic manufacturers, military electronic manufacturers, Department of Defense, the Institute for Interconnecting and Packaging Electronic Circuits (IPC), the EPA, and other interested organizations (see Table I). Utilizing this broad participation, the ad hoc group has worked (and continues to work) at an extraordinary pace to provide a uniform and timely evaluation of materials/processes which reduce the level of CFC-113 used in electronic cleaning processes.

One of the first and most critical tasks for the ad hoc group was to define the scope of effort. When choosing an alternate material and/or process for cleaning of electronic assemblies, there are a number of technical and economic issues that an electronics manufacturer should consider. These include environmental, worker safety and health, performance, and cost issues. As Figure I shows, the list of issues that require definition for each alternative is extensive. In addition, the cleaning process, although critical, is just one of a number of interconnected processes that affect the reliability of an electronic assembly.

In order to be effective in the shortest time possible, the group decided to tackle the issues that could not be addressed by other interested parties, and which required the broader review that the ad hoc group could provide. These items are denoted with the crosshatches in Figure I. The ad hoc group chose to focus on issues related directly to the ability of a process and its associated materials to produce an assembly that was "as clean as" or "cleaner than" an assembly produced using CFC-113. In addition, the group chose to tie the cleanliness levels to the effect on electrical performance. In this way the test program addressed the questions "how clean is clean" and "how clean is clean enough."

The ad hoc group was assured by the EPA and the suppliers of the alternate materials that they would answer questions concerning environmental issues, such as the ozone depletion potential (ODP), the global warming potential (GWP), energy efficiency, regulation as a volatile organic compound (VOC), and waste handling and disposal. In addition, the EPA assured the group that they would work with other government agencies to expedite the evaluation of alternatives with regard to their effect on worker safety and health. This information is supplied by materials vendors on the Materials Safety Data Sheets that are required by law.

The electronics manufacturer, with help from material and equipment suppliers, was left the task of deciding which materials and processes make the most sense for their designs, processes, materials, equipment, corporate culture, specifications, facility, and budget. The ad hoc group did not attempt to answer each and every manufacturer's question of "what should I switch to?" The answer to this question is as varied as the differences between manufacturers, and even between an individual company's plants and production lines. Instead, the ad hoc group developed the standard test that can be used by electronics manufacturers and by the military for deciding which alternatives are capable of cleaning "as well as" or "better than" CFC-113.

Once the focus had been established, the ad hoc group proceeded to develop the objective test protocols that are tied to actual manufacturing conditions and answer the questions "how clean is clean?" and "how clean is clean enough?" Beginning in the spring of 1988, the ad hoc Group developed a three phase test program. The goal of Phase I/Benchmark is to produce the cleanliness reference data for a typical CFC-113 process. The objective of Phase II/Alternate Cleaning Materials and Processes is to evaluate a wide variety of cleaning materials and processes, compare them to the cleanliness reference data obtained in Phase I, and to make recommendations with regard to their suitability for military use. The goal of Phase III/Alternate Manufacturing Media and Processes is to evaluate alternate soldering materials and processes in combination with alternate cleaning materials and processes. Replacement of soldering materials/processes could eliminate the need to clean with CFC-113, and might even eliminate the need to clean at all. Current Phase III efforts include the Water Soluble Flux Evaluation, No-Clean Fluxes, and Controlled Atmosphere Soldering. The ultimate goal for each of these phases is to reduce the level of ozone-depleting chemicals used in the manufacture of electronic assemblies.

The link between all three phases of the test program is the standard printed wiring board (the IPC-B-36) developed by the ad hoc group. The IPC-B-36 incorporates both through-hole and surface mount features with ten electrical circuits that can be used for evaluating cleanliness using surface insulation resistance (SIR). The assembly design also provides a reasonably tough cleaning challenge with its 0.050" pitch leadless chip carriers mounted on 0.005" stand-offs and 0.020" vias that permit liquid flux to flow up the via holes and under the carriers.

In addition, the ad hoc group developed a second standard printed wiring board (the IPC-B-24), to be used in the evaluation of alternate soldering processes and materials in Phase III. A relatively simple design, the IPC-B-24 has four SIR circuits that can be used for evaluating the interaction between soldering fluxes or pastes, metals, and the cleaning materials.

The Cleaning and Cleanliness Test Program for Phases I and II[3], and the Phase III Water Soluble Fluxes Test Program[4], define the materials and processes to be used in each phase. (Note: Test plans for the controlled atmosphere and no-clean flux portions of Phase III are in the process of development.) Not only does the IPC-B-36 link the three phases, but the materials and processes are also kept as uniform as possible. The IPC-B-36 and IPC-B-24 test sequences simulate surface mount assembly processes utilizing solder pastes, infrared or vapor phase reflow, liquid fluxes, and wavesoldering. The test results include data for boards which have **Not Been Exposed** to the typical electronic assembly residues and which represent the cleanest condition. The test results also include data for assemblies which have been exposed to the typical electronic assembly residues and have **Not Been Cleaned**. These unclean assemblies represent the dirtiest conditions and indicate what would happen if an electronic assembly were not properly cleaned. The final two sets of test results include the data for assemblies which have been exposed to typical electronic assembly residues and which have **Been Cleaned**. These cleaned assemblies represent the relative performance of a cleaning material/process, and indicate what effect the cleaning material/process and any residues would have on the electrical performance of an electronic assembly.

Standard cleanliness tests were developed and defined for each of the three phases. For Phases I and II, four standard cleanliness tests were defined. Ionic cleanliness levels are measured using the ionic conductivity test and the commercially-available Omegameter 600SMD. Non-ionic cleanliness levels are measured using two methods: the residual rosin test, which utilizes an ultraviolet/visible spectrophotometer, and high performance liquid chromatography (HPLC). The final test is surface insulation resistance (SIR), which is a measurement of electrical performance after prolonged exposure to elevated temperature and humidity. Phase III utilizes three of the four standard tests used in Phases I and II: ionic conductivity, HPLC, and SIR. In addition, Phase III requires tests using ion chromatography. Using both ionic and non-ionic tests, the ad hoc group ensured a full picture or "how clean is clean." Inclusion of SIR provides a good indication of "how clean is clean enough."

Although not included as pass/fail criteria, visual examination is also required. Each of the test plans requires photographs of representative assemblies and of all anomalies.

The only differences between Phase I and Phase II are the cleaning material and the cleaning process used. Differences between Phase II data and the Phase I/Benchmark data are attributable to the alternate cleaning material/process. The Phase II data are evaluated to determine if it is "worse than," "as good as," or "better than" the Phase I/Benchmark data.

Phase III processes are significantly different from both Phases I and II; however, the common test methods and the use of the IPC-B-36 provide a link to the Benchmark.

TEST MONITORING AND VALIDATION COMMITTEE

Citing the need to expedite the evaluation of alternatives, the ad hoc group decided that testing should not be confined to a single site. In order to ensure that the test plans were strictly followed and that tests were comparable between sites, the ad hoc group established the Test Monitoring and Validation Committee (TMVC). The TMVC is a subset of the ad hoc group. Members of the TMVC (called the Test Monitoring and Validation Team) attend all official tests, monitor all official testing, review all reports, and determine if an alternate material/process is "worse than", "as good as," or "better than" the Benchmark. The sponsor of the alternate material/process is required to publish all test results, even in the case of "worse than results", and provide copies of the report upon request. The Phase II TMVC is chaired by Dr. Leslie Guth, AT&T. The Phase III TMVC is chaired by Dr. Laura Turbini, Georgia Tech.

PROGRESS/PHASE I

The Electronics Manufacturing Productivity Facility (EMPF) and Naval Avionics Center (NAC) completed the Phase I tests in January and February 1989 using a nitromethane-stabilized azeotrope of CFC-113 and methanol. The tests were made possible by significant contributions of labor, materials, and equipment by members of the ad hoc Solvents Working Group. EMPF and NAC estimate that they alone spent two and one half man years of effort preparing for and actually performing the tests.

Results of the Phase I/Benchmark tests were fully reviewed by the ad hoc group and are reported in IPC-TR-580, "Cleaning and Cleanliness Test Program Phase I Test Results,"[5] which is available from the IPC. The primary result of Phase I/Benchmark is a benchmark data set for comparison of alternate materials and processes to the performance of a typical CFC-based cleaning process. In addition, many of the processes used in Phases II and III were defined during Phase I.

PROGRESS/PHASE II

David Bergman, IPC, has coordinated the Phase II tests. As of 25 January 1991, seven materials and their associated processes have been tested and received Test Monitoring and Validation Team approval as part of Phase II. The results of these tests are summarized in Table II. Three of the alternate materials are hydrogenated chlorofluorocarbons (HCFCs); three of the alternate materials are semi-aqueous; and the remaining material is a CFC-113 based solution which uses less CFC-113 than is used in the Benchmark solvent. Copies of the test reports are available from the sponsors of the tests (see Table III). In addition, for those interested in performing Phase II evaluations, David Bergman's address and phone number are listed. Additional Phase II tests are anticipated.

Results of the Phase II tests have been forwarded with recommendations for inclusion into military specifications to an extensive list of technical contacts within Department of Defense and industry, including members of the Military Electronics Technical Advisory Group (METAG) CFC Subcommittee. In addition, members of the ad hoc group have been involved in numerous briefings at a variety of levels within the Department of Defense (DOD), including the DOD CFC

Advisory Committee which was established by Congress to provide feasibility and cost estimates of CFC chemical substitutes and alternative technologies, and assist in technology transfer.[6] The METAG CFC Subcommittee, a group established by the Deputy Assistant Secretary of Defense (Total Quality Management) and the Deputy Assistant Secretary of Defense (Environment), and chaired by Harold Rife, Crane Naval Weapons Support Center, has recommended the following:

- "...that Phase I Benchmark become the acceptance criteria for all electronic assemblies"[7]
- "...any environmentally safe and compatible process proven capable of cleaning materials to the benchmark be allowed for military electronic assemblies, even if a contract modification is required."[7]
- "...that U.S. military standards officially recognize and cite the benchmark test procedure ... and the test results ... as the full confirmation of the capability of cleaning as well or better than CFC-113"[7]

PROGRESS/PHASE III

The Phase III Water Soluble Flux Test Plan was issued in August 1990 after months of development. Naval Avionics Center was chosen as the primary test site for the Water Soluble Flux (WSF) Evaluation, which was designed to demonstrate the performance of water cleaning in combination with water soluble fluxes and pastes. The WSF Evaluation consists of two parts: evaluation of the flux/substrate interaction using the IPC-B-24 assembly and the evaluation of cleanliness using the IPC-B-36 assembly (same assembly used in Phase I/Benchmark and Phase II testing). Process development has been completed at Naval Avionics Center. The evaluation of three fluxes and three pastes using the IPC-B-24 test is scheduled for 1-3 February 1991. The "best" paste and the "best" flux from the IPC-B-24 test will undergo the IPC-B-36 test at NAC, tentatively scheduled for April 1991.

The Phase III No-Clean Flux Test Plan is being developed by a subset of the ad hoc group, primarily through the efforts of the IPC. Compared to the traditional rosin-based fluxes and CFC-113 cleaning, no-clean fluxes and pastes (which require no cleaning at all) promise the greatest environmental improvements. However, they also raise the most questions with regard to performance and long-term reliability. Dr. Laura Turbini, Georgia Tech, is leading this formidable effort.

The Phase III Controlled Atmosphere Test Plan is also being developed by a subset of the ad hoc group and through the efforts of the IPC. Using a variety of inert or reactive atmospheres, new soldering equipment and processes, which do not require traditional fluxes, have been developed. Larry Lichtenberg, Motorola, is leading this effort.

CONCLUSIONS

The EPA/DOD/IPC/Industry ad hoc Solvents Working Group has worked effectively and efficiently to identify viable alternatives for ozone-depleting CFC-113 for use in cleaning electronic assemblies. Thus far, the ad hoc group has:

(1) developed a set of standard tests for evaluating alternate materials and processes.

(2) established a group and method for monitoring and recognizing all official tests of alternatives.

(3) produced a benchmark data set for comparison of alternate materials and processes to the performance of existing CFC-based cleaning materials.

(4) monitored and validated the testing of seven alternate cleaning materials and processes.

(5) worked with Department of Defense representatives to modify military specifications and existing contracts.

More than ever, the industry/government cooperation demonstrated by the ad hoc Solvents Working Group needs to continue. The Clean Air Act of 1990 includes legislation concerning CFCs, methyl chloroform, and volatile organic compounds (VOCs). This legislation challenges the U.S. electronics industry to reduce the use and release of ozone-depleting and global warming chemicals. In addition, the stringency, scope, and timetable for the Montreal Protocol were recently reviewed by United Nations Environment Programme Committees and the protocol was modified on 29 June 1990 in London, England (see Table IV).

The Protocol now calls for a 100% phase-out of CFC-113 by the year 2000. Methyl chloroform (otherwise known as 1,1,1-trichloroethane) was also added to the list of regulated substances. 100% phase-out of methyl chloroform, which was at one time considered a viable alternative for CFC-113 in electronics cleaning, is required by 2005. A resolution calling for the use of hydrogenated chlorofluorocarbons (HCFCs) only where other alternatives are not feasible was also included, with a notice that 100% phase-out of HCFCs should be expected no later than the year 2040 (Note that the Clean Air Act calls for phase-out of HCFCs by 2030).

At the recent International Conference on CFC & Halon Alternatives held 27-29 November 1990 in Baltimore, Maryland, Eileen Claussen, Director of the Office of Atmospheric and Indoor Air Programs at the U.S. EPA, announced that dependent on further scientific and technical evaluations, the phase-out of CFCs may be moved to 1997. Obviously, the challenge to the electronics industry remains just as imperative today as it did in the spring of 1988.

ACKNOWLEDGEMENTS

The author would like to recognize the members of the ad hoc Solvents Working Group who donated the necessary labor, materials, equipment, and technical support to expedite the transition to non-ozone-depleting alternatives within the electronics industry.

REFERENCES

1. United Nations Environmental Program (UNEP), "Montreal Protocol on Substances that Deplete the Ozone Layer", 1987.

2. UNEP Solvents Technical Options Committee, "Electronics Cleaning, Degreasing, and Dry Cleaning Solvents Technical Options Report", 30 June 1989, p. 13.

3. Ad hoc Solvents Working Group, "Cleaning and Cleanliness Testing Program, a Joint Industry/Military/EPA Program to Evaluate Alternatives to Chlorofluorocarbons (CFCs) for Printed Board Assembly Cleaning", published by the IPC, 1 September 1990 revision.

4. Ad hoc Solvents Working Group, "Phase 3/Water Soluble Fluxes, Cleaning and Cleanliness Testing Program, a Joint Industry/Military/EPA Program to Evaluate Alternatives to Chlorofluorocarbons (CFCs) for Printed Board Assembly Cleaning", published by the IPC, 1 September 1990.

5. Ad hoc Solvents Working Group, "Cleaning and Cleanliness Test Program Phase 1 Test Results", IPC-TR-580, published by the IPC, October 1989.

6. "Charter for the Department of Defense Chlorofluorocarbons (CFC) Advisory Committee" established by Section 356 of the National Defense Authorization Act for Fiscal Years 1990 and 1991, Public Law 101-189, 29 November 1989.

7. Harold Rife, "Department of Defense Chlorofluorocarbon Subcommittee Meeting of 14 June 1990", Memorandum 5050/648, 603A, dated 9 July 1990.

TABLE I

PARTICIPANTS / AD HOC SOLVENTS WORK GROUP

Alternative Producers

Allied Signal
DuPont
ICI Chemicals
Petroferm
DuBois Chemicals
By Pas of Toledo
Advanced Chemical Tech.
Advanced Chemical Tech.
Alpha Metals
Chem-Tech International
Dow Chemical
GAF Chemical
Kester Solder
London Chemical
Martin Marietta Labs
Mirachem
Pennwalt Corp.
Van Waters & Rogers

Flux/Equipment Manufacturers

Alpha Metals
Baron Blakeslee
ECD
Electrovert
Kester Solder
London Chemical
Unique Industries
Unique Industries
Forward Technologies
Stoelting
Branson
Exxon Chemical
Branson
Detrex
Gram Corporation
Hollis Automation
Multicore

Defense Contractors

Boeing
General Dynamics
Honeywell
Hughes Aircraft
IBM
Litton
Magnavox
Martin Marietta
Motorola
Texas Instruments
Lockheed
McDonnell Aircraft
Raytheon
Sunstrand
General Electric
Grumman Aerospace

Commercial Manufacturers

AT&T
Digital Equipment
Northern Telecom
Ericsson
Ford
Apple Computer
Delco

Industry Associations

Institute for Interconnecting and Packaging Electronic Circuits (IPC)
Halogenated Solvent Industry Association (HSIA)

Government Agencies/Other

EPA
U.S. Navy (EMPF)
U.S. Navy (NAC)
Sandia National Laboratories
Underwriters Laboratory
Robisan Laboratory
U.S. Air Force
U.S. Army
D.O.D.
Georgia Tech.
NASA
DESC
International Conservation Center Foundation (ICF)
Naval Weapons Support Center (Crane)

TABLE II

TESTED PHASE II ALTERNATIVES

MATERIAL	TYPE	MATERIAL DATE	TEST RESULTS
Allied Signal Genesolv[R] 2010	HCFC *	Sept. 89	Passed
Martin Marietta Marclean™-R	S/A **	Jan. 90	Passed
Petroferm Bioact[R] EC-7	S/A	Feb. 90	Passed
DuPont Axarel[R] 38	S/A	Feb. 90 June 90	Passed
Allied Genesolv[R] 2004	HCFC	Apr. 90	Passed
DuPont Freon[R] SMT	CFC	June 90	Passed
DuPont KCD 9434	HCFC	July 90	Passed

Results reported are for specific material, equipment, and operating parameters.

* HCFC = hydrochlorofluorocarbon

** S/A = semi-aqueous material

TABLE III

CONTACTS FOR PHASE II

Institute for Interconnecting and
Packaging Electronic Circuits (IPC)
David Bergman
7380 North Lincoln Avenue
Lincolnwood, IL 60646
708-677-2850

Allied-Signal, Genesolv/Baron-Blakeslee
Dr. Kirk Bonner
2001 N. Janice Avenue
Melrose Park, IL 60160
708-450-3880

Martin Marietta
Dr. Maher Tadros
Systems, Baltimore Division
103 Chesapeake Park Plaza
Baltimore, MD 21220
301-247-0700

Petroferm
Dr. Mike Hayes
5400 First Coast Highway
Fernandia Beach, FL 32034
904-261-8286

E.I. duPont de Nemours & Co., Inc.
Carroll Smiley
Chestnut Run Plaza
P.O. Box 80711
Wilmington, DE 19880-0711
302-999-2629

TABLE IV

SUMMARY OF LONDON AMMENDMENTS TO MONTREAL PROTOCOL

(Percent of Reduction)

	1993	1995	1997	2000	2005	2020	2040
CHLOROFLUOROCARBONS Study by 1992 to see if earlier phase-out is possible	20%	50%	85%	100%			
HALONS [1]Exemption for essential uses		50%		100%[1]			
OTHER FULLY HALOGENATED CFCs	20%		85%	100%			
CARBON TETRACHLORIDE		85%		100%			
METHYL CHLOROFORM	freeze	30%		70%	100%		
HCFCs Transitional Substances for use only where other alternatives are not feasible						100% Target for Phase-out	100% Required Phase-out

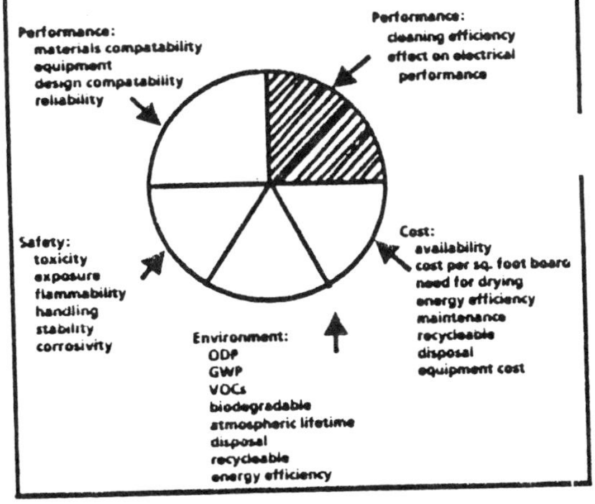

FIGURE I

CRITERIA FOR CHOOSING A CFC ALTERNATIVE

SPRAY FORMING AS A NEW PROCESSING TECHNIQUE

Scott A. Ploger and Lloyd D. Watson
Custom Spray Technologies, Inc.
Rigby, Idaho

INTRODUCTION

A versatile process has been discovered for spraying solid materials in various forms from liquid feedstock. As displayed in Figure 1, the basic principle involves aspirating liquid into an inert gas nozzle, where incoming streams are efficiently nebulized into a directed mist of fine droplets. The liquid is typically in a molten state, such that droplets cool rapidly in flight before collecting against a substrate where solidification is completed.

Although seemingly similar to other "gas-atomization" techniques, the **Controlled Aspiration Process** (CAP) offers unique advantages for spray forming numerous materials. As discussed below, the CAP approach emphasizes tailoring plume characteristics for individual applications. Critical aspects involve designing spray systems for specific purposes, using sophisticated control over component temperatures, and employing comprehensive monitoring of performance parameters. Consequently, free-standing objects and adherent coatings have been sprayed successfully. Whereas most activities thus far have been aimed at rapidly solidified metals, this technology has also been extended to polymer-based materials.

Spray forming normally offers economic and environmental benefits to complement superior product properties. By comparison to traditional processing techniques, spray-forming technology is highly efficient in use of energy and conversion of feedstock. It is safe for personnel, and it produces virtually no hazardous solid, liquid, and airborne wastes. Spray forming can be marketed successfully to industries interested in the broad realms of waste minimization and environmental compliance, where the attractive "bottom line" considerations on product properties and process economics can readily offset capital costs of retrofitting. In particular, spray forming could conceivably eliminate solvent emissions in several commercial situations of immediate concern.

BACKGROUND

The CAP version of spray forming was originally developed at the Idaho National Engineering Laboratory (INEL). As outlined in Reference 1, EG&G Idaho, Inc., devoted internal seed funds toward elaborating upon pioneering research at the University College of Swansea and the Massachusetts Institute of Technology (MIT). These two institutions essentially adapted the standard nozzle used to fabricate metal powder[2] into devices for spraying consolidated metal deposits. In both cases the driving forces were reducing metal processing costs and improving mechanical properties.

Nozzles designed for powder metallurgy were primarily aimed at convenience. As such, they consist of an open crucible with a plugged hole in the base. When a stopper rod is pulled, molten metal drains down the hole, whereupon the metal stream is nebulized by a ring of inert gas jets. Metal on the outside of the stream is nebulized more efficiently than at the core, yielding droplets with spatially varying size distributions. Consequentiy, most of the mass remains near the plume centerline, producing a Gaussian-shaped deposit when the droplets impact onto a substrate before completely solidifying. Microstuctures and mechanical properties change over the deposit because of differing thermal fluxes and cooling rates.

The rate at which molten metal is nebulized in the standard design is largely determined by the liquid level in the crucible. Instead, the INEL chose to investigate an approach where the liquid feed rate is governed by aspiration. Here, over a certain range of gas pressure, liquid metal is drawn by suction into the throat of a converging/diverging nozzle, much as gasoline enters a venturi carburetor. The nebulization process is very efficient in the physically confined ineraction zone, resulting in unusually fine droplets. Furthermore, operating pressures are significantly lower than in conventional metal-spraying nozzles, producing a gentle, low-velocity plume that resists entrapping gas bubbles in deposits. The U.S. Department of Energy obtained a patent that recognized the unique features of the INEL version of spray forming[3].

EG&G Idaho employees also developed a nozzle prototype for spraying wide, flat deposits, as detailed in[1]. This rectangular nebulizer contained a slot-shaped gas throat fed by multiple liquid orifices from a heated trough (tundish). This design was critical to attracting a DOE program under the Steel Industry/Federal Laboratories Research Initiative, in conjunction with MIT, Oak Ridge National Laboratory, and eight industrial participants. The main objective is reducing the energy consumed in manufacturing steel strip by eliminating most rolling steps.

Another environmentally beneficial project was initiated at EG&G Idaho at the same time, where funding from the Engineering and Services Center at Tyndall Air Force Base was directed at spraying high-performance coatings. The ultimate intention is eliminating hazardous wastes from the chromium electroplating process, which demands coatings of high-melting-point metals with established resistance to wear and corrosion. First, however, the feasibility had to be confirmed at low temperature with tin[4]. In this low-budget study, a bench-scale spray system readily produced dense adherent coatings, and rapid solidification also strengthened the coating layer. Over 99 percent of the metal sprayed was consolidated into each deposit and virtually no wastes were generated. The low heat flux delivered to the base material shows that CAP systems are ideal for depositing metal coatings on plastics, cellulose fibers, and heat-sensitive alloys. These novel coating attributes yielded a pending DOE patent[5].

ACTIVITIES OF CUSTOM SPRAY TECHNOLOGIES, INC.

The two principal investigators technically responsible for spray-forming growth at the INEL (Lloyd Watson and Scott Ploger) elected to form their own research company under DOE's "technology transfer" auspices. **Custom Spray Technologies** (CST) is thus a two-person private venture created to explore possibilities for rapidly commercializing certain spray-forming applications outside the domain of direct INEL interests. To this end, CST constructed its own 4000 square-foot laboratory in 1990 to offer timely, economical design and testing investigations. Consulting support will also be available during scale-up and pilot-plant stages to properly escort particularly promising applications into full production.

As noted, CSTs primary objective is conducting industrially oriented research and development. Nevertheless, CSTs first two years will be occupied honoring previous commitments to federal agencies. Successful Phase I efforts on the USAF coatings project naturally led to approved funding for a high-temperature Phase II demonstration with hardfacing alloys, where CST provided full-time consulting support to EG&G Idaho. Phase II was also successful, yielding microcrystalline cobalt-chromium coatings with hardness values among the highest ever measured on metallic materials--in the realm of cemented carbide cutting tools[6]. More CST support was thus requested by Mountain States Energy, Inc., to assist on Phase III pilot-plant experiments.

Because CSTs two-person staff might otherwise limit the commercial implementation of spray-forming technology, CST has entered into a teaming agreement with SCIENTECH, Inc., of Idaho Falls. This established engineering firm will be responsible for scaling up prototype sytems developed by CST, as well as providing procurement services, contract management, human factors input, approved drawings, system operating procedures, and other pedigreed documentation required for production purposes. The first example of this joint arrangement is a net-shape-forming project for Martin Marietta Energy Systems, Inc., which is intended to minimize generation of contaminated wastes at Oak Ridge National Laboratory.

DEVELOPMENT METHODOLOGY

Successfully confirming the feasibility of spray forming on diverse applications depends upon several inherent features of a CAP nebulizing system. Of particular importance is the ability to control temperatures of all major components. Figure 1 reveals that heating is employed at the melt furnace, tundish reservoir, nozzle body, and incoming inert gas plenum. Auxiliary heating and/or cooling is also frequently implemented at the substrate to influence deposit wetting and adhesion, along with solidification rates. For any fixed nebulizer geometry, these independent thermal controls offer considerable flexibility for optimizing both plume characteristics and properties in the consolidated droplets.

Performing meaningful experiments further requires carefully monitoring and recording critical operating variables. Figure 1 illustrates the typical positions of thermocouples for temperature measurements, plus the pressure and flow transducers necessary to control the nebulizing gas. Measurement signals are displayed dynamically on computer screens for on-line monitoring through the use of versatile data-acquisition software. Menu-driven screens are configured for all operating modes desired, including detailed characterizations of individual components prior to assembly of the integrated system. Thoroughly understanding subsystem performance has proven essential to later data interpretation from spraying experiments. Dynamic computer displays of key system parameters aid manual adjustments of regulated gas pressure, along with visual observations of droplet spraying and deposition behavior.

Figure 1 also shows that spraying is normally conducted in a sealed, inert environment to eliminate interactions with reactive gases. Additional benefits are protecting personnel from heated components, electrical hazards, and any ingestible unconsolidated particulates. As indicated, particle concentrations are measured on both sides of the filter to determine filter efficiency and to absolutely guarantee against significant atmospheric discharges. Comparing chamber concentrations to spraying rates further assists in calculating melt-to-deposit consolidation efficiencies, as well as estimating needs for respiratory protection on portable open spray systems that would be designed with inert sheathing gas flows.

SOLVENT SUBSTITUTION APPLICATIONS

Ozone non-attainment and air toxic problems in the United States are heavily influenced by emissions of volatile organic compounds from stationary area sources[7]. Such small area sources include auto body shops and cabinet makers, where applying polymer-based paints and protective films invariably frees large amounts of solvents to the atmosphere. Fortunately, the multi-component temperature control already discussed for spray forming metals enables polymer films to be sprayed in a molten state with little or no solvent emission and no later curing step. This new approach is straightforward for linear polymers, but the kinetics of crosslinking reactions can also be accommodated.

In fact, feasibility of spray forming linear and crosslinked polymers has already been demonstrated at EG&G Idaho. Here the objective was fabricating polyphosphazene membranes for extracting sulfur compounds. Despite the narrow focus of this exploratory exercise, no major impediments were encountered that would prevent spraying other polymer-based materials. Furthermore, performance of the rectangular nebulizer geometry was verified with regard to depositing films of uniform thickness. These results were embodied in another DOE patent idea record[8]. As co-inventors and a qualified small business, CST has applied to the DOE Office of Patent Counsel for associated technology transfer rights.

Another potential application for spray-formed polymers relates to the semiconductor industry, where protective films are used to package printed circuit boards and other electronic modules against environmental degradation. Here spray forming would have two benefits beyond eliminating solvent emissions. In current situations, difficulty occurs frequently in obtaining a consistent film thickness, because the dissolved polymer tends to "run" prior to curing. Since sprayed molten polymers could set up instantly by solidification, deposited layers would resist later relocation.

Protective packaging must also be kept from covering isolated areas such as electrical contacts, which often requires masking during the application process. This would be the case as well with spray forming, but a highly collimated plume can be achieved by optimizing the exit contour of a converging/diverging nozzle. Consequently, spraying through a template would leave the polymer film only where desired. Provisions must be made, however, for periodic template cleaning, lest accumulation narrow the openings.

Hazardous airborne emissions from the semiconductor industry are not confined to polymer solvents, and attention must also be paid to cleaning agents. The most notorious of these agents are the ozone-depleting chlorinated fluorocarbons (CFCs), which are especially important for removing flux from soldered connections on printed circuit boards. It is conceivable (albeit ambitious) that conductors could be sprayed onto printed circuit boards in a manner that altogether eliminates the soldering process.

The concept for solderless PC boards is presented in assembly-line fashion in Figure 2. Beginning on the left side, a PC board enters the process upside-down with electronic components already mounted underneath. As envisioned, component connections would be flat tabs bent over both to secure the components in place and to present a perpendicular surface to the spray plume. The first step shown is placement of a template over the PC board base, followed by a cleaning and roughening operation such as grit-blasting. Not only would this step remove surface contaminants from the exposed connectors, but it would also roughen both them and the board base for mechanical bonding to conducting metal. Once remaining grit has been blown free, a conductive coating is sprayed through the template. At this point, the template is detached for removal of accumulated metal (probably by scraping), recycling of surplus conductor, and subsequent reuse of the template.

In evaluating this concept, it should be recognized that spray-formed mechanical bonds can be quite strong. References 4 and 6 describe coating bonds exceeding 3000 pounds per square inch, with negligible interfacial porosity. Another attractive feature is that this approach opens the choice of conductors to a wide range of alloys, rather than merely plated copper. And, of course, the photo-resist techniques now required to etch conducting paths on PC board backings would no longer be necessary.

An additional method for reducing solvent emissions by spraying metals concerns "coating-free" materials. The Environmental

Protection Agency is seeking new ways to cover exterior surfaces that do not demand an initial layer of paint, as well as periodic reapplication of protective coatings[7]. Although beyond the scope of this paper, a tandem nozzle concept exists whereby a corrosion- and wear-inhibiting layer could be co-deposited while spraying items such as aluminum and vinyl siding.

CONCLUSIONS

Spray forming has been introduced as a versatile process for fabricating materials, including both high-performance metals and polymer coatings. With further research and dedicated development, this technology has considerable potential for reducing solvent emissions in several areas of immediate national concern.

REFERENCES

1. Watson, L.D. et al., "Nozzle-Aspirated Metal Forming," Paper presented at the Metallurgical Society's International Symposium on Casting of Near-Net-Shape Products, Honolulu, HI, November 13-17, 1988.

2. Hall, E.J., "Process for Disintegrating Metal," U.S. Patent 1,659,291, February 14, 1928.

3. Alvarez, J.L. and Watson, L.D., "Apparatus and Method for Spraying Liquid Materials," U.S. Patent 4,919,853, April 24, 1990.

4. Ploger, S.A. et al., Spray Coating of Metals, Phase I: Feasibility of Concept, ESL-TR-89-61, U.S. Air Force Engineering and Services Center, Tyndall AFB, FL, May 1990.

5. Ploger, S.A. and Watson, L.D., "Low Temperature Process of Applying High Strength Coatings to a Substrate and Article Produced Thereby," U.S. Patent Application No. 7,599,773, October 18, 1990.

6. Ploger, S.A. et al., Spray Coating of Metals, Phase II: Proof of Concept, U.S. Air Force Engineering and Services Center, Tyndall AFB, FL, to be published.

7. Kosusko, M., "Demonstration of Emerging Area Source Prevention Options for Volatile Organics," U.S. Environmental Protection Agency, Air and Energy Engineering Research Laboratory, Proceedings of the AIChE 1990 Summer Meeting, San Diego, CA, August 19-22, 1990.

8. McHugh, K.M. et al., "Spray-Forming Process for Polymer Films," DOE Case Number S-71,998, assigned August 20, 1990.

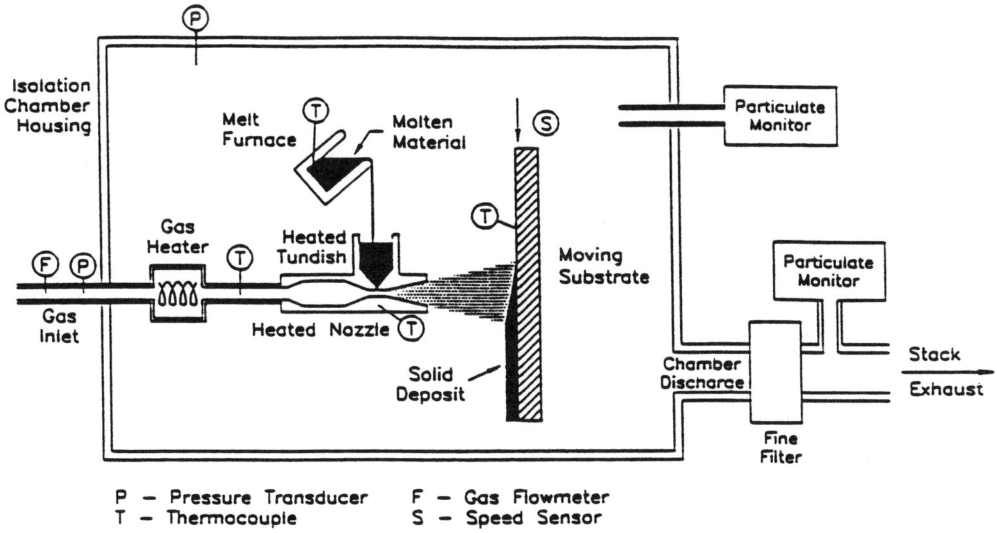

Figure 1. Basic Components of a CAP System.

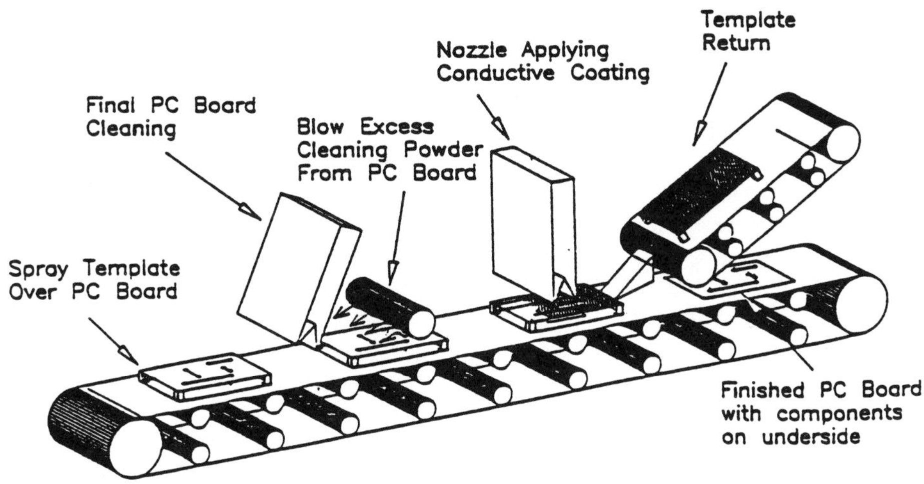

Figure 2. Depositing Conductors onto Printed Circuit Boards.

REDUCTION OF SOLVENT USE THROUGH FLUXLESS SOLDERING*

F. Michael Hosking
Sandia National Laboratories
Albuquerque, New Mexico

ABSTRACT

Conventional soldering typically requires fluxing to promote wetting. Halogenated solvents must then be used to remove the flux residues. While such practice has been routinely accepted throughout the DOE weapons complex, new environmental laws and agreements will eventually phaseout the use of these solvents. Solvent substitution or alternative technologies must be developed to meet these restrictions. SNL, Albuquerque is characterizing and developing alternative fluxless soldering technologies that will reduce solvent use and be compatible with prototypic packaging materials. The program is focusing on controlled atmosphere (vacuum, inert/reducing gas, reactive plasma, and activated acid vapor) soldering, metallization and inhibitor technology, and thermomechanical surface activation (laser, infrared, solid state diffusion, and ultrasonic) soldering. Since there is no universal method that can be applied to every electronic application, the study is defining technological options and limitations. Fluxless soldering would reduce the number of cleaning steps and the subsequent volume of mixed solvent waste. This paper will present an overview of the effects of atmosphere, materials, and processing conditions on attaining a fluxless operation. Examples of applying these technologies to electronic packaging will be given.

----

*This work is supported by the U.S. Department of Energy under Contract DE-AC04-76DP00789.

INTRODUCTION

There has been increasing concern about the environmental effects of chlorofluorocarbons (CFCs) by the scientific and political community over the past decade. CFCs have been identified as a source of the depletion of stratospheric ozone. Continued ozone depletion would seriously affect both the environment and human health. The evidence for this scenario is well documented (1-3). The most celebrated example of ozone depletion is the ozone hole discovered over the Antarctic. It is being extensively studied and monitored. The hole has been associated with the emissions of fully halogenated CFCs and halons. The Montreal Protocol, which is an international agreement that was originally drafted and submitted for signature in 1987 and has gone through several revisions, attempts to reverse this depletion problem. The current Protocol schedule requires a complete phaseout of controlled CFCs by 2000. Other fully halogenated CFCs, carbon tetrachloride, and methyl chloroform will be also affected by the Protocol controls (4).

The international restrictions on CFCs will significantly impact the electronics industry. Cleaning is a major element in electronic manufacturing, especially as a part of solder processing. An electronic package is typically populated with many devices (surface mount devices, capacitors, resistors, chips carriers, leaded devices, etc.) that are attached to the host board by one of several soldering methods. Whether the operation is manually performed with a soldering iron or batch processed with a wave soldering machine, each method has the common feature of using a flux to help the molten solder alloy wet the base material (5). The flux has three functions. The first is to chemically remove surface oxides and provide a protective layer over the cleaned surface while solder wetting occurs. The second is to assist heat transfer to the joining surfaces. The third is to assist the removal of the reaction products. The reaction products and flux residue must be removed after soldering. Although the residues are generally nonconductive, they are corrosive and could create a reliability problem, especially for applications where extended storage in uncontrolled environments is expected. These conditions make mandatory their complete removal from the assembly.

Most military electronic applications use rosin fluxes when soldering. The flux residues are typically removed with halogenated solvents. This practice is changing because of the impact of the Montreal Protocol. New solvents, fluxes, and cleaning methods are being consequently developed to satisfy the CFC phaseout. Terpene solvents, aqueous based cleaning, water soluble fluxes, and low solids ("no-clean") fluxes are showing promise. Alternative technologies, such as fluxless soldering, must also be developed to supplement these other activities.

Fluxless soldering is not intended to eliminate all cleaning during electronic manufacturing, but it will reduce the total number of cleaning steps and the subsequent quantities of mixed solvent waste that must be handled. An example of this is step soldering where two or more solder alloys with different melting temperatures are used to attach more than one components in multiple processing sequences. If fluxing could be reduced or eliminated during these multiple steps, the need for cleaning could also be reduced and a significant quantity of solvent saved. This quantity is dependent on the number and size of parts processed, but for a typical hybrid microcircuit, up to 250 ml of mixed solvent waste can be generated from one cleaning cycle.

The purpose of this paper is to present an overview of the Fluxless Soldering activities that Sandia National Laboratories, Albuquerque (SNL) has in progress. SNL is characterizing and developing several alternative technologies that will be applied to waste minimization in the Department of Energy (DOE) weapons complex. The work is being funded by the DOE Office of Technology Development (DOE/OTD) which is committed to developing faster, better, cheaper, and safer processes and materials which can achieve and sustain environmental restoration and waste management compliance. The project objective is to safely integrate these alternative technologies onto the production floor.

FLUXLESS SOLDERING TECHNOLOGY OVERVIEW

SNL's Fluxless Soldering effort covers a broad range of technologies (6) that either reduce surface oxides or prevent surface oxidation prior to and during soldering. Most of the technology currently exists but has not been fully applied to soldering. Fluxless soldering is consequently not well understood and must be better characterized and developed if it is to succeed in reducing solvent use at the manufacturing level. There are four key elements to the SNL task. They involve the characterization and development of Controlled Atmosphere Soldering, Thermomechanical Surface Activation Soldering, Metallization Technology, and Inhibitor Technology, Figure 1. These activities are being supported by thermodynamic and kinetic analyses and wetting experiments. Figure 2 lists the principal contacts in SNL's Metallurgy Department 1830 who are involved in the investigation.

Controlled atmosphere soldering utilizes various "clean" or reducing atmospheres to maintain or produce a solderable base surface. These atmospheres typically depend on a vacuum, inert or reducing gas, reactive plasma, or dilute acid vapor-inert gas mixture (eg. formic acid and nitrogen). More will be said on the use of these controlled atmospheres for fluxless soldering in the next section.

Thermomechanical surface activation soldering depends on kinetic or directed thermomechanical energy to spall or ablate the surface oxide and facilitate wetting of the underlying, pristine metal. Laser, solid state diffusion, and ultrasonic soldering are typical ways in which this can be accomplished. These processes can be done in air or in a controlled atmosphere. An example of fluxless laser soldering, coupled with metallization and controlled atmosphere technology, will be given in the next section on controlled atmospheres.

Ultrasonic soldering uses an ultrasonic probe which is immersed in a solder bath and generates ultrasonic vibrations that reduce thin oxide layers through cavitation. It is difficult, however, to accurately direct these ultrasonic waves. There is also a lack of fundamentally understanding the interaction effects between the process parameters and materials. SNL is conducting experiments to characterize these fundamental properties. Cu and Al substrates are being fluxlessly and ultrasonically tinned with elemental Sn. The effects of tinning temperature, probe separation, probe power, probe angle/position in reference to the base surface, and vibration time on wetting are being studied. Preliminary results on flat Cu substrates have shown that excellent wetting can be achieved on both sides of the immersed substrate, but cavitation is very sensitive to sample thickness.

Protective coating technology also shows promise as a compliment to fluxless soldering. Nonoxidizing surfaces, such as Au, have a long history of being readily wettable without fluxing. Their deposition, however, must be closely controlled. A thick layer of Au is generally required to guarantee complete coverage and wettability of the underlying metal. However, the metallization must not be too thick or the extra Au will produce a brittle solder joint. Au metallizations should generally not exceed 3-5 wt. % in a 63Sn-37Pb solder joint. The resulting compromise in thickness typically results in a thinner, porous layer of Au that exposes the underling metallic surface, usually Ni, and degrades subsequent wettability under oxidizing conditions, Figure 3. These porous metallizations can be protected by applying organic inhibitors, especially if stored in an uncontrolled environment before soldering. SNL is working with the University of California at Berkeley to characterize the microstructures and the fluxless wettability of Ni-Au platings and the State University of New York at Stony Brook to study the bonding behavior of organic inhibitors on metallic surfaces and their effect on subsequent solder wetting. Work is underway to examine the effects of Au thickness and porosity on the degradation of wetting under fluxless soldering conditions. These protective coatings generally work in both air or a controlled atmosphere, although a dry, nonoxidizing cover gas is more effective.

The above fluxless soldering technologies must be compatible with not only the base and filler metals, but also with any neighboring materials that might be exposed to the same process during soldering. Sensitivity to lasers, infrared heating, or reactive plasmas is of special concern since they could effect the functional performance of an electronic component. Materials such as alumina, glass frits, epoxy, polyester, phenolic, polyimide, plastics, and conformal coatings could be degraded by exposure to these processes.

CONTROLLED ATMOSPHERE SOLDERING TECHNOLOGIES

Controlled atmosphere soldering (7) utilizes vacuum, inert or reducing gas, reactive plasma, or acid vapor-inert gas mixtures that function as either a protective or reducing cover during processing. Vacuum and inert/reducing atmospheres restrict the supply of oxygen to the workpiece with oxygen levels as low as 5 ppm. Although thermodynamic data suggests that the reduction of metallic oxides in hydrogen or vacuum is feasible, the kinetics for it to occur at typical soldering temperatures, 200-300°C, is negligible and the oxide remains relatively stable requiring the use of a flux. If fluxless vacuum or inert gas soldering is to succeed, therefore, the base metal must be oxide-free throughout the heating and wetting cycle. Gas flow rates are important because volatile contaminants must be removed from the work area with a dynamic flow of "clean" process gas. Metallizations, whether they are plated or tinned, provide an added margin for fluxless soldering in vacuum or inert atmospheres. Controlling the time at which the solder alloy is molten is also critical since extended soldering times could cause excessive solder alloy and base metal reaction and produce a new surface (eg. intermetallic) that could dewet. Infrared heating helps to minimizes thermal gradients and heating times and is consequently being applied to various controlled atmosphere soldering systems.

A Controlled Atmosphere Solder Wettability System is being constructed at SNL to study the effects of vacuum, inert gas, and dilute hydrogen reducing gas atmospheres on fluxless wetting. A schematic of the system is shown in Figure 4. The system uses an electrobalance to measure wetting force as a function of time. An auxiliary video system can record the wetting event and analyze the wetting images to determine the effects of processing conditions (pretreatment, atmosphere type, soldering temperature, immersion time, flow rates, etc.) and materials (base metal, metallization, solder alloy, inhibitor, etc.) on achieving fluxless wetting. A second system is available to perform area-of-spread (sessile drop) experiments in activated acid vapor-inert gas atmospheres.

SNL has developed fluxless laser soldering to fabricate the closure joints on an electronic radar package. The process utilizes the combined features of controlled atmosphere, metallization, and thermomechanical surface activation soldering. The application joins Ni-Au plated Kovar pieces with Sn-Pb, In-Pb-Ag, or In-Pb solder alloys and a 100 watt CW Nd:YAG laser, Figure 5. The laser beam is directed on the solder preform because of the reflection properties of the Au plating that would inhibit the absorption of the laser energy. Satisfactory hermetic joints were achieved with a 90 watt, 0.4 mm spot size, and 5 mm/s travel speed laser setting in a forming gas cover of 5 vol. % hydrogen in argon. Figure 6 shows a cross-section of a typical laser soldered joint from the parametric study. Although the process is being developed for closure joints, it can be readily applied to attaching discrete leaded devices.

Reactive gases or plasmas are also being investigated. Reducing plasmas can be used in a two step process that cleans the base metal and solder alloy during the first step and uses an auxiliary heat source, such as a heated platen, laser, or infrared heater, to make the solder joint in the second step, Figure 7. Compatibility between the plasma and the packaging materials is an important consideration. Reactive gases have the potential for effectively reducing surface oxides. For example, thermodynamic data suggests that atomic and ionic hydrogen have a higher copper oxide reduction potential than molecular hydrogen at 250°C, Figure 8. Ionic hydrogen appears especially effective. Preliminary cathodic plasma cleaning experiments on heavily oxidized Cu have resulted in oxide-free solderable surfaces, Figure 9. Experiments are underway to comprehensively characterize the effect of ionic hydrogen on Cu and Ni oxide reduction and fluxless wetting.

The final element of SNL's controlled atmosphere soldering effort is focused on acid vapor-inert gas mixtures. Although there are commercial systems available that use variations of this process, the fundamentals of their operation are not well characterized. The process is readily applicable to wave or batch furnace soldering, Figure 10. Dilute additions of formic or acetic acid vapor are added to argon or nitrogen to promote fluxless wetting through the reduction of metallic oxides:

$$MO + 2HCOOH \rightarrow M + 2CO_2 + H_2 + H_2O$$

In actual practice, it is difficult to reduce most surface oxides and an adipic acid additive is generally required to achieve wetting. Adipic acid is a major constituent of low solids or "no-clean" fluxes. As with the low solids fluxes, the adipic acid residues left on a soldered component must be completely removed to satisfy the long term reliability requirements imposed on most military electronic applications. SNL is characterizing the effect of these acid vapor additions on oxide reduction and fluxless soldering. The acid-gas mixture, gas flow rate, activating additives, soldering temperature, time, and materials are important parameters that influence wetting and their interaction effects are being determined. The objective is to develop a scientific understanding of how the process works and what must be done to attain true fluxless wetting.

SUMMARY

Fluxless soldering is a viable and supplemental technology to solvent substitution for the electronics industry. It has a high potential for reducing the storage and handling of hazardous fluxes, environmentally harmful solvents, and the subsequent mixed solvent waste generated by flux residue removal. Since there is no universal method that can be applied to every soldering application, technologies must be identified, characterized, and developed to satisfy the increasing number of environmental, safety, and health regulations. The objective is to quickly integrate these fluxless processes into full scale manufacturing.

SNL, Albuquerque has an active program that is evaluating various fluxless soldering technologies. It includes controlled atmosphere soldering, thermomechanical surface activation soldering, metallization technology, and inhibitor technology. These processes offer a wide range of fluxless soldering options which can be combined to enhance the reliability of the final product. Laser and atmosphere soldering are excellent examples of this dual technology concept. The key to fluxless soldering is to maintain a "clean" surface that the molten solder will directly wet or to reduce surface oxides that the solder will not wet. Materials compatibility must also be considered. Otherwise, the functional performance of the final product could suffer if the selected process degrades sensitive components near the solder joint.

ACKNOWLEDGEMENTS

The author would like to acknowledge the work of members from the SNL Fluxless Soldering Task group. Charlie Robino, Paul Vianco, Darrel Frear, Dave Keicher, Mark Smith, and Rob Sorensen were especially helpful in providing background and experimental information. I would like to also acknowledge the work of Rusty Cinque, Choong-Un Kim, and Bill Morris of UC-Berkeley and Clive Clayton of SUNY-Stony Brook. I also appreciate the program support of Joan Woodard, SNL, Pam Saxman, DOE/AL, and Clyde Frank, DOE/OTD.

REFERENCES

1. Molina, M. J. and Rowland, F. S., "Stratospheric Sink for Chlorofluoromethanes: Chlorine Atom - Catalyzed Destruction of Ozone," Nature, 249, 810-812 (1974).

2. Derra, S., "CFCs, No Easy Solutions," R&D Magazine, 56-66 (May 1990).

3. Anderson, S. O., "Progress by the Electronics Industry on Protection of Stratospheric Ozone," 40th Electronic

Components & Technology Conference Proceedings, IEEE, 1, 222-227 (1990).

4. Schuessler, P., "CFC Alternatives: Examining the Emerging Technologies," SUNY-Binghamton Soldering Technology for Electronic Packaging Symposium Program, November 12-13, 1990.

5. Wassink, R. J. K., "Soldering in Electronics," Electrochemical Publications, Scotland, 2nd Edition, 204-262 (1989).

6. Hosking, F. M. (Principal Investigator), "Fluxless Soldering to Reduce Solvent Use," DOE/OTD Technical Task Plan, revised January 30, 1991.

7. Hosking, F. M., "Fluxless Soldering with Controlled Atmospheres," SUNY-Binghamton Soldering Technology for Electronic Packaging Symposium Program, November 12-13, 1990.

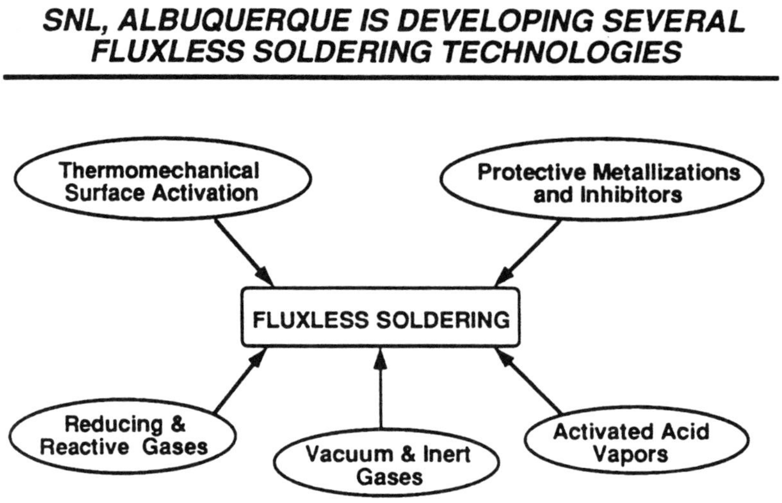

Figure 1. Flow chart for the SNL DOE/OTD Fluxless Soldering Task.

SNL METALLURGY DEPARTMENT 1830
FLUXLESS SOLDERING INVESTIGATORS

- *Thermodynamic/Kinetic Analyses* - *Charlie Robino, 1831*
- *Controlled Atmosphere Soldering* - *Darrel Frear, 1832*
  Mike Hosking, 1833
  Jim Jellison, 1833
  Dave Keicher, 1833
  Mark Smith, 1833
  Janda Panitz, 1834
- *Thermomechanical Surface Activation (Ultrasonic & Laser) Soldering* - *Paul Vianco, 1831*
  Dave Keicher, 1833
  Mike Hosking, 1833
- *Metallization Technology* - *Darrel Frear, 1832*
  Mike Hosking, 1833
  (UC-Berkeley)
- *Inhibitor Technology* - *Rob Sorensen, 1834*
  Mike Hosking, 1833
  (SUNY-Stony Brook)

Figure 2. Fluxless Soldering group responsibilities in SNL's Metallurgy Department 1830.

Oxide Free Surfaces Are Necessary If Vacuum or Inert Gas Soldering Is to Succeed

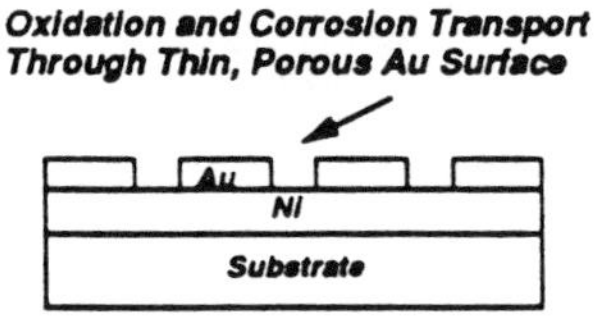

- *Au metallizations can provide an oxide free, solderable (fluxless) surface.*
- *Control of the Au thickness is critical; too thick and Au intermetallics will embrittle the joint; too thin and porous Au will allow oxidation of the underlying metal.*
- *Recommended Au thickness is 50-75 μin. (1.3-1.9 μm).*
- *Fraction of Au in a 63Sn-37Pb solder joint should not exceed 3-5 wt. %.*

Figure 3. Fluxless soldering can be achieved in a controlled atmosphere by overlaying the base metal with Au.

Solder Wettability In Controlled Atmospheres Can Be Determined With A Wetting Balance

Controlled Atmosphere Solder Wettability System

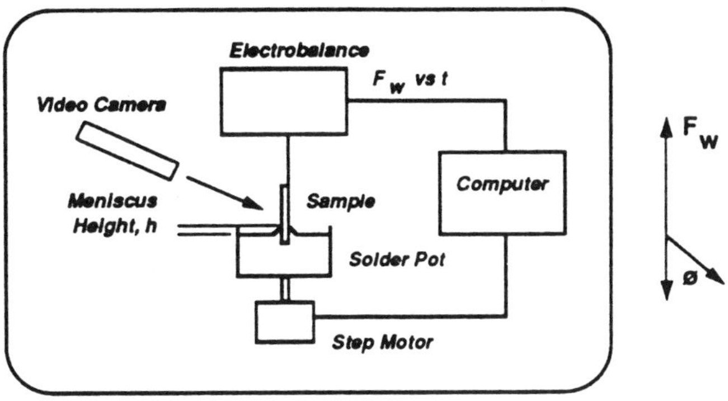

Figure 4. Diagram of the Controlled Atmosphere Solder Wettability System which is under construction and will characterize the effect of controlled atmospheres on fluxless wetting.

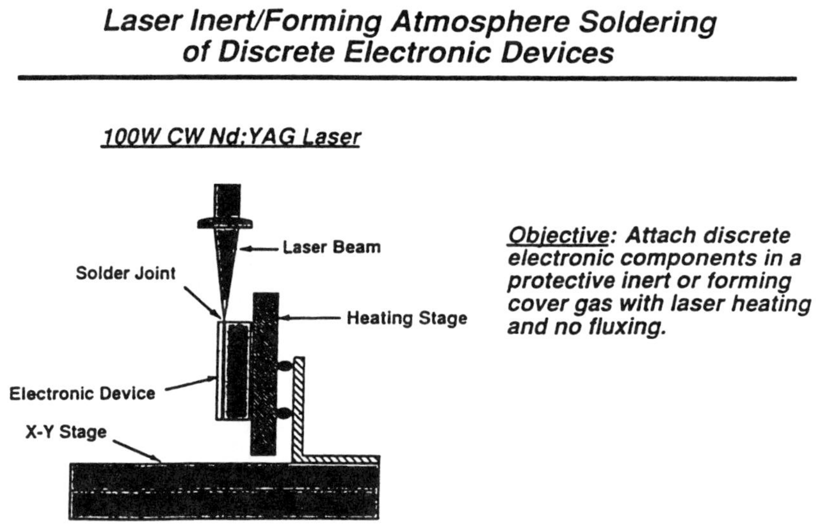

Figure 5. Laser and controlled atmosphere soldering can be combined to produce a fluxless soldering operation.

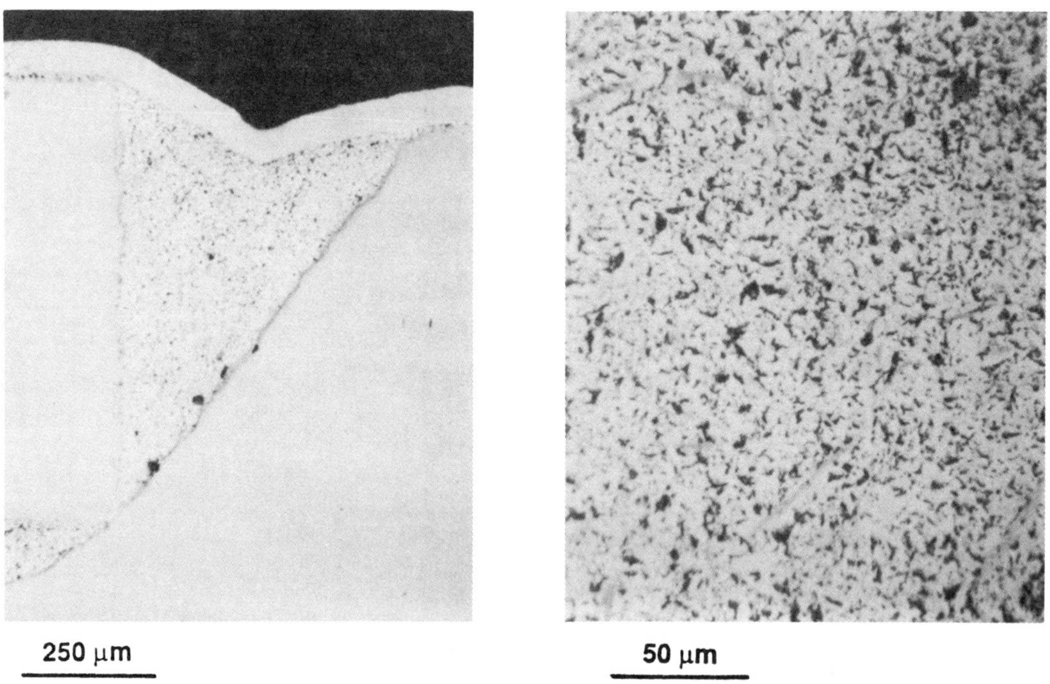

Figure 6. Optical micrographs of a fluxlessly laser soldered joint (Ni-Au plated Kovar with 60Sn-40Pb solder) showing excellent solder wetting and flow.

Two Step Plasma Cleaning and Soldering Is Best Suited for Batch Processing

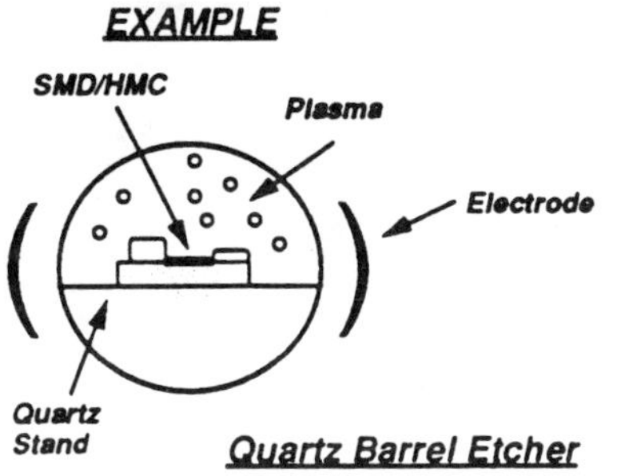

- *Cleaning Variables -*
  - *a) power*
  - *b) chamber pressure*
  - *c) time*
- *Soldering (fluxless) assisted with auxiliary heating (hot stage, laser, infrared)*
- *Reducing plasma produced by an RF electric field.*

Figure 7. Fluxless soldering can be accomplished in a two step operation: plasma cleaning immediately followed by soldering with an auxiliary heat source.

Thermodynamics of the Reduction of Cu_2O at 250°C Suggests That Ionic Hydrogen Has the Best Potential

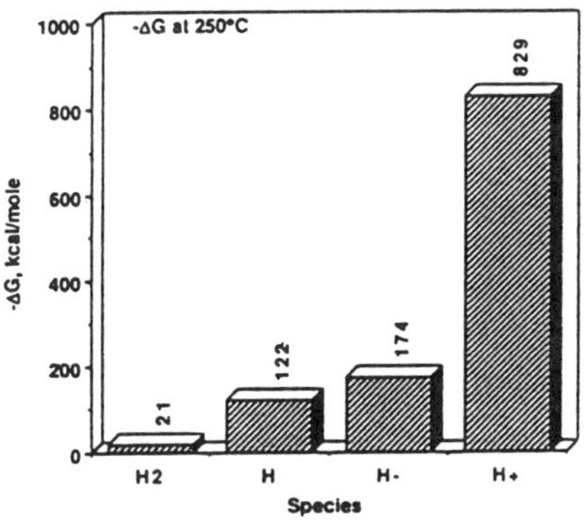

Gas Species	*Reduction Factor*
H_2	1
H	6
H-	8
H+	39

Figure 8. Thermodynamic data demonstrating the oxide reduction potential of ionic hydrogen.

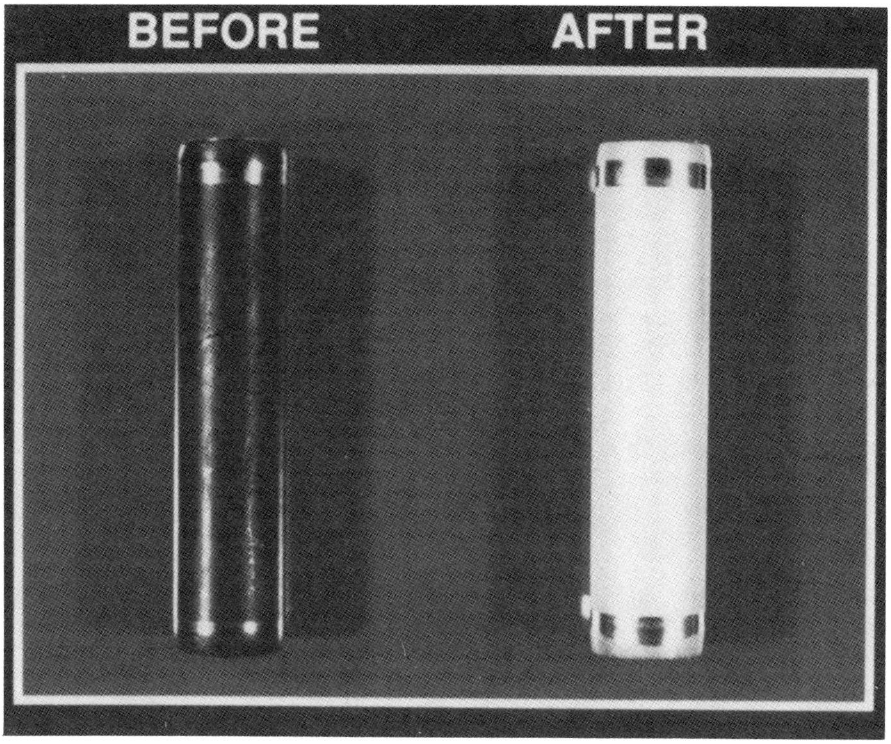

Figure 9. Heavily oxidized copper tube cathodically cleaned with a reducing plasma.

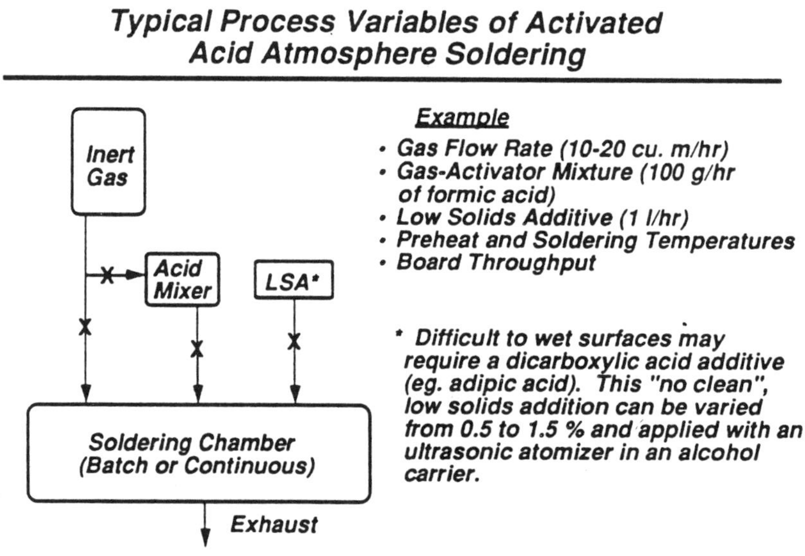

Figure 10. Dilute additions of formic or acetic acid vapor to an argon or nitrogen atmosphere have the potential for fluxless soldering.

PLASMA STRIPPING OF MAGNETIC COMPONENTS

T.J. Gillespie and T. Mehrhoff
GE Neutron Devices*
Largo, Florida

ABSTRACT

A plasma-stripping process and its associated product fixtures and equipment has been developed and evaulated for stripping wire insulation on both coils and magnetic assemblies. The stripping process uses tetrafluoromethane in oxygen as the active plasma gas and a metal plasma containment system. The system provides residue-free stripped leads when inspected at 200X magnification using Scanning Electron Microscopy (SEM). An evaluation of the plasma-stripping process was conducted on a flyspec inductor and a converter assembly. The results showed that the parts met all drawing requirements. In all cases, the fixtures provided a sufficient shield to prevent the plasma from attacking areas of the product which were not to be stripped. Material specifications for the plasma gas mixture and a Manufacturing/Engineering Equipment Instruction (MEEI) have been issued to support the plasma stripping activity in production.

INTRODUCTION

In 1985, the magnetics production assignment was assumed by GE Neutron Devices (GEND). One of the processes required by drawing as a part of this transfer involved the use of Iso Verre™ chemical stripper to strip wire insulation on the many types of products. Iso Verre is formulated in France and contains hydrofluoric and formic acid, phenol and methylene chloride. The material is provided as a kit with a bottle of thinner supplied with each four bottles of active stripper. In production, the basic stripper must be thinned to a viscosity that can easily be applied to the wire. The stripper is a very active chemical agent and causes frequent burning of operators and severe fume problems. One of the main objectives of this study was to eliminate the use of the Iso Verre chemical stripper in magnetics production.

In 1986, GEND and Sandia National Laboratories (SNL), Albuquerque, initiated a joint study on the development of a process to plasma-strip wire insulation. The initial study was designed around the flyspec inductor, which was the most difficult product to strip, since it required the leads to be cleaned such that they could be thermosonically bonded and required stripping within .010" of the core (see Figure 1).

In addition to the flyspec inductor, there were many magnetic assemblies where up to 4 coils with a total of 18 leads were mounted in diallyl phthalate (DAP) contact assemblies. Each lead on these assemblies requires that the insulation be stripped so that the lead can be soldered to a specific contact on the contact assembly. Some examples of these assemblies are shown in Figure 2.

PLASMA PROCESS DEVELOPMENT

Initial plasma stripping studies were performed in plasma cleaners with quartz-barrel reactors using various types of plasma gases including argon, oxygen, helium, pure air, sulfur

----

*GE Neutron Devices operates the Pinellas Plant for the U.S. Department of Energy under Contract No. DE-AC04-DP00656.

hexaflouride and combinations of these gases. The studies were performed on polyester-coated copper and gold wire. Results of these studies indicated that complete removal of all insulation residue and oxicdation of the gold-coated wire would be major problems. An additional major problem that surfaced at this time was development of a masking material that could survive the plasma treatment and provide a tight seal around the leads, since any small opening in the seal would cause etching of the product. These problems were solved by switching to a primary-type plasma cleaner in an all-metal enclosure (shown in Figure 3) using tetrafluoromethane in oxygen as the active plasma gas and using unfilled Sylgard™ as the sealing material.

PRODUCT PROCESS DEVELOPMENT

Ferrite Core Inductor

The inductor is wound on a ferrite core with an outside diameter of .050" and an inside diameter of .020". The wire is American Wire Gauge (AWG) No. 40 (.003"). The conductor is gold plated and is covered by a polyester insulation. A typical ferrite core inductor is shown in Figure 4.

Winding Development

Early in development of the processes for this inductor, the intent was to develop both new stripping and winding processes, since manual winding of the inductor was very slow. GEND purchased two winders for this product that were fabricated by the Jovil Manufacturing Company of Danbury, Connecticut. Initial evaluation of the winder showed that the polyester-coated wire was too weak to withstand the winding process; it cracked at each core/wire interface.

The General Electric Research Laboratory recommended that GEND evaluate the winder using other insulations. Following this lead, GEND evaluated polyester amid imide, polyimide and polyester imide nylon. Evaluation of these insulations showed that all of the new insulations resisted cracking during the winding operation. However, a new problem emerged; all wires were found to stretch during the winding process. The resulting reduction in cross section of the wire caused the product to violate the product specification. No further winding activity has been performed to date.

Having established that the combination of the metal plasma cleaner and the tetrafluoromethane/oxygen gas would provide acceptable stripping of the wire insulation without oxidation of the gold-plated coating, it became a matter of designing the right fixture to achieve an acceptable product. The fixture shown in Figure 5 is designed to hold the wound inductors in a Sylgard pocket with the wires extending through the seam of the pocket. Twenty inductors can be processed in each fixture, and up to three fixtures can be processed at one time in the plasma cleaner.

Two evaluations of the plasma stripping process have been performed in production to date, with 200 products being made in each evaluation. All products met the specifications. The stripping process was performed at .5 Torr, 400 watts for 45 minutes.

PLASMA STRIPPING OF MAGNETIC ASSEMBLIES

When an operator is preparing to finish-solder a magnetics assembly, he or she routes the wire in the specified path to the contact to which the wire is to be soldered. The wire is then bent around the contact, and the bend is used as a stripping guide with all the insulation being removed beyond the bend in the wire. The stripped wire is then looped around the contact twice and soldered in place.

In designing a fixture to strip a magnetics assembly, the outside edge of the stripping mask defines where the stripping action starts. A spring-loaded retainer is positioned in the fixture to hold the wire rigid during stripping. After stripping has been completed, the wire should be attached to the designated contact with the wire following the normal wire routing. Figure 6 shows a photograph of two converter assemblies in position in the stripping fixture and a diagram of the assembly after wiring has been completed.

ALTERNATE METHOD OF STRIPPING ASSEMBLIES

Many of the magnetic assemblies, such as the one shown in Figure 6, consist of one or more coils mounted into a DAP contact assembly. In these assemblies, the coils are secured in the contact assembly with eposxy prior to being stripped. The alternate method of stripping an assembly is to strip the coil to dimension prior to securing the coil into the contact assembly, the advantage being that many more products can be stripped at one time and the number of fixtures required could be significantly reduced. A fixture has been designed to strip the coil on the magnetic assembly previously mentioned and is presently being fabricated.

CONCLUSIONS

A mixture of 8% tetrafluoromethane in oxygen is effective in stripping wire insulation such that no residue can be detected at 200X using SEM. The plasma stripping process is very fixture dependent, but it appears from these evaluations that this process can be effective in stripping wire insulation in production.

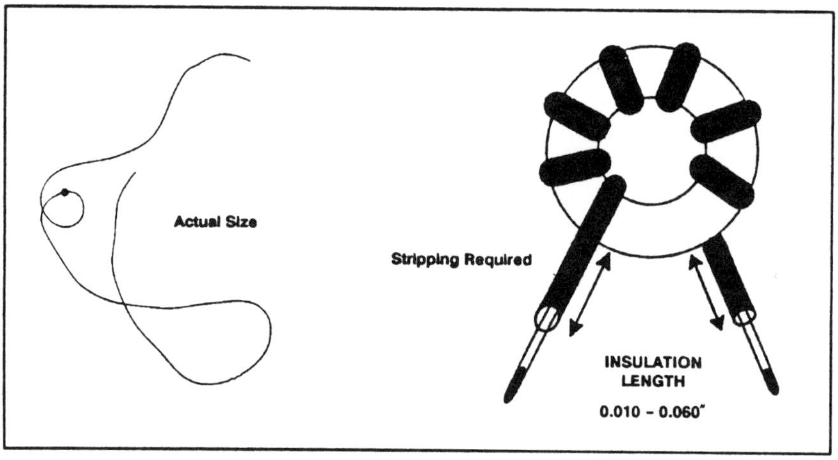

Figure 1. Flyspec Inductor

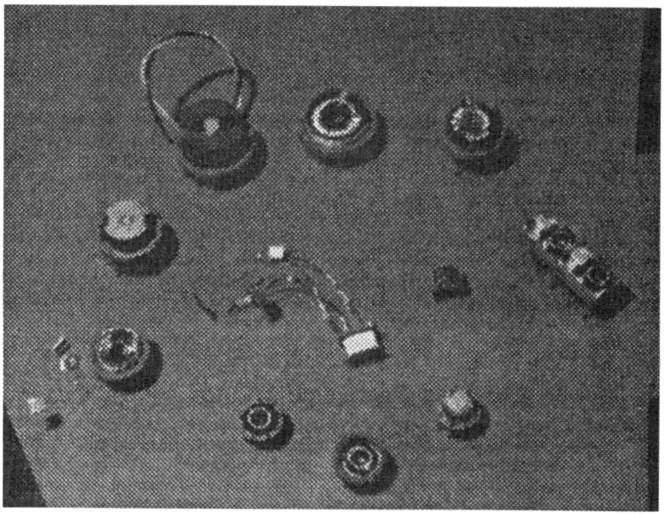

Figure 2. Examples of Magnetic Assemblies Requiring Stripping

Figure 3. Metal Containment Plasma Cleaner

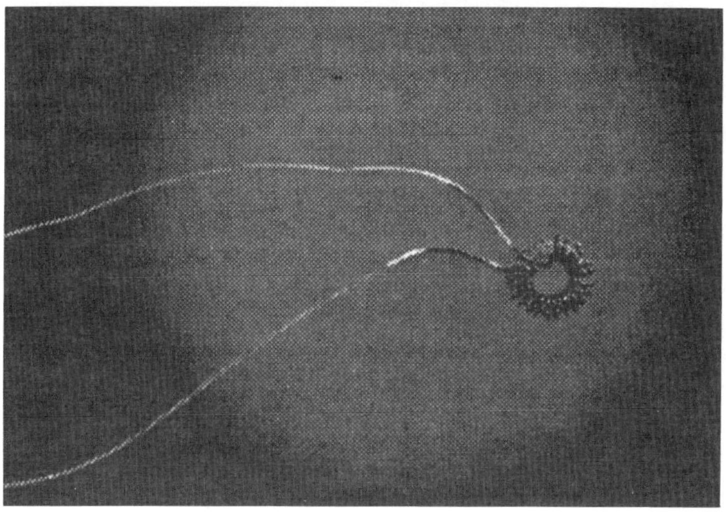

Figure 4. Wound Flyspec Inductor

Figure 5. Stripping Fixture for Flyspec Inductor

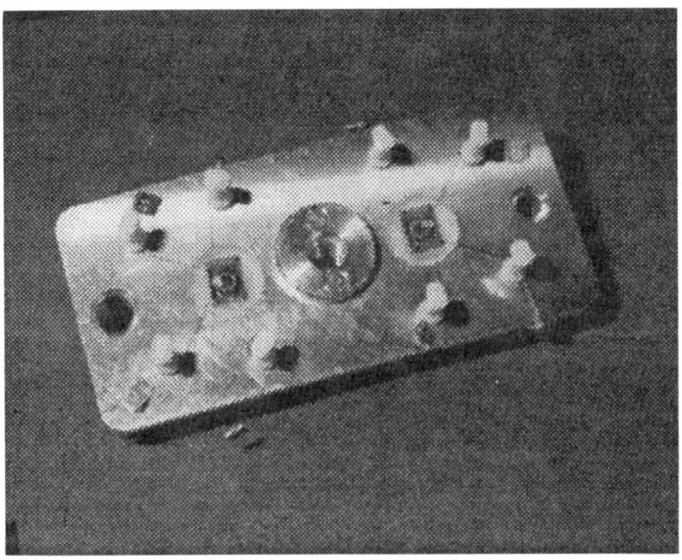

a. Assemblies in Position in Stripping Fixture

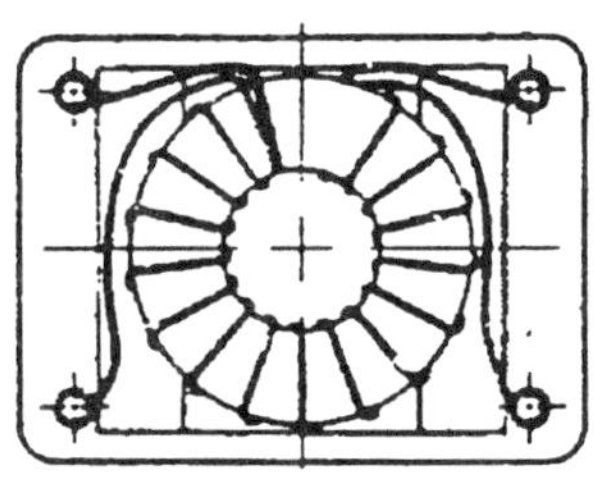

b. Assembly After Wiring

Figure 6. Converter Assemblies in Position in the Stripping Fixture and Assembly After Wiring Has Been Completed

SODIUM BICARBONATE BLASTING FOR PAINT STRIPPING

N.E. Wasson Jr. and Michael N. Haas
U.S. Air Force
Kelly Air Force Base, Texas

ABSTRACT

The San Antonio Air Logistics Center (SA-ALC) is one of five industrial activities in the United States that supports the maintenance and modification requirements for the United States Air Force. SA-ALC is especially concerned about the corrosion control requirements necessary to support their mission, since they remove the coatings, inspect, and repaint over 1.6 million square feet of delicate aerospace structures each year. Recent legislation and public awareness has encouraged the pursuit of alternatives to the costly chemical stripping operation. Plastic Media Blasting (PMB) has come forth as the most viable alternative in recent years. The primary weakness of the PMB process, however, is the generation of a substantial hazardous waste stream. SA-ALC decided to investigate the use of Sodium Bicarbonate as a coatings removal technique, because the process offered the potential to eliminate over 90% of this waste stream. The Sodium Bicarbonate, better known as the Bicarbonate of Soda Stripping (BOSS) process, is similar to the PMB process except that a small volume of water is injected into the blast stream at the nozzle to eliminate nuisance dust. SA-ALC decided to perform a thorough evaluation of the process to determine where it could be used in their industrial environment. This paper outlines the work initiated for evaluation of the BOSS process; another report will summarize the results.

BACKGROUND

Four primary areas of study will be conducted to determine whether the BOSS process is viable for use on aerospace structures. Initial work will be centered around aluminum alloys such as 2024-T3 and 7075-T6 clad and bare materials. First, corrosion tests will be conducted to determine what effects any residual sodium bicarbonate or its byproducts may have on an aerospace structure when exposed to the temperature and humidity conditions of a military operating environment aircraft. Second, the waste stream will be examined to determine the most suitable waste handling and disposal methods. Third, an optimization of the process, varying the many parameters affecting material degradation and production rates, will be analyzed to develop the most suitable parameters for thin-skinned, aluminum aerospace structures. The final work will be a material characterization of the optimized parameters to determine exactly what long term effects the process may have on the life of the structure. Most of this work will be performed by independent research laboratories under contract with the government. The corrosion testing and the waste stream evaluation are under contract already and work has begun. The optimization and material characterization test should begin by September, 1991.

CORROSION TESTING

Research has been performed over the last two years by J.H. Van Sciver Associates, who specialize in materials and corrosion engineering. The corrosion work they performed has demonstrated there is no adverse corrosion effects on aluminum alloys tested in sodium bicarbonate solutions, especially when compared to chemicals presently used in aircraft paint stripping. The Air Force Corrosion Program Office is presently managing a Corrosion Study contracted to Battelle Laboratories. The purpose of the study is to evaluate the BOSS process for military weapon systems. The program consist of two phases. Phase I will address the thermal stability of potentially entrapped BOSS media when exposed to aircraft operating environments. Phase II will determine the relative corrosivity of the BOSS media and byproducts vs. existing paint stripping methods (Chemical and PMB) used on aerospace materials. Phase I will be completed in March, 1991, and Phase II will be completed by October, 1991.

WASTE CHARACTERIZATION

A contract has been awarded to EG&G Idaho Inc. to perform an evaluation of the waste stream of the BOSS process and how to treat it. The work to be performed by EG&G will include:

1) Development of a test plan for characterization studies.

2) Characterization of new and spent BOSS media.

3) Identification and evaluation of material separation technologies.

4) Prepartion of a final report on materials characterization and separation technologies.

This work should be completed by the end of 1991. The primary objective of the study will be to develop an efficient method of separating the paint particles from the waste stream. Then the bulk of the waste could be disposed of as a non-hazardous component in the sanitary sewer system; the hazardous component consists primarily of paint chips, will be reduced to an absolute minimum. The BOSS process will help the Department of Defense work towards the goal of reducing its hazardous waste stream.

PROCESS OPTIMIZATION

Test specimens will be primed, painted and artificially aged. The specimens will be blasted to determine the optimum parameters to produce the best combination of stripping rate with the least amount of damage to the substrate. Parameters to be varied are: traverse speed, standoff distance, angle of impingement, nozzle pressure, media flow rate, and water pressure. Once these parameters have been established, they will be used to perform a material characterization study, quantifying the damage imparted by the abrasive effects of the sodium bicarbonate on clad and bare aluminum alloys. This work will probably be contracted out to an independent test laboratory. This will be (Phase I) with the follow-up material characterization to be performed by the same laboratory. The award of a contract is expected to occur by September, 1991 with work completed by January, 1992.

MATERIAL CHARACTERIZATION

Phase II of the contract will be to characterize the effects of the optimized process parameters on 2024-T3 bare aluminum. We feel the data generated through the evaluation of the PMB process has adequately characterized the effects of abrasive paint removal processes. A substantial amount of time and money will be saved by limiting the material characterization studies without affecting the accuracy of the assessment of the BOSS process. Stress saturation curves will be generated by using almen strip data. This data can be correlated to the existing PMB data without having to do extensive fatigue, tension, crack growth, and surface flaw fatigue testing. Some X-Ray Defraction testing with some minor fatigue data will be generated for accurate correlation to the existing PMB data. This work will be completed by March, 1992.

SUMMARY

Coatings removal, utilizing sodium bicarbonate as the abrasive, appears to be a viable alternative. Concerns of long-term corrosion effects of any media left in an aircraft or component will be resolved in this study. The material characterization is expected to mirror that of PMB and other abrasive blasting processes. The waste characterization will determine the true potential of this process. The promise of the elimination of one of the Department of Defense's largest waste streams offers the most benefit. The results of this work will be published upon final completion of all tests discussed in this paper.

LOW TOXICITY PAINT STRIPPING OF ALUMINUM AND COMPOSITE SUBSTRATES

Nona Larson
Boeing Aerospace
Seattle, Washington

INTRODUCTION
Part 1

The effective chemical paint strippers for aerospace coatings are toxic and present a hazard to both personnel and the environment. They contain phenol, methylene chloride and hexavalent chromium materials which have been targeted by governmental regulations for future elimination. These strippers are also detrimental to composites. This report focuses on impact damage inflicted on composite substrates by some of the mechanical methods meant to replace chemicals.

OBJECTIVE

The objective of this program is to find benign alternates to hazardous chemical paint strippers.

APPROACH

The approach with the least impact on production was taken for this project; i.e., identify and evaluate commercial chemical strippers which did not contain the targeted materials; if none were available, develop a chemical stripper; and finally evaluate non-chemical methods. To accomplish this work, the following three tasks were identified:

Task 1: Evaluate Chemical Strippers

Identify coatings and substrates to be evaluated; contact commercial sources of strippers; evaluate strippers based upon their ability to strip the coating, length of time it takes, and damage sustained by the substrate. If necessary, blend Boeing proprietary formulas.

Task 2: Evaluate Non-Chemical Stripping Methods

Investigate the mechanical, radiation and other stripping methods being tested throughout industry and evaluate only those which appear to offer solutions to aerospace problems.

Task 3: Specification Coverage

Change Boeing specifications for abrasive blasting to include effective and non-damaging paint removal methods. This will include Process Specification Departures for interim solutions and specific applications.

CHEMICAL STRIPPER ALTERNATIVES

Conclusions from Boeing Document D180-30690-4, Comprehensive Chemical Reduction Research Projects Final Report 1989, summarize the work from Tasks 1 and 2. No low toxicity chemical paint strippers were found in this program which can be recommended to replace those presently used; however, the four strippers: Brulin EXP 2187 mod., ManGill LP 4566, and Turco 6088 may be used effectively on selected coatings. For example, EXP 2187 mod. is extremely effective on removing melamine enamel and military epoxy primer; Turco 6088 is effective in removing phenolic resin varnish. The mechanical methods appear to be the best overall approach to removing the high performance aerospace and commercial coatings. Results are shown in Tables 1 and 2.

Recommendations made in D180-30690-4 state that we should monitor (rather than duplicate) data on alternate stripped methods, such as plastic media blasting and laser paint removal.

TECHNICAL ACCOMPLISHMENTS

For 1990 this program was to continue on a monitoring basis. Processes included were: plastic media, sodium bicarbonate, and carbon dioxide blasting, high pressure water jet, xenon lamp and laser paint removal. Envirostrip wheat starch media was not available in 1989.

Wheat Starch Media

In April 1990, a new paint stripped media was introduced[1]. This media, made of 100% crystallized wheat starch, is non-toxic, biodegradable in the true sense of the word, and made out of a renewable resource (as opposed to petroleum based plastics). Since this product was new, independent research for us to monitor, we started our own evaluation program. The first step in the study of this new process was to send samples to the vendor (Ogilvie Mills). When they came back, these samples looked good enough to pursue this process more actively than program planning had anticipated. Media characteristics and results are summarized in Table 3.

The coatings we tested were chosen because they were the most difficult to remove of the military and commercial airplane specification coatings. Some of these were brought to us on scrapped or test parts as a challenge because the owners did not believe that the wheat starch media would remove the coating. Table 4 lists the coatings removed, and Table 5 lists the effects.

Composite Testing

Wheat starch blasting caused the least. damage of any other blasting method we tested directly. Comparisons were made by blasting identical panels (composites laid up at the same time by the same person), or by blasting different parts of a single large panel. Figures 1 to 18 show these comparisons. These cross-sections were photographed at 200x.

Aluminum Testing

No damage was detected on 0.020-inch thick aluminum, with the exception of some deformation of the clad surface. Figures 19 and 20 show results of stripping the thin aluminum panels. Sandwich corrosion tests were performed per D6-17487, Certification Testing of Aircraft Maintenance Materials. As expected, the media passed both in the crystalline state and dissolved in deionized water.

SUMMARY

Discussion of Envirostrip Wheat Starch Blasting

It is not difficult to explain the differences in results obtained with this and other blasting media. Figures 21 & 22 show the fracture surfaces of new and used media. A particle hitting the paint surface at a pressure above approximately 30 psi will break. The glassy fractures shown in the figures explain why the media remains effective cycle after cycle. It becomes more effective as the particles get smaller simply because there are more sharp edges per pound of media in the blast stream. The limiting factor for size of this media is drag. When the particles break down enough that dust hangs in the air, the operator cannot see through it. A dust separation system easily removes this.

The fact that the particle breaks at around 30 psi makes the media very forgiving. Turning up the pressure will increase the flow rate, but will not make the media itself more aggressive.

One common concern when using "wheat" in a blasting system is the so called "silo effect," where a high concentration of wheat dust will spontaneously combust. This is not expected to be a problem because crystallized starch molecules are different in structure than the naturally occurring polymer. Nevertheless, we asked the vendor to supply data on the explosive limits of the dust. The following table summarizes the finds.

Envirostrip Properties: Explosive Limits

Media Size Mesh (US Std)	Ignition Temperature °C	Min. Explosive Concentration oz per ft^3
12/30	>850	Non-explosive
30/50	>850	Non-explosive
50/250	530	Non-explosive
Dust Bin	460	0.060

There are many interesting and useful properties to this media which we discovered during our study.

a. Envirostrip does not remove alodine. The aloding remains after depainting passes 7-day salt spray exposure, but it is dehydrated so paint adhesion suffers. This can be remedied by a quick dip in an alodine tank. Ten seconds should be sufficient.

b. New media is less aggressive than used media. When stripping Kevlar, new media gives the operator more control than the used media.

c. Large media works better on the elastomeric coatings, while small media works better on the more brittle coatings, i.e., epoxy primers and topcoats.

d. The media works best by breaking through the paint surface and peeling the paint up from the edges of the stripped area.

e. Some of the harder coatings (i.e., BMS 10-11 type 1 epoxy primer) will break the media down faster than others. This is not true of Mil-P-23377.

Discussion/Results of Other Methods

There are a number of paint stripping methods currently commercially available. Each of these has good and bad points. The information listed herein has been gathered from a variety of sources in addition to the data generated by this project.

Xenon Flashlamp Depainting

This method does not lend itself well to production use. Coated surfaces are exposed to high intensity pulses of light. A special head to focus the light must be used for each different part configuration. The three primary drawbacks are that it is not very effective on light colored coatings, it treats composite substrates the same as paints, and acutely toxic gases are released when polyurethane paints are broken down without adequate oxygen flow. An oil smut is left behind the flashlamp, so a cleaning step such as carbon dioxide pellet blasting is required. Dark, low-gloss topcoats (i.e., MIL-C-83286) can be removed at rates up to one square foot per minute. Light colors are removed much more slowly, and high gloss white is not efficiently removed. No damage to metal substrates has been found with this process; the metallic surface completely reflects the xenon flash. Composite surfaces do not reflect the flash, and are therefore removed in layers analogous to paint.

Laser Depainting

This method is similar to the Xenon flashlamp method, but is more easily controlled. Laser depainting is geared toward robotic applications where capital cost is high, but some applications justify the cost. Optimization is difficult due to the uncontrollable variations in paint thickness. Light colored and high-gloss topcoats are removed less efficiently than darker coatings, but they can be removed in a reasonable amount of time. Two companies are in the process of making this method commercially available.

Sodium Bicarbonate Blasting

This method, more than any other mentioned in this report, is a case of trading off pros and cons. The media is very effective without causing much impact damage to the substrate. Since rinsing is very difficult, there is a high potential for corrosion problems. The sodium bicarbonate breaks down in water, forming sodium sesquicarbonate, which has a pH of approximately 10. Leaving this on aluminum parts will be very detrimental to the part. The usual solution to this problem is to use a dilute acid rinse. Small parts are well suited to this, but large parts or structures are not. The dilute acid rinse also creates more hazardous waste, driving up the cost of the total paint removal process.

The media itself is inexpensive at about $0.50 per pound. Unfortunately, it is not recyclable. Enormous amounts of water are required to dissolve the spent media before sewering. To keep the dust down, it is blasted with water which contributes to the corrosion potential. The positive aspect of non-recyclable media is that a dedicated facility is not required.

Carbon Dioxide Pellet Blasting

CO_2 pellet blasting has been the subject of an enormous amount of testing. This process is ideal in the sense that no toxic substances are generated or released, a dedicated facility is not required, and precleaning or surface preparation of the painted surface is not needed. However, there are some very serious areas of concern with this process. Some of which limit the use of CO_2 blasting to steel or very thick aluminum parts. Very high pressures are used to accelerate the particles. The particles impact the surface at extremely high velocities, high enough to leave dents in 0.020 inch thick aluminum.

There are two systems commercially available to produce CO_2 pellets for blasting. The more widely used system extrudes the pellets, producing small cyclinders. They are not optimally shaped for removing paint, and unfortunately the optimization of the system that has been performed to date has consisted of changing pressure, impingement angle and stand-off distance. There has been some planning to optimize the process from the front end, beginning with producing pellets with different shapes, sizes, and/or hardness. This new approach to optimization should increase the strip rates. Previous rates have been slower than acceptable for production operations.

Plastic Media Blasting

Plastic media blasting has been studied in great detail. There are a number of complete programs, both military and commercial, which report widely varying results (Ref. 2 to 12). The process is partly accepted (one cycle only) by the FAA, and it is widely used by the military (Ref. 13 & 14). There are seven types of media, varying from soft to hard and mild to aggressive. The soft media requires a longer dwell time, so it does not necessarily cause less damage than the harder types. Plastic media is quite aggressive on composites, causing an unacceptable amount of erosion and/or fiber damage, with the exceptions of graphite or boron epoxy.

Ice Crystal Blasting

The Canadian Navy funded a program to devise a way to remove coatings from the interiors of submarines. This method must be safe in a confined environment with minimal air flow. This method is very effective on the interior coatings it was designed to remove. The company which invented the process is now optimizing to remove aerospace and other high performance exterior coatings. Progress is being made in this area. The coatings can be removed with very little damage to any substrate tested, including composites and thin aluminum. This process will be watched closely during this program, In addition to being effective, it has the attractive properties of requiring no precleaning of the painted surface, minimizes the waste generated, does not require a dedicated facility, and is completely non-hazardous to personnel.

General Concerns

Two primary concerns have been repeatedly expressed about any blasting method for paint removal of structures: crack closure and media entrapment.

Crack closure has been a concern ever since the early days of plastic media blasting. Research has shown (15) that different plastic media types do not close cracks, but the media can become lodged in the crack and prevent it from showing up during dye penetrant inspection. This is more likely to happen with plastic media than with other types that are water soluble.

Media entrapment is of serious concern to some, while others do not consider it a major problem. Any accelerated particles will find ways to enter openings. This can add weight to an aircraft and interfere with moving parts. Ideally, these areas would all be masked before stripping, but in reality, this is not always done properly. The other side of this issue was expressed very well by a shop foreman (16) who said, "I can live with that," comparing media entrapment to residual chemical paint stripper that was oozing out of faying surfaces on the repainted airplane parked behind him.

CONCLUSIONS

During 1990 this program investigated a number of mechanical paint stripping methods. Most of these methods were evaluated by monitoring research done by others. Due to the newness of the Ogilvie Mills Envirostrip media, there was no outside work to monitor. Therefore, we did our own evaluation. Overall, this media proved to be the best currently available technology. Clearly there is no panacea, but this method is more widely applicable than any other. Recommendations for using this method on aircraft are awaiting fatigue testing, which the Boeing Commercial Airplane Group plans to do in 1991.

On the other hand, chemical strippers will remain a problem, especially when pH is a consideration. There is simply no low toxicity, low vapor pressure analogy for methylene chloride and phenol. When tank stripping is possible, the N-methyl-2-pyrrolidone based formulations can be used at elevated temperatures. However, these cases can usually be mechanically stripped, making preferential the wheat starch media. When pH is not a factor, concentrated, low molecular weight organic acids will work, but they are relatively slow and unpleasant to use. Therefore, it can be concluded that a highly innovative approach will be required for developing a good neutral pH, room temperature and chemical aerospace coating stripper.

Part 2

FLUIDIZED BED ABSORBENT CLEANING

After demonstrating the feasibility of non-solvent substitutes during 1989, the objective in 1990 was to develop methods for their use that were suitable for industrial scale-up. The most promising materials at that time were absorbents such as starch and uncalcined diatomaceous earth. This could lead to the conclusion that very fine particulate matter is best, but developments in 1989 dealt primarily with substitution materials usable in the same approximate manner as a wipe solvent. While it can be generalized that a finer particle provides more surface area per volume, density and other properties become major considerations with other delivery systems. When fluidized bed technology was investigated as a possible means to scale-up and automate absorbent cleaning, the absorbent media had to be completely re-evaluated.

While it was theorized that the particles would aggregate with oil absorption and gain sufficient bulk and density to drop off the test panels, in reality the adhesion of the Zyglo test oil to the panel was the dominant force in the system. Aggregated particle mass never reached the point where it could exert the tensile and sheer stresses necessary to overcome the oil-panel bond. This precipitated a search for denser, larger absorbent materials which were also sewerable. Flours, cornmeal, and finally grain cereals were tried. Whole grains cereals proved to possess the right balance of particle size, absorbency and density. Specifically, Bear Mush, branch whole wheat cereal was used as the active absorbent in our most successful trials.

The actual fluidized bed material used was "Envirostrip," a modified wheat starch resembling large sugar crystals. The Envirostrip crystals provided the abrasiveness and bulk necessary to remove the oil laden wheat particles without manual assistance. This is a critical consideration for scale-up and prospective process automation.

The media selection thus dictated the procedure which evolved to: 1) Immersion of racked parts or sheet stock in the wheat cereal; 2) Removal of the wheat/oil conglomerate in the fluidized bed turbulent phase. Zyglo penetrant inspection oil was used for artificial contamination because it could be observed during bed operation; traditional cleaning tests indicated comparable properties with the MIL-L-7870 protective oil commonly used by our sheet stock suppliers.

The lab scale bed diameter was 14 cm with a total height of 43 cm. 301 CRES screening was used as the air spreader (funnel). Oil/wheat removal occurred almost instantaneously when the bed was turbulent.

Operating parameters were determined experimentally, and it was found that the most efficient cleaning was in a region referred to as "rapid bubbling." Some spouting occurs in this region. Combined with the rapid rolling of the bed, this spouting results in the maximum contact of the media and soiled surface. Figure 23 shows the characterization of pressure as a function of flow rate for the system. Figure 24 shows pressure as a function of velocity.

Scale-up of the fluidized bed can be calculated from the best fit curves of the plots shown in Figures 25 through 29. These plots are based on experimental values, with the exception of ▲P. ▲P was calculated using Ergun's equation. (4) The more commonly used equation established by Baeyens and Geldart (5) was determined to be inappropriate for this situation. Their equation does not hold true when the bed volume is small, or when the fluid and particle densities are orders of magnitude different, both of which were true of our system.

Figure 25 shows the operating air velocities at varying bed heights. Using the best fit curve for this plot, a bed height of one meter will require an air velocity of approximately 0.28 meters per second for efficient operation.

Scale-up to a bed height of one meter should be achievable using available plant air. Multiple ports may be necessary to achieve the operating flow rate. This size bed will approximate the size of many vapor degreasers.

Another non-solvent substitute investigated in 1990 was an oil absorbing cloth made by 3M called "Oil Sorbent, Type T-151." This cloth is made of 1/4 inch thick coarsely felted polypropylene. 3M markets this material for cleaning oil spills, and it is available to our shops for that purpose.

Our interest in this material was precipitated by an inquiry to Environmental Affairs from the Everett Production Drawing area. This area uses large quantitities of Freon to remove finger prints from the mylar drawing film. T-151 samples were tested in a variety of oil removing situations including the coated mylar film. It performed very well demonstrating a high degree of competitiveness for any oil accretion. On the basis of the initial work, it is concluded that these cloths could replace the naphtha/petroleum distillate preclean now employed in a two-step solvent cleaning process involving a non-polar precleaning solvent followed by a polar final cleaning solvent.

REFERENCES

1. Lenz, Ruben, "Envirostrip, A Non-Petroleum Based Natural Dry Blast Media Engineered for the Aerospace Industry," DOD/Industry Advanced Coatings Removal Conference, Atlanta, GA, May, 1990

2. D204-14436-1, "Impact Study of Plastic Bead Paint Removal on Corrosion Prevention and Control," The Boeing Company, Boeing Aerospace, Seattle, WA, March, 1987.

3. Panciera, H., "Surface Finish Removal from Advanced Composites Prior to Repair and Refinishing, Naval Air Rework Facility, Alameda, CA.

4. "Coating Removal via Plastic Media Blasting, "NAVAIR Engineering Support Office, Materials Engineering Division, Naval Air Rework Facility, Pensacola, FL.

5. NESO Code 34132, "Preliminary Report on Plastic Media Paint Stripping from Graphite Epoxy Surfaces, "Materials Engineering Lab, March, 1984.

6. 86-E3B2-19, "Impact Study of Plastic Bead Paint Removal on Corrosion Prevention and Control, "The Boeing Company, Boeing Aerospace, Seattle, WA.

7. Project No. 00-143, "Paint Stripping of F-4 Aircraft and Component Parts Using Mechanical Methods, "ALC/MABEB Ogden, UT.

8. Kelley, Stephen, "Methods for Mechanically Removing Paint from Aircraft Structures," Robotic Solutions in Aerospace Manufacturing Roboics International/S.M.E., Orlando, FL, March, 1986.

9. N00019-83-G-0049 IMP Contract No., Project WBS-235, "Automated Painting and Stripping Project, Sixth Quarterly Report," Department of the Navy, prepared by Grumman Aerospace Corp. (Confidential).

10. 00-143 Stage 2 PRAM Project, Roberts, R.A., "Mechanical Paint Removal Process," Interim report on stripping paint from the first F-4E prototype at Hill AFB, UT, July, 1984, and May, 1985.

11. AFWAL-TR-85-4138, Childers, Sidney, et al., "Evaluation of the Effects of a Plastic Bead Paint Removal Process on Properties of Aircraft Structural Materials," December, 1985.

12. Bullington, J.B., Williams, D.R., "Organic Coating Removal via Multiple Plastic Media Blast Cycles on Clad Aluminum Airframe Skins," Corpus Christi Army Depot Chemical Branch, Engineering Branch, Corpus Christi, TX.

13. A.21 Process Standard, "Plastic Media Blast Cleaning and Paint Removal," Training Guide for Corpus Christi Army Depot.

14. Manufacturing Operating Instruction 8-1951985937, "Plastic Media Blast Cleaning for CH-47D Modification Program," July, 1985.

15. 8516 8-427, "Flourescent Penetrant Detection of Fatique Cracks After Plastic Bead Paint Removal," Boeing Vertol, Philadelphia, PA, December 1985.

16. MDSR 330036-1, "PMB Stripping of Aircraft," Manufacturing Development, Boeing Vertol Company, Philadelphia, PA, May 1986

Table 1. Formulated Strippers

COMPONENTS	MIL-P-23377	MIL-C-81352	BMS10-11 Tp1 & 2	MIL-E-15090	COMMENTS
Acetic acid (Glacial)	4	4	4	2	Slower than formic acid.
Hydrochloric Acid	1	1	1		
DiBasic Esters (DBE)			0		
N-methyl pyrrolidone (NMP)			2		
Dimethylsulfoxide (DMSO)/NMP	2	2			Detergent helps with wetting, speeds up effects of stripping.
Ethyl 3-ethoxypropionate (EEP)	3				Thickened with Knox gelatin, it still works. Used mostly as an additive in the commercial mixtures.
50%Acetic Acid/50% Triacitin	1				Paint removed in vapor phase. Percents varied with little change. The less A.A., the slower it works.
Acetic Acid 50%/ MOK 50%	3				
50% DBE/50% NMP			1		
ETHANOL/ KOH	1	0	0		pH > 14

0 = no effect
5 = immediate effect

Table 2. Coatings and Commercial Products Tested

NAME	PRIMARY COMPONENTS	MIL-P-23377	MIL-C-83286	MIL-P-23377/ MIL-C-83286	BMS10-11Tp1	BMS10-11 Tp2	BMS 10-11 Tp 1&2	MIL-P-85582	MIL-E-15090	BMS-10-101 (MIL-C-27725)	TT-P-645	MIL-L-8937	MIL-L-81352	BMS 5-95G	TT-V-119	COMMENTS
Brulin EXP 888	Diethylene glycol; n-butyl ether; N-methyl pyrrolidone	0	/	/	/	/	/	/	/	/	/	/	/	/	/	
Brulin EXP 2187	Diethylene glycol n-butyl ether; N-methyl pyrrolidone (elevated temp.)	2	/	/	/	/	/	/	/	/	/	/	/	/	/	
Brulin EXP 2187 mod	Diethylene glycol n-butyl ether; N-methyl pyrrolidone (elevated temp.)	4	4	/	3	3	/	4	3	1	4	5	5	2	5	Stronger version of EXP 2187.
Brulin Safety Strip 1000	Ethyl 3-Ethoxypropionate; 2-Butoxyethanol	1	/	/	/	/	/	/	/	/	/	/	/	/	/	
DuPont B60988	Dibasic esters; N-methyl pyrrolidone; Exxon Aromatic 200	3	/	/	/	/	/	/	/	/	/	/	2	/	/	
IRCL J-7040	Dimethyl glutarate; N-methyl pyrrolidone; Aromatic petroleum solvent; Dodecyl benzene; Dimethyl formamide; Sulfonic acid; Formic acid	4	/	0	/	/	1	/	/	/	/	/	4	/	/	Too slow on topcoats.
ManGill LP4566 thin	Formic acid; Dibutylthiourea	4	4	/	3	4	/	5	2	0	3	/	3	/	3	
ManGill LP4566 thick	Formic acid; Dibutylthiourea	4	4	/	3	4	/	5	3	0	3	/	3	3	2	
ManGill CF20	Propionic acid	4	/	/	/	4	2	/	/	/	4	/	/	/	/	99% acid.
Oakite 108-PT-14	Aromatic hydrocarbons; Formic acid; Dodecylbenzenesulfonic acid; Nonylphenoxy polyethoxy ethanol	3	/	/	/	/	/	/	/	/	/	/	4	/	/	Difficult to work with.
Oakite 108-PA-126/8035PD	Aromatic hydrocarbons; Formic acid; Dodecylbenzenesulfonic acid; Nonylphenoxy polyethoxy ethanol	3	/	/	/	/	/	/	/	/	/	/	/	/	/	
Oakite 108-NF-2	Aromatic hydrocarbons; Benzyl phenoxy polyethoxy ethanol; Monoethanolamine; Diethylene glycol methyl ether; Tetrahydro furfuryl alcohol; Tributyl phosphate	2	/	/	/	/	/	/	/	/	/	/	/	/	/	PH too high.
Oakite Flexsolve	Ethylene glycol phenyl ether; Sodium dodecylbenzenesulfonate; Sodium hydroxide; Dipropylene glycol methyl ether	1	/	/	/	/	/	/	/	/	/	/	/	/	/	
Turco EXP789	N-methyl pyrrolidone; OXO-Decyl Acetate (elevated temp.)	4	/	4	2	/	2	/	/	0	3	/	4	/	5	Removes topcoat but not primer.
Turco 6088A thin	Hydroxyacetic acid	4	/	/	/	/	/	/	2	/	/	/	3	/	3	
Turco 6088A thick	Hydroxyacetic acid	4	/	/	/	/	/	5	3	0	3	/	3	/	/	
Westcoast Paper SR3000	Cyclic amide; Diethylene glycol-ether blend	0	/	/	/	/	/	/	/	/	/	/	/	/	/	

Rating:
0 = No effect to 5 = Immediately effective.
/ = Not tested.

Table 3. Media Characteristics and Blasting Parameters

MEDIA CHARACTERISTICS

Hardness	3.0 moh
Size	12-30, 30-50, and 50+mesh
Specific gravity	1.45
Chemical characteristics	crystalline wheat starch

BLASTING PARAMETERS

Pressure	24-45 psi
Flow rate	6-10 pounds per minute
Angle	dependent on coating and substrate
Nozzle size	3/8 inch

Table 4. Coatings Removed

Coating	Description
BMS 10-11 Tp1	Epoxy primer
BMS 10-11 Tp 11	Epoxy topcoat
BMS 10-20	Integral fuel tank coating
BMS 10-60	Protective enamel
BMS 10-79	Urethane compatible primer
BMS 10-21	Anti-static coating
BMS 5-95	Sealant
BMS 10-86	Teflon-filled coating
BMS 10-101	Urethane for integral fuel tank
BMS 5-89	Bonding primer
Mil-P-23377	Epoxy primer
Mil-C-83286	Polyurethane topcoat
Mil-C-27725	Corrosion preventive coating
Mil-C-22750	Epoxy polyamide
Mil-C-24441	Epoxy polyamide
Mil-P-85582	Water-based epoxy primer
Mil-C-85285	High solids topcoat
Mil-P-53030	Lead & chromate free water reducible epoxy primer
Mil-C-53039	Aliphatic polyurethane, chemical resistant
TT-P-1757	Zinc chromate primer
TT-E-489	Alkyd enamel
AMS 3138	Rain erosion resistant coating

Table 5. Blasting Effects

	Substrate Material	Damage	Comments	Corres. Figure
Substrate effects Metals	Aluminum and aluminum alloys	no effect	pending fatigue data, see figures	19, 20
	Ferrous Alloys	no effect		
	Chrome plate	no effect		
	Nickel plate	no effect		
	Cadmium plate	no effect		
Substrate effects Composites	Fiberglass	minimal or no effect when done properly	Possibly less damaging than current method, see figures	13 to 16
	Epoxy/E-Glass	minimal or no effect when done properly		17, 18
	Graphite	no effect		1, 2
	Kevlar	minimal damage	fibers exposed, very delicate operation	
	Cyanate Ester	minimal or no effect when done properly		2 to 12

Figure 1. Graphite Surface Before Depainting with Envirostrip

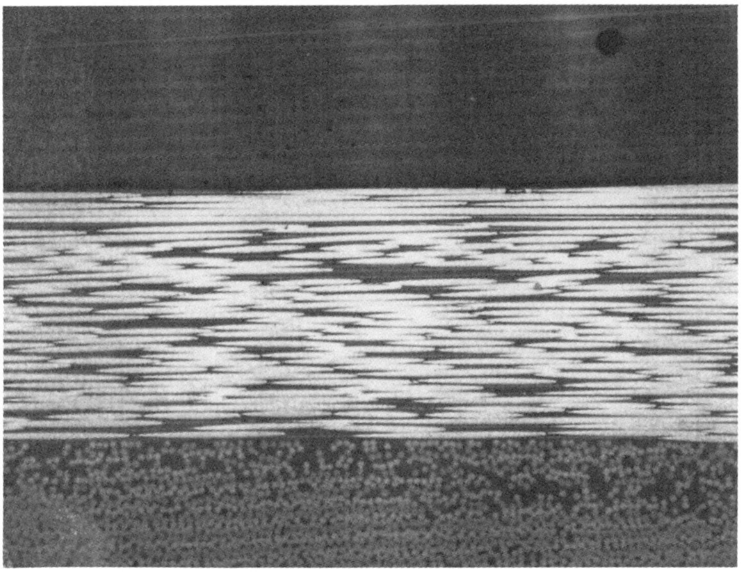

Figure 2. Graphite Composite After Depainting with Envirostrip
(No Noticeable Fiber Damage or Delamination)

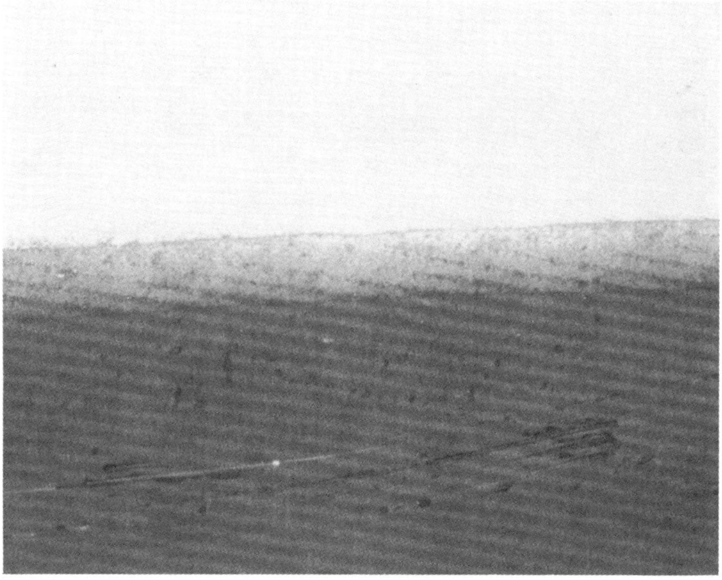

Figure 3. Cyanate Ester Panel Before Depainting

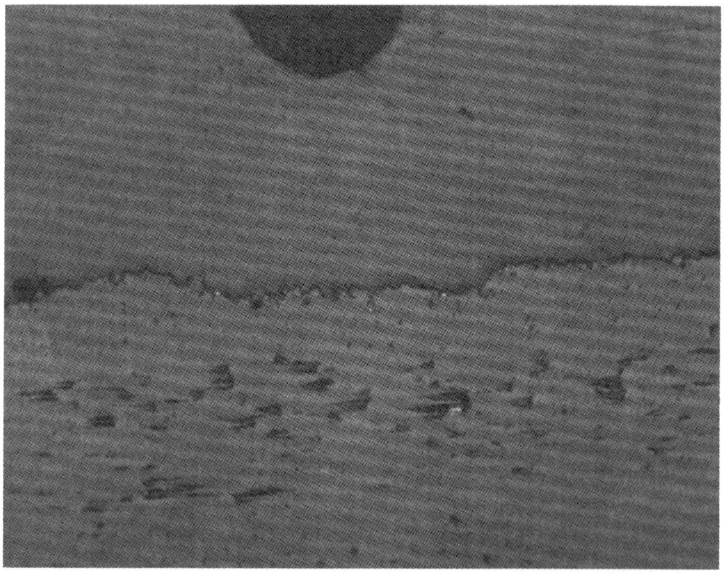

Figure 4. Cyanate Ester Panel After Depainting with Plastic Media (Type V)

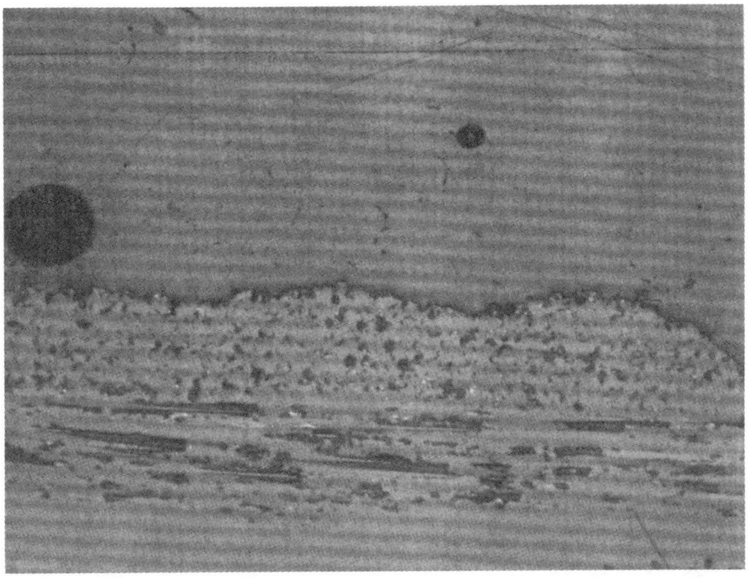

Figure 5. Cyanate Ester Panel After Depainting with Plastic Media (Type V)

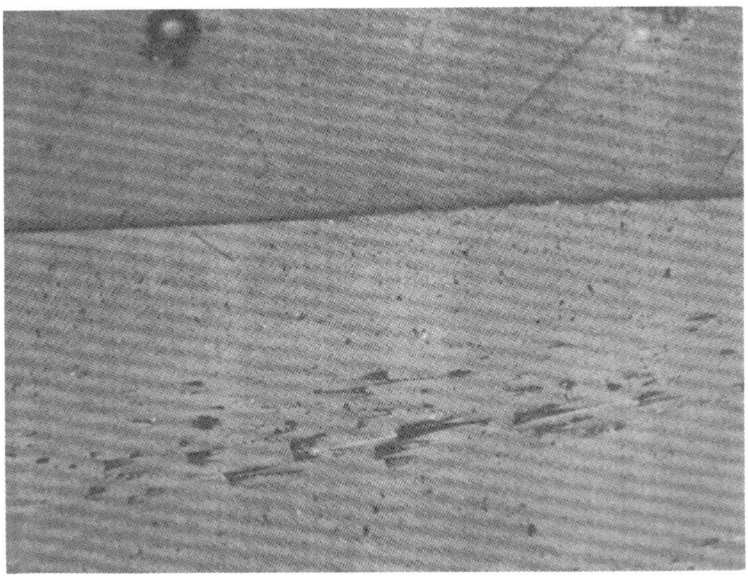

Figure 6. Cyanate Ester Depainted Using High Pressure Water at 10-15K psi

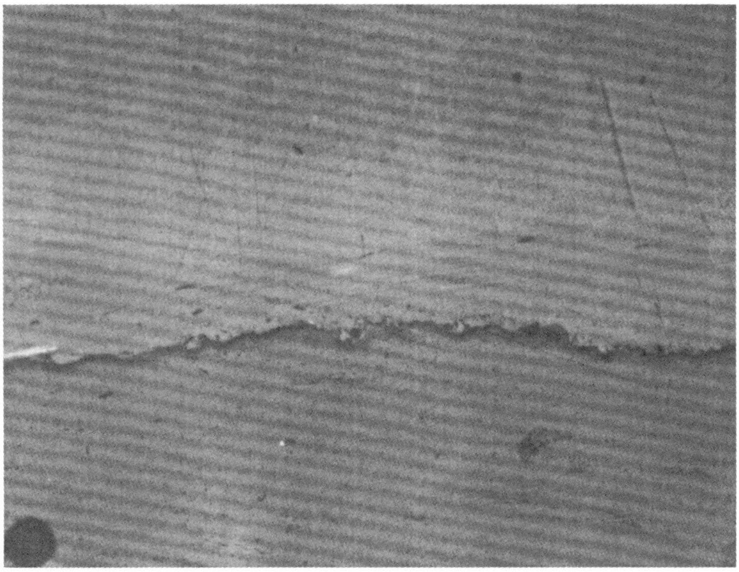

Figure 7. Cyanate Ester Depainted Using High Pressure Water at 18K psi

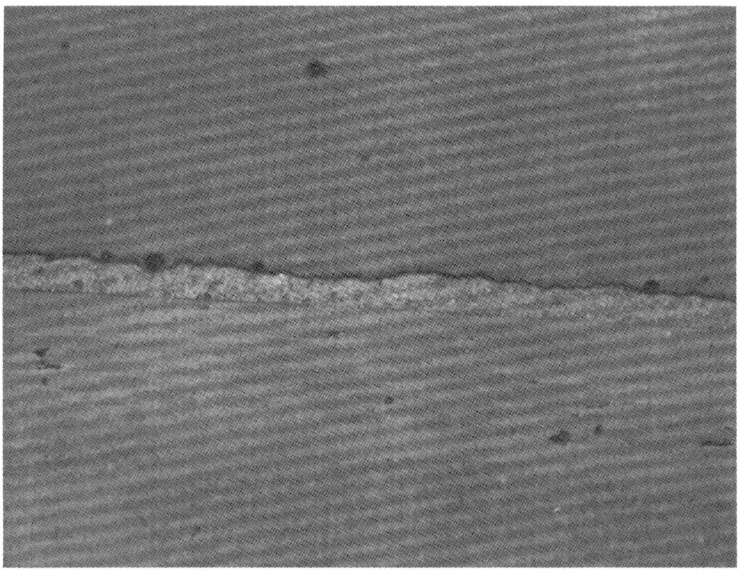

Figure 8. Cyanate Ester Stripped to Primer Only with Envirostrip

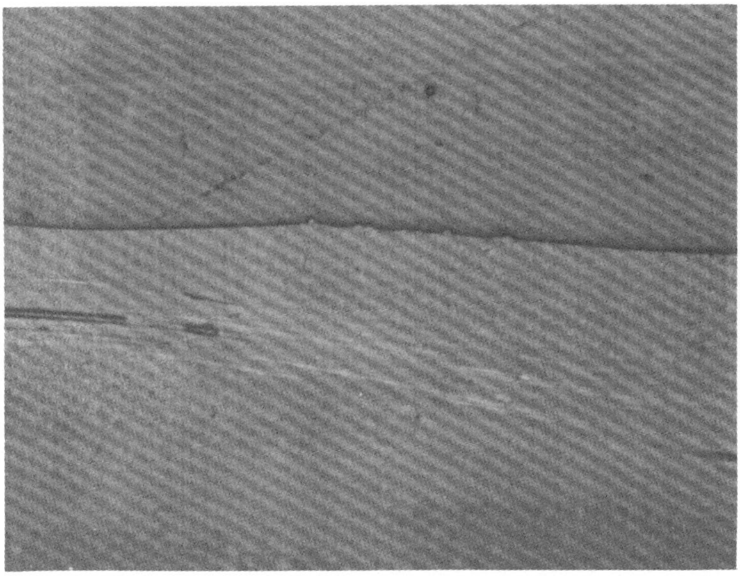

Figure 9. Cyanate Ester Depainted with Envirostrip

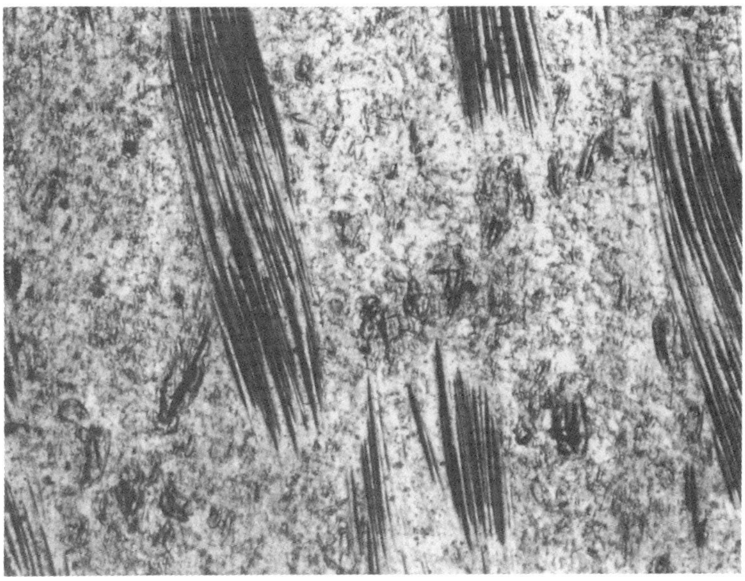

Figure 10. Cyanate Ester Panel Top View After Depainting with Envirostrip

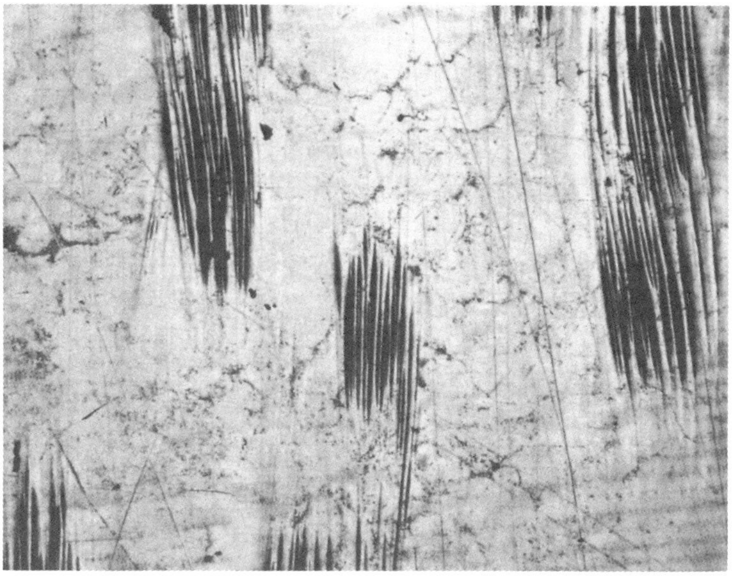

Figure 11. Cyanate Ester Panel After Depainting with High Pressure Water

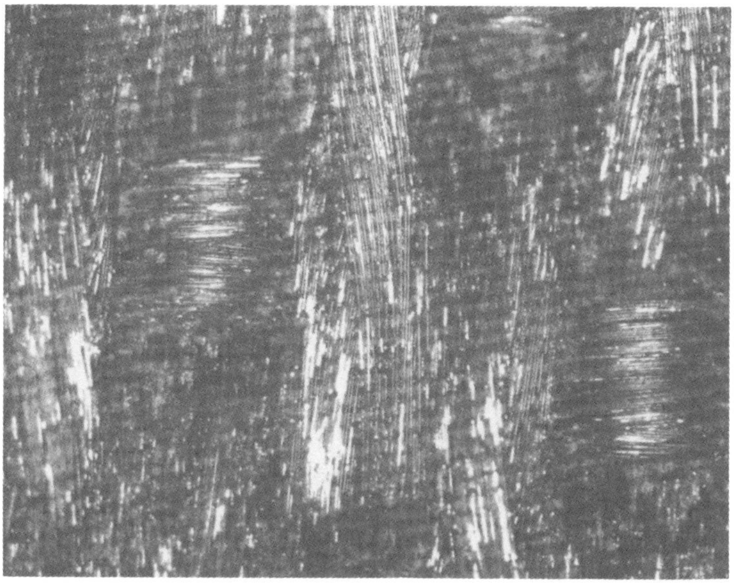

Figure 12. Cyanate Ester Panel After Depainting with Type V Plastic Media (Note: All of the fibers are exposed. Plastic media removes the resin.)

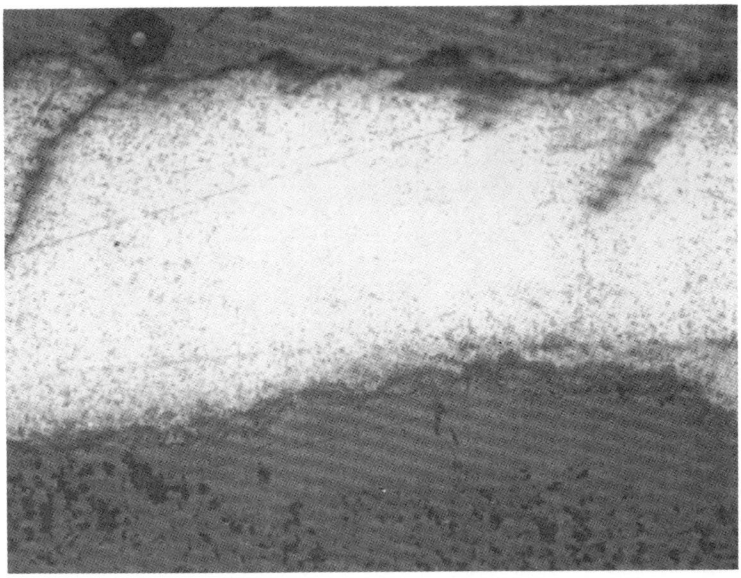

Figure 13. Fiberglass--Piece of Scrapped AWACS Rotodome Before Depainting

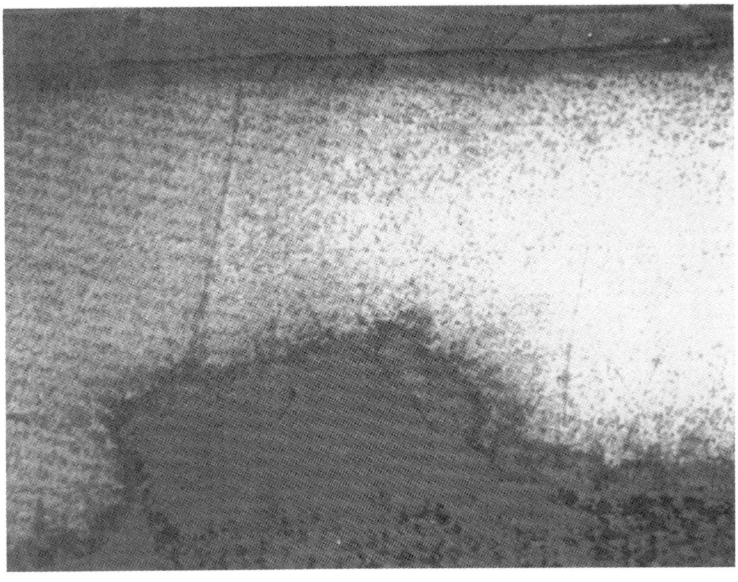

Figure 14. Fiberglass--Piece of Scrapped AWACS Rotodome Before Depainting

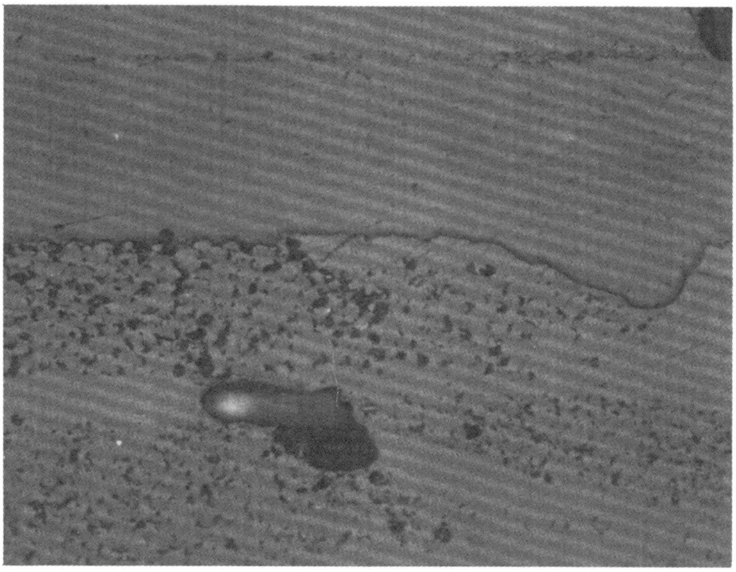

Figure 15. Fiberglass--AWACS Rotodome After Depainting with Envirostrip

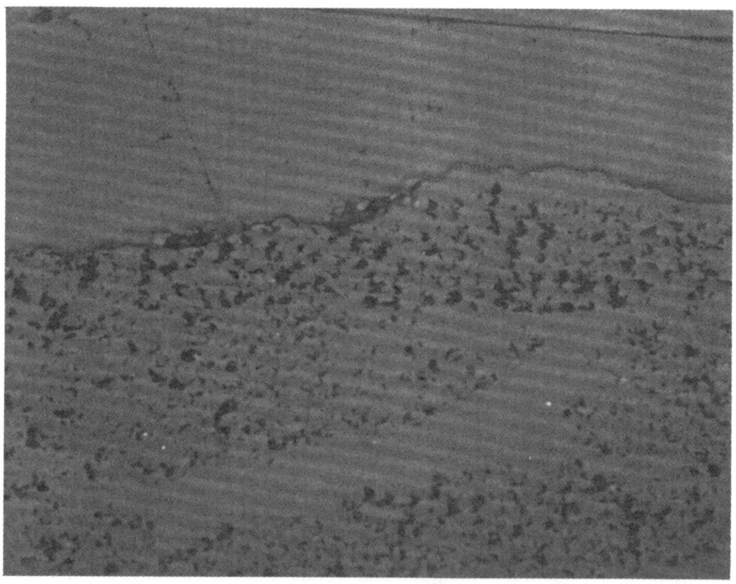

Figure 16. Fiberglass-AWACS Rotodome After Depainting with Envirostrip (Impossible to determine if damage was caused by the Envirostrip depainting or the previous rework.)

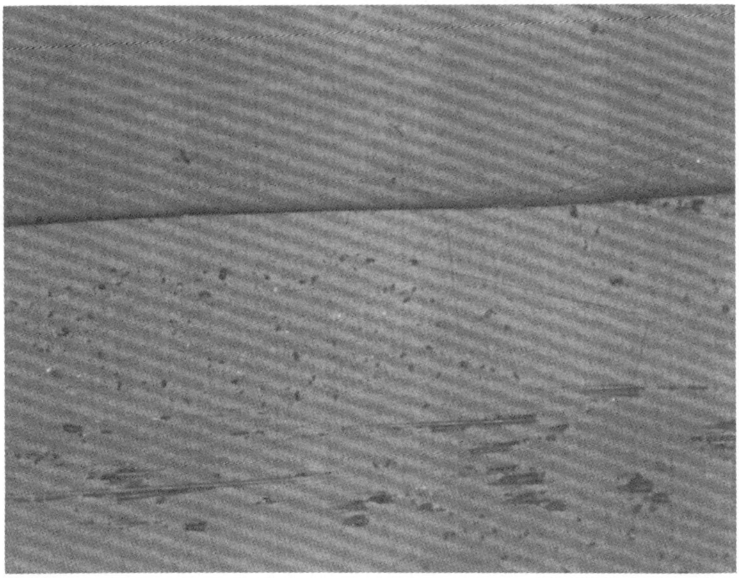

Figure 17. Epoxy E-Glass After Depainting with Wheat Starch

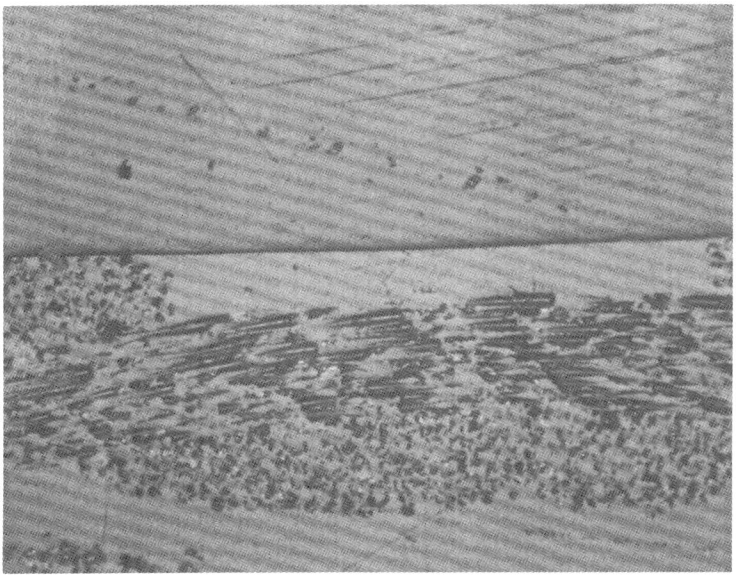

Figure 18. Epoxy E-Glass After Depainting with Wheat Starch

Figure 19. 2024 Aluminum Panel Before and After Depainting with Envirostrip (This photomicrograph is representative of all the aluminum photos. No damage is visible.)

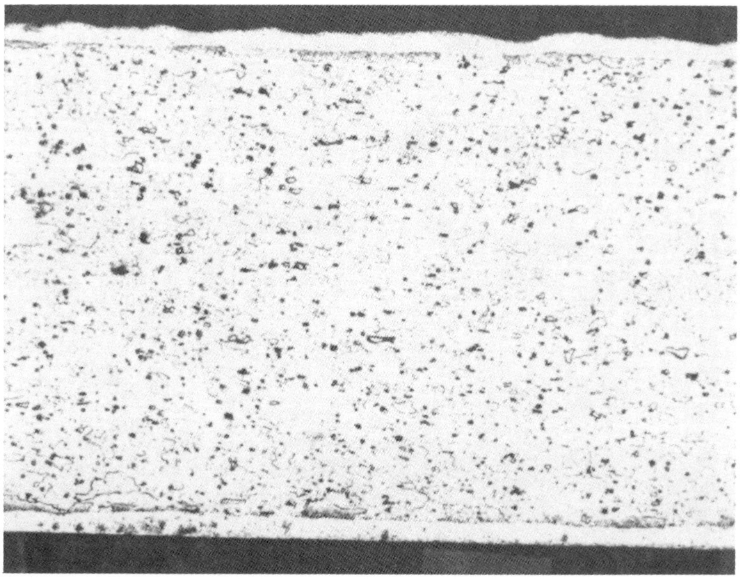

Figure 20. .020 Inch Clad Aluminum Panel (150X) Upper Surface Depainted with Envirostrip (Very limited damage testing was performed on aluminum clad; therefore, quantitative data is not reported.)

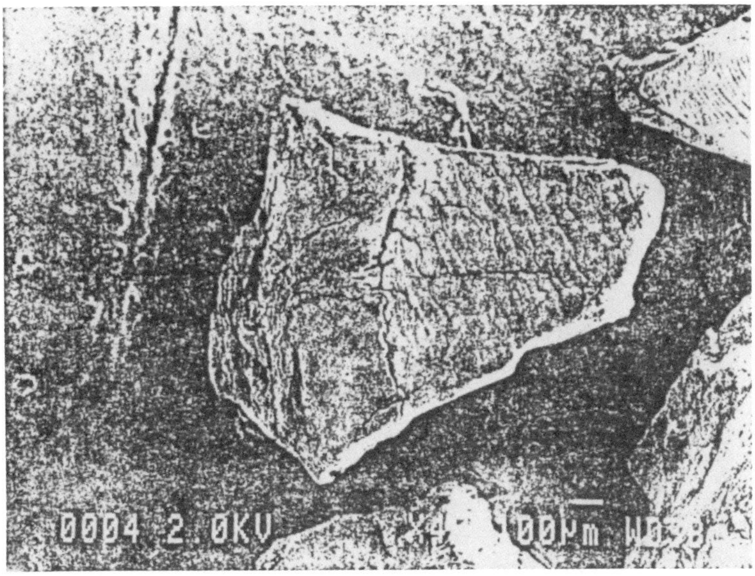

Figure 21. New Envirostrip Media

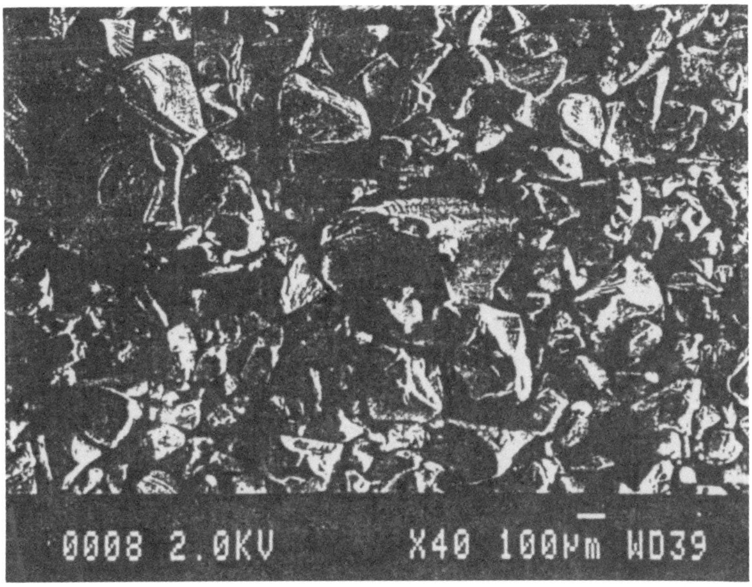

Figure 22. Used Envirostrip Media
(Note the particles have the same sharp angles as the new media.)

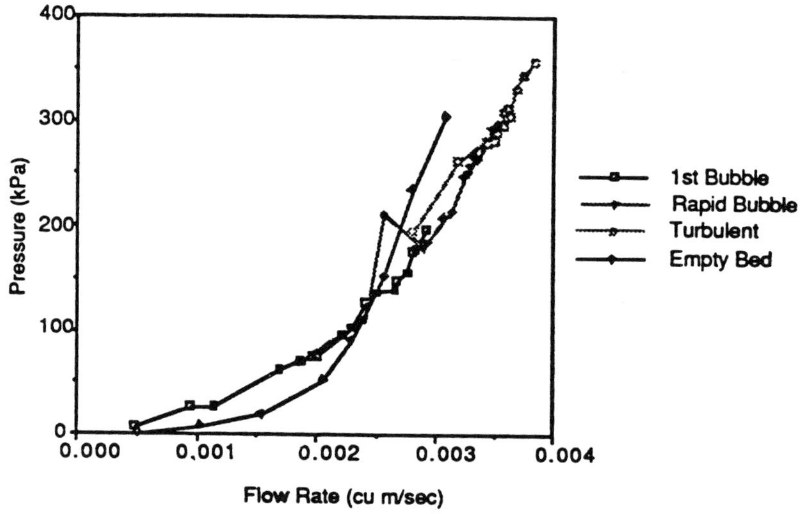

Figure 23. Flow Rate vs Pressure of Empty Bed at Bed Height 20 cm

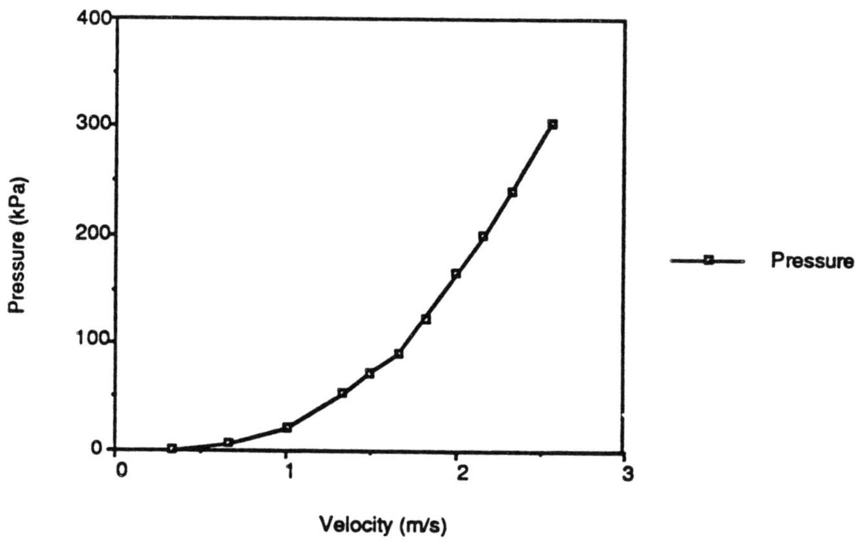

Figure 24. Velocity vs Pressure at 20 cm Bed Height

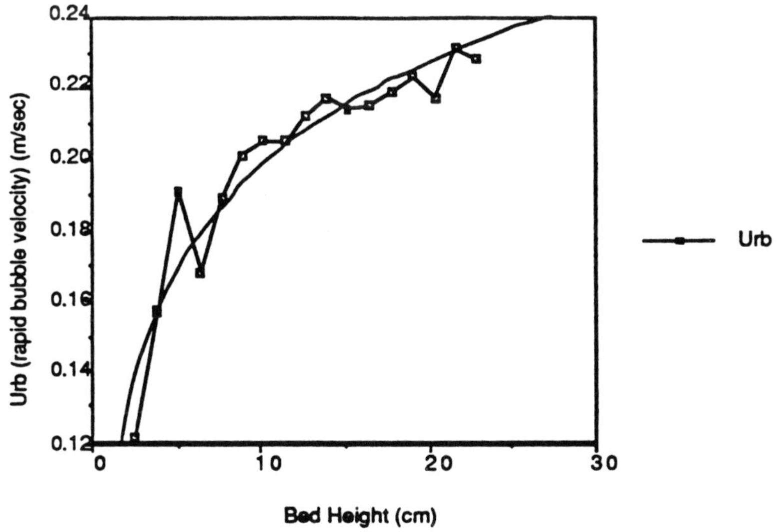

Figure 25. Bed Height vs Rapid Bubble Velocity

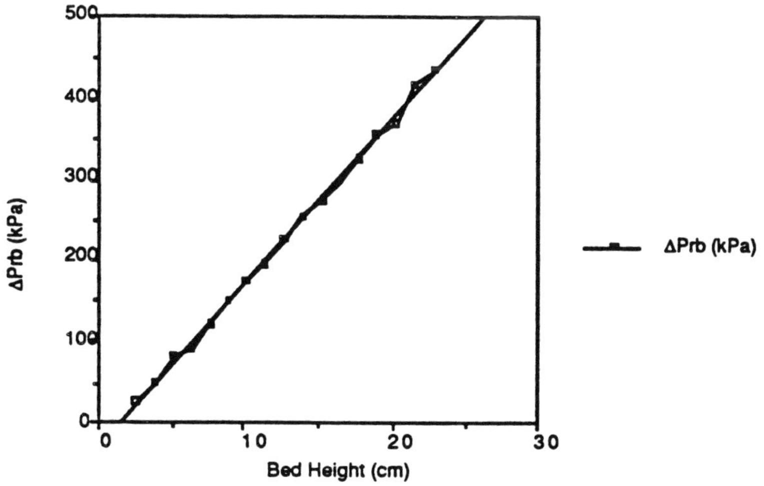

Figure 26. Bed Height v. ΔPrb (rapid bubble)

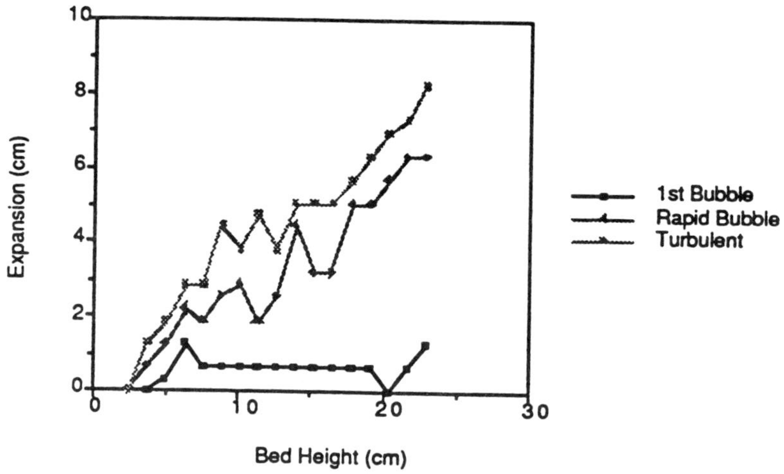

Figure 27. Bed Height vs Expansion

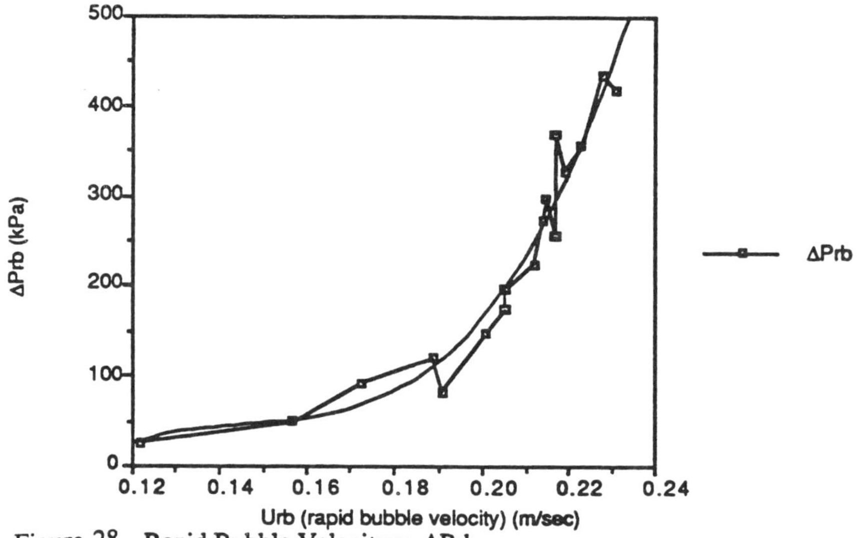

Figure 28. **Rapid Bubble Velocity v. ΔPrb**

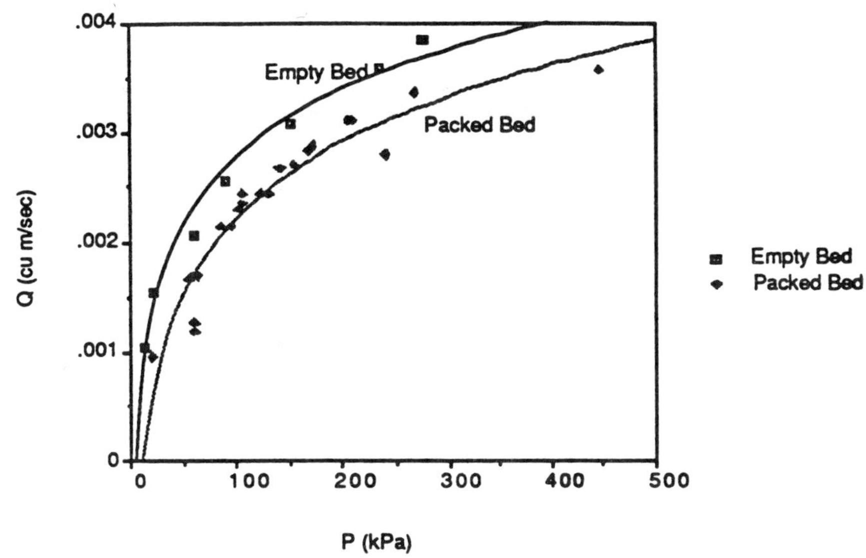

Figure 29. Pressure vs Flow

PRECISION PARTS CLEANING WITH SUPERCRITICAL CARBON DIOXIDE

Paula M. Gallagher and Val J. Krukonis
Phasex Corporation
Lawrence, Massachusetts

INTRODUCTION

Over the past two decades, supercritical fluids have been utilized as solvents for accomplishing separations of materials as diverse as foods, polymers, pharmaceuticals, petrochemicals, natural products, and explosives. More recently they have been used for non-extractive applications, such as recrystallization, deposition, impregnation, and surface modification. Today, supercritical fluid extraction is being practiced in the foods and beverage industries; there are commercial plants for decaffeinating coffee and tea, extracting beer flavoring agents from hops, and separating essential oils and oleoresins from spices. The use of supercritical fluids, especially carbon dioxide, for cleaning metal, ceramic, or composite parts is an almost natural outgrowth of these previous developments.

With the recent scrutiny of chlorofluorocarbons (CFCs) and the new regulations concerning the phaseout of these and other ozone-depleting chemicals, the search for alternative solvents and technologies is becoming widespread. Phasex Corporation has been developing a process called DriClean™ in which carbon dioxide is used as a CFC replacement for cleaning intricate parts such as gyroscopes, laser optics components, accelerometers, nuclear valve seals, and thermal switches. Carbon dioxide has the advantage of environmental acceptability, is non-flammable, non-corrosive, and is "worker friendly." Additionally, carbon dioxide has no ozone-depletion potential, and while it does have some global warming potential, its use in cleaning operations would contribute insignificantly to global warming in comparison with, for example, automobile emissions.

This paper will describe the application of carbon dioxide to the cleaning of precision parts, specifically on the removal of organic-based contaminants from these parts; the contaminants are actually dissolved by the carbon dioxide and not simply dislodged by flow or abrasion. Included will be background information on supercritical fluids and their behavior as solvents, as well as preliminary data which demonstrates the effectiveness of the DriClean™ process.

BACKGROUND

The "simple" supercritical fluids, such as carbon dioxide and the light hydrocarbons, are typically gases at room temperature and pressure. Above their respective critical points (carbon dioxide, for example, has a critical temperature of 31°C and a critical pressure of 1072 psi), these fluids can be used as solvents; even liquid carbon dioxide has the ability to dissolve some materials. SCFs have high density properties and attractive transport properties such as gas-like viscosities and diffusivities, which render these fluids capable of penetrating very small pores and interstices of complex parts. Perhaps the most unique property of an SCF is that its "dissolving power" is pressure-dependent, such that, simply put, at higher pressures more material will dissolve in the SCF than at lower pressures.

The phenomenon of supercritical fluid solubility was first reported over 100 years ago by Hannay and Hogarth. As an example of the pressure-dependent dissolving power of a supercritical fluid solvent, Figure 1 shows the solubility of a simple solid, naphthalene, in carbon dioxide and ethylene. Above the critical pressure of each gas, it is clear that relatively small increases in pressure result in large increases in the solubility of naphthalene. The directed line between Points 1 and 2 shows the very large change in solubility that results when the pressure of a saturated solution at 200 atm is lowered to, for example, 100 atm. Because of such dissolving characteristics, it is possible to design a process to extract, purify, or fractionate materials based on changes in pressure of a supercritical fluid solvent. At high pressure an extraction (i.e., a dissolution) can be carried out, and by lowering the pressure, a separation of the dissolved material can be made to occur. The solute-free solvent can then be recycled to the extractor. The process can be carried out in either a batch or a continuous mode, depending upon the nature of the feed and the nature of the extraction, i.e., whether it be a purification (or topping), fractionation, or extraction from a reaction mass. Most of the work that is being carried out industrially usually involves an extraction of one material from a mixture. The operation of an extraction process will be described.

A schematic diagram of a process which uses a supercritical fluid as a pressure-dependent solvent to extract an organic substance is given in Figure 2a. Four basic elements of the process are shown, viz., an extraction vessel, a pressure reduction valve, a separator for collecting the material dissolved in the extractor, and a compressor for recompressing and recycling fluid. (Ancillary pumps, valving, facilities for fluid makeup, heat exchange equipment are omitted from the figure for clarity and ease of presentation.) Figure 2b shows extensive data on the solubility of naphthalene in carbon dioxide as a function of temperature and pressure. Reference to Figures 2a and 2b is made in explaining how a supercritical fluid process operates. Some process operating parameters are indicated on two solubility isobars in Figure 2b: Point 1 represents conditions in the extractor, e.g., 300 atm, 55°C, and Point 2 the conditions which exist in the separator, 90 atm, 32°C. The extractor vessel is assumed to be filled with naphthalene in admixture with some other material, which for the purpose of this discussion is assumed to be insoluble in carbon dioxide. Gas at condition 1 is passed through the extraction vessel, wherein it dissolves (and extracts) the naphthalene from the insoluble material. Leaving the extractor, the carbon dioxide-naphthalene solution is expanded to 90 atm through the pressure reduction valve as indicated by the directed path in Figure 2b. During the pressure reduction step, naphthalene precipitates from the solution, because as Figures 1 and 2b show, the dissolving power of carbon dioxide is low at low pressure. The precipitated naphthalene is collected in the separator, and the carbon dioxide leaving the separator is recompressed and returned to the extractor. This recycle process continues until all the naphthalene is dissolved and extracted, the directed line segment 1-2 in Figure 2b and its reverse on the solubility diagram representing approximately the cyclic process. (The other directed lines, e.g., 1-3, 4-5, etc., designate other extraction/separation paths; they show that isobaric conditions can also be used to separate a material and a case-by-case evaluation will dictate the appropriate operation of any specific process.)

Many materials including other SCF soluble solids, polymers, and oils will exhibit the general behavior shown in Figures 1 and 2b, and thus the principles involved in the extraction process described by these figures are directly applicable to precision parts cleaning. The next section describes the results of prior solubility studies on which the effectiveness of the DriClean™ process is based.

PREVIOUS RESULTS

There has been a considerable amount of research devoted to the study of the behavior of oils and polymers in various supercritical fluids (SCFs). Previous studies by Phasex Corporation have demonstrated that SC carbon dioxide is an excellent solvent for oils such as hydrocarbons, esters, silicones, perfluoropolyethers, halocarbon-substituted triazines, and organosilicones with various reactive functionalities; many of these oils are associated with the manufacture of precision components such as gyroscopes and accelerometers. The ability to dissolve a particular oil or polymer at any given pressure will greatly depend on the molecular weight and structure of the material. The implications of these characteristics are dramatic, particularly for polymers which are inherently composed of a range of molecular weight chains that give rise to a wide polydispersity. For example, by conducting a gradually increasing pressure profile, a polymer can be fractionated into increasingly higher molecular weight fractions, each fraction having a much narrower polydispersity than the parent polymer. Likewise, for both synthetic and natural oils, which generally contain chains of varying molecular weight, separation based on chain length has been demonstrated. The work described subsequently was carried out at the Phasex laboratories; much of it was reported nearly six years ago. Since that time, there has been a significant amount of investigations involving solubility studies of various other oils and polymers in SCF's. Much of this prior research was the basis for subsequent gyroscope cleaning work done for Draper Laboratories, Naval Avionics Center, and Honeywell Avionics.

As an example of the separation properties of supercritical fluid solvents, a high molecular weight silicone oil (M_w = 90,000) was fractionated with supercritical carbon dioxide. This oil was analyzed by size exclusion (gel permeation) chromatography to determine the molecular weights of the fractions. Table I gives the molecular weight values (both M_n and M_w) of the parent silicone oil and of the supercritical fluid fractions. Carbon dioxide over a pressure range of 5500 to 1800 psi at a temperature of 80°C was used to carry out this fractionation. The data in the table show that supercritical fluid fractionation can produce cuts with very narrow molecular weight range and of course, they show that the silicone was completely dissolved.

As is frequently found with synthetic oils, this particular silicone (which was a commercial 8000 centistoke oil) contained a small fraction of very low molecular weight material which could be isolated easily; analogously, the data in the table show that a small fraction of very high molecular weight material of about 150,000 (M_w) was also present in the silicone oil. Figure 3a shows the gel permeation chromatogram (GPC) of the parent silicone oil; the small peak of low molecular weight silicone is evident. Figure 3b reproduces the parent GPC and onto which are superposed the GPC's of the highest and the lowest molecular weight fractions. Thus, from an examination of the GPC traces shown in Figure 3b, it is seen that the dissolving power of carbon dioxide can be adjusted to selectively extract the "lows"; additionally, it is interesting to point out here that supercritical carbon dioxide can dissolve a silicone polymer of 150,000 molecular weight, as the data for Fraction 6 in Table I and the superposed trace in Figure 3b show.

As another especially pertinent example, a fractionation was made on a commercially-available, high molecular weight perfluoroalkyl-polyether oil which is being used in specialty lubricant and barrier fluid applications; the oil had previously undergone processing in three sequential molecular distillation steps to obtain as narrow a molecular weight range and to remove as many lower molecular weight species as possible. Using supercritical carbon dioxide at a temperature of 80°C and over a decreasing pressure range of 4000 psi to 1500 psi, five fractions (of an approximately equal weight) were obtained from this parent perfluoroether oil. The vapor pressure of the perfluoroether oil was too low to permit characterization of the fractions to be made by gas chromatography, and the viscosity of each fraction was measured as a means of indicating molecular weight.

The viscosity-temperature curve of each fraction and of the parent oil are shown in Figure 4. The room temperature viscosities of the fractions range from 800 cps to 9000 cps; the viscosity of the parent oil is 2000 cps. Published data give the molecular weight (M_w) of the perfluoroether at about 5000; if an eight-tenths power relationship between viscosity and molecular weight is assumed for the oil, the average molecular weight of the fractions is calculated to range from about 1600 to 13,000. Thus, although the particular commercial perfluoroether oil has been molecularly distilled in three steps, the data in Figure 5 show that supercritical fluid extraction can separate the oil into still narrower fractions and that there is still a large component of low molecular weight fluoroether that molecular distillation cannot "extract."

One final example of the solvent capabilities of supercritical fluids is given here. Although the use of supercritical ethylene is described, it has recently been demonstrated for Naval Avionics Center that supercritical CO_2 also has the ability to dissolve and fractionate chlorotrifluoroethylene.

Chlorotrifluoroethylene (CTFE) oligomer oils are currently being evaluated as flotation fluids in several applications. Narrow molecular weight ranges, or single oligomers, are desirable for these applications. Currently, only preparative GC or HPLC is satisfactory for achieving the narrow distributions desired. Supercritical fluids were able to process this chemical species also; the results of a fractionation test with CTFE are described. Supercritical ethylene at a temperature of 80°C over a pressure range of 4800 psi to 1900 psi was used to fractionate the oligomers of CTFE. Eight fractions were (arbitrarily) obtained, and GC-MS analysis of the fractions provided the information on the separations achieved. There were primarily four oligomers in the oil, with molecular weights ranging from about 700 to 1048; based upon the structure of the molecule and the measured molecular weight, it can be calculated that the mixture consists of species with six, -seven, -eight and -nine-mers, respectively. The composition of mers in each of the eight fractions obtained from the test and in the parent are given in Table II. (The fractions were not of equal weight, and the right hand column gives the weight contribution of each fraction.)

The composition data in Table II show that substantial changes in the ratio of mers in a fraction can be achieved by supercritical fluid fractionation; in Fractions 3 and 6, for example, the predominant mer represents about 90% of the fraction.

The examples detailed above are only a few of the many systems that have been studied.
The ability of supercritical fluids to dissolve many types of oils and organic materials, coupled with the ability to penetrate minuscule pores and interstices of metal, ceramic, and composite parts, suggest that these fluids could partially replace CFC. Although SCF's are tailored for the cleaning of intricate parts, they are by no means recommended as a general purpose CFC alternative.

While each potential cleaning application must be evaluated on a case-by-case basis, it is of value to point out here that there are some situations which are not amenable to cleaning with SC carbon dioxide. In general, contaminants which do not dissolve in carbon dioxide cannot be removed by this process; these contaminants include rust, scale, lint or dust, ionic species, metal salts, and many (but not all) fluxes. Removal of rosin flux residue, for example, is one of the more commonly encountered cleaning problems where replacements for CFCs are being sought. Rosin is mainly composed of isomers of abietic acid which, in itself, is soluble in carbon dioxide but which tends to polymerize during solder reflowing when heat is applied; the solder process renders the flux residue insoluble. Water soluble fluxes which typically contain polyglycols as carrier fluids may have the potential for removal by SC carbon dioxide; parts contaminated with this type of flux residue are being evaluated. Certain polymers, such as high density polyethylene and cross-linked polymers, (e.g. epoxies and phenolics), will not dissolve in carbon dioxide; this fact may be beneficial for those situations where it is desirable to leave the polymer untouched during cleaning procedures, however.

There are many misconceptions (perhaps, more accurately stated, exaggerations) being circulated through the precision parts cleaning community that, for example, "very high velocity" streams of CO_2 can shear off the particulates; this is not true except in some fortuitously placed large adherent particle on the outside surface of the gyroscope part. (Note CO_2 pellet blasting, one trade name being "Cold Blast", can remove only line of sight particulates; pellet blasting cannot remove the oils from interstices or pores.) This abrasive procedure may be suitable for stripping large surface area components, ranging from circuit boards to airplane wings, but it is important to distinguish this process from DriClean™ in which the contaminants to be removed are actually dissolved by the CO_2, not simply dislodged by flow.

PRELIMINARY CLEANING DATA

Much of the information regarding the effectiveness of the DriClean™ process was obtained through preliminary feasibility testing for several private companies that provided the parts to be cleaned. All of the pre- and post-cleaning analyses was conducted by the respective companies. A brief summary of the operation of the DriClean™ process as well as some of the results of the research is presented in this section.

Process Operation

In concept, precision parts cleaning is exactly analogous to the extraction procedure described in the background section. Parts to be cleaned are placed into a vessel; carbon dioxide, at some pressure and temperature, is passed over the parts to dissolve the oils; the parts, now free of contaminating oils, are removed from the vessel. The contaminant-laden stream of CO_2 leaving the vessel is directed to a second chamber in which the pressure is lowered and the contaminant oil that was dissolved in the CO_2 drops out of solution. If the gas is to recycled, it may be passed through a bed of activated carbon to remove residual contaminants and then recompressed. Typically, a batch-continuous mode of operation would be employed. Since carbon dioxide is a gas at one atmosphere, there is no danger of solvent residue remaining on the parts. The dissolving and transporting properties of carbon dioxide described earlier, plus the integrated industrial experience in large scale (e.g., 60,000,000 lbs/yr of decaffeinated coffee at one plant location) operations combine to point out that precision parts cleaning can be scaled to essentially any manufacturing level.

RESULTS

As an example of the capability of SC CO_2 to replace CFC-113, a perfluoropolyether (Krytox 143 AD) contaminated beryllium oxide part which typically requires 20 cycles of CFC-113 rinsing, was cleaned with supercritical fluid: one pass with CO_2 resulted in a degree of cleanliness much better than 20 rinses with CFC-113. Figure 6 shows the analytical results using FTIR. This work was carried out for Draper Laboratories. In late 1990 Phasex carried out tests for the Naval Avionics Center, Indianapolis, Indiana on the declassification of two Pulsed Integrating Pendulums (Mod B - Poseidon). Carbon dioxide extracted literally all the CTFE as determined by visual inspection. The instrumental analytical data was not available for this publication.

A series of cleaning tests was conducted for a

communications company; the samples were ceramic substrates contaminated with fingerprints. It was anticipated that the oils would be removed but that the ionics would not; the results confirmed these expectations. The samples were intentionally contaminated with a solution of fatty acids and salts that would approximate the composition of fingerprint residue. Prior to cleaning, surface analysis using XPS (x-ray photoelectron spectroscopy) was conducted. The low levels of ionic species on the original sample as indicated in Table 3 are strictly an artifact of the analytical technique; XPS samples only the top ~30Å and the ionics were essentially "hidden" beneath the fatty acid layer. Table 3 gives the results of the cleaning tests in comparison to using Freon TA (a mixture of CFC-113 and methanol) and peroxide. Supercritical CO_2 was able to clean the parts to a level comparable to that using Freon TA.

SUMMARY

It has been demonstrated that the DriClean™ process is indeed an effective replacement for CFCs in the cleaning of precision parts. The process can be tailored for many types of cleaning operations by modifying process conditions, and it can be employed as a complementary second-stage cleaning procedure. Although a case-by-case evaluation of each potential application must be conducted, there exists an extensive knowledge base from which to derive the information necessary to determine the potential for success. Issues such as ozone depletion and global warming potentials, flammability and toxicity, safety and health, process compatibility and ease of installability need to be addressed, as well as economic considerations, such as operating and capital costs, throughput, process flexibility, floor space requirements, and waste treatment/disposal/recycle/reclamation issues. However, for the particular case of gyroscopes, accelerometers, and other intricate parts that are contaminated with the oils enumerated above, carbon dioxide presents itself, at least at this period in time, as the only potential solvent to replace CFCs for the problem of removing every oil that is associated with machining, assembly, filling, or declassification of precsion parts.

Table I. Molecular Weight of Silicone Oil Fractions

Fraction	Molecular Weight M_n	M_w	WT % of Parent
Parent			
1	42,500	90,000	100%
2	428	789	4.0
3	3,310	11,500	5.1
4	27,100	53,200	27.7
5	43,000	57,500	27.9
6	58,900	91,500	28.3
	112,600	149,900	7.0

Table II. Composition of Chlorotrifluoroethylene Fractions

Fraction	Composition,				Wt% of Parent
	6	7	8	9	
Parent	4.5	49.5	39.5	6.5	100%
1	51.1	37.6	12.3	----	5.8
2	24.4	73.5	2.0	----	8.6
3	7.6	86.4	6.0	----	20.2
4	1.8	75.3	22.9	----	15.2
5	0.7	38.7	58.1	2.5	22.8
6	----	5.7	94.1	0.1	15.6
7	----	0.6	66.4	33.0	10.5
8	----	3.0	55.5	41.5	1.3

Table III. XPS Data

Sample	C	O	Al	Na	Cl	F	Ta	Cu	Si	S	N	Sn
Contaminated	88	10	0.2	0.2	-	0.2	-	-	-	1.3	-	-
Freon TA	45	30	14	4	3	0.7	0.1	0.4	0.3	-	0.2	.2
Peroxide	67	21	9	0.5	-	-	-	-	-	1	1	-
supercritical-CO_2	43	35	16	2	1	-	1.5	1	-	-	0.5	-

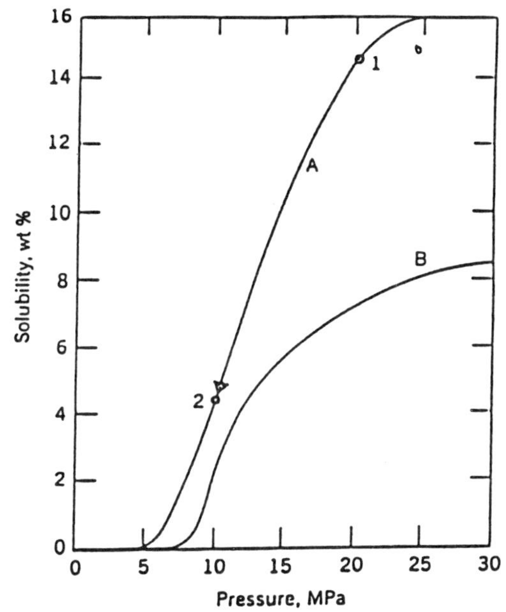

Figure 1. Solubility of Naphthalene. A, in Ethylene at 35°C, and B, CO_2 at 45°C. 1 and 2 are Explained in the Text (To Convert MPa to Bar, Multiply by 10).

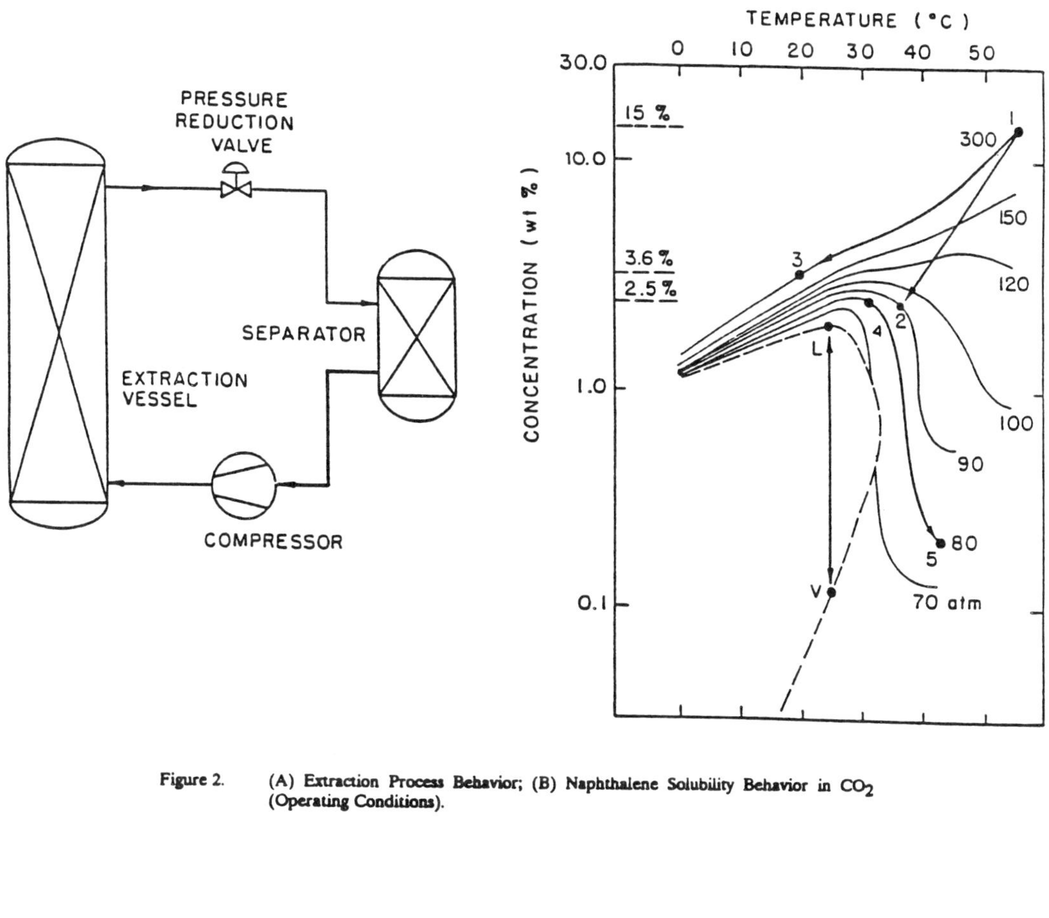

Figure 2. (A) Extraction Process Behavior; (B) Naphthalene Solubility Behavior in CO_2 (Operating Conditions).

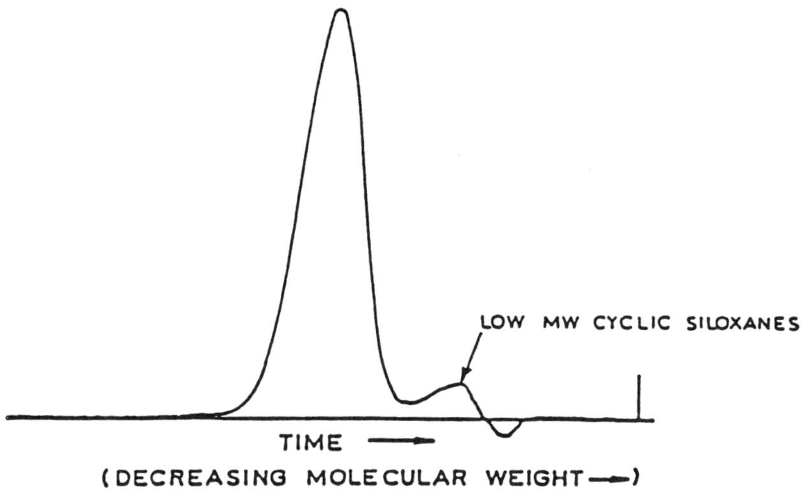

Figure 3a. GPC of Silicone Oil ($M_W = 90,000$).

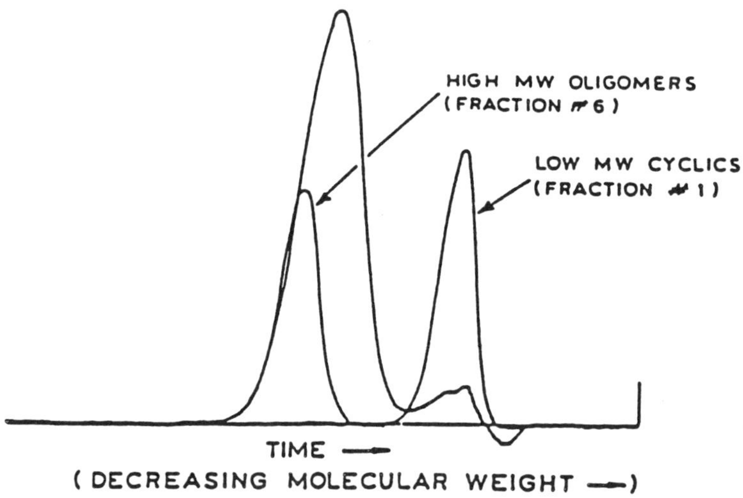

Figure 3b. GPC of Parent Oil and High and Low Molecular Weight Fractions.

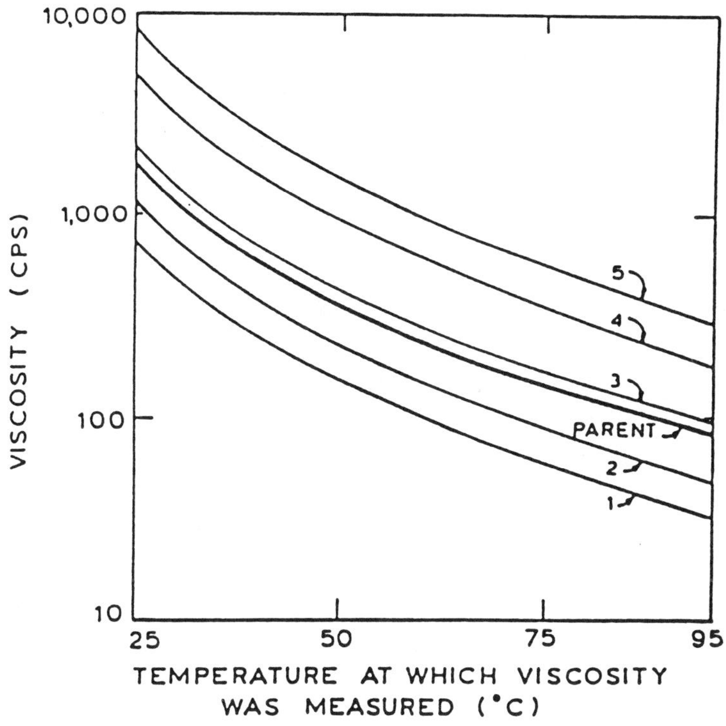

Figure 4. Viscosity of Perfluoroether and of Fractions Extracted with Supercritical CO_2.

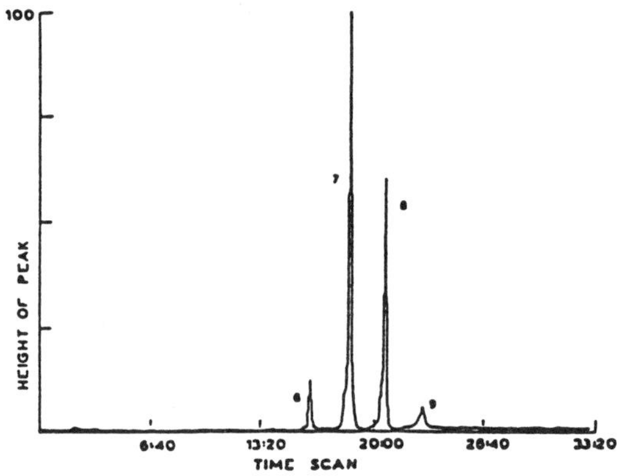

Figure 5a. GC-MS Trace of Parent Chlorotrifluoroethylene (Numbers by Peaks Indicate MERS).

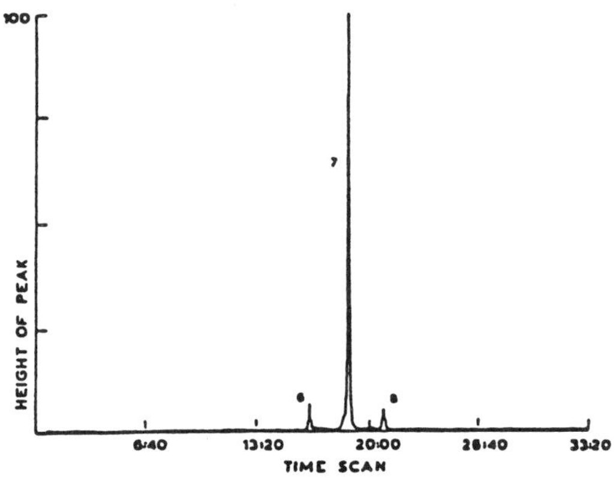

Figure 5b. GC-MS Trace of Fraction 3 of Chlorofluoroethylene (Numbers by Peaks Indicate MERS).

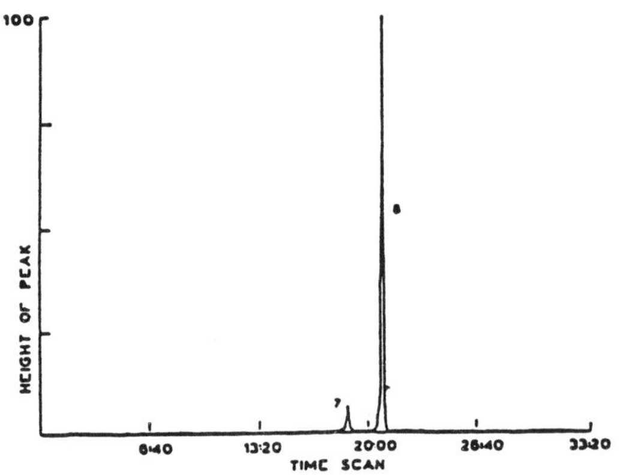

Figure 5c. GC-MS Trace of Fraction 6 of Chlorofluoroethylene (Numbers by Peaks Indicate MERS).

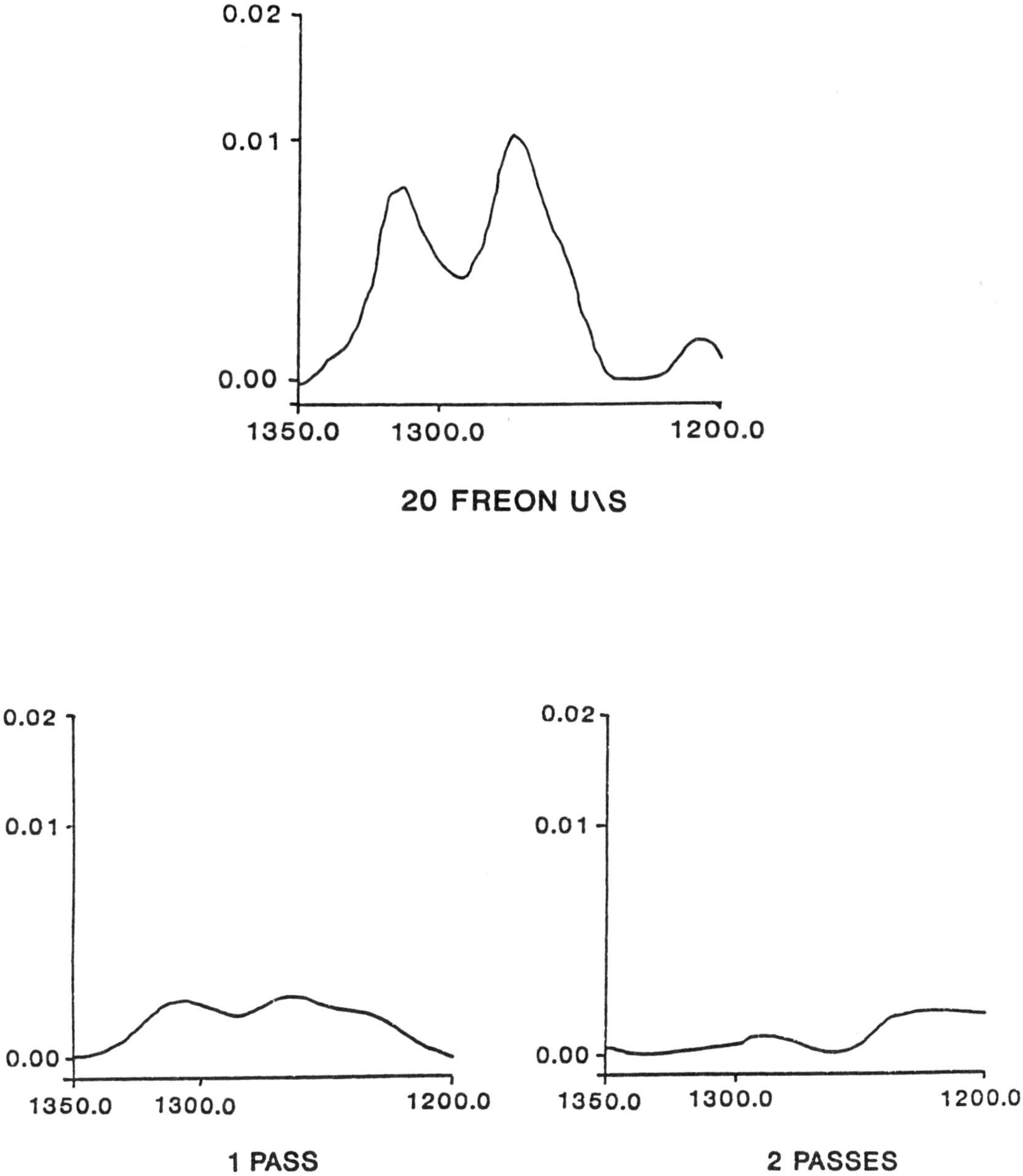

Figure 6. Krytox Removal Using (a) Freon-113, (b and c) SC CO_2.

CARBON DIOXIDE PELLET BLASTING PAINT REMOVAL FOR POTENTIAL APPLICATION ON WARNER ROBINS MANAGED AIR FORCE AIRCRAFT

Randall B. Ivey
Air Force Corrosion Program Office
Robins Air Force Base, Georgia

INTRODUCTION

WR-ALC has been striving toward the reduction or elimination of environmentally unacceptable waste produced from aircraft repair processes. One of the largest sources of hazardous waste comes from the aircraft paint removal operations. In this operation, the aircraft will go through several processes, all of which produce some form of waste product. This paper will discuss the investigation of the carbon dioxide pellet blasting (CDPB) process as a potential environmentally compliant replacement for current coatings removal technologies.

CHEMICAL PAINT STRIPPING DEFINED

Prior to chemical paint stripping, the aircraft is washed to remove oil, dirt, and grime which may prevent the chemical strippers from working properly. The wash and rinse water requires industrial sewage treatment because of the soils and, in many cases, the contents of the soap. The chemical strippers are then applied to the aircraft.

Chemical strippers have several ingredients which make them hazardous and dangerous to use and discard. Chemical strippers typically contain up to 50 percent methylene chloride solvent, which the Occupational Safety and Health Administration (OSHA) has listed as a known carcinogen. OSHA has limited worker exposure to levels that are not achievable in an aircraft stripping environment. In addition, methylene chloride may soon be classified as a volatile organic compound (VOC), further restricting its use. Chemical paint removers may also contain phenolic compounds in concentrations up to 20 percent, a concentration which cannot be treated effectively in industrial sewage treatment plants. Therefore, the phenolic sludge must be collected, dried, stored, and discarded as a "hazardous waste."

Chemical strippers are applied to the aircraft and scraped off. Because the paints on U.S. Air Force aircraft are chemically resistant, they are difficult to remove with this process. Therefore, the application and scraping of the stripper may be repeated up to nine times until all of the paint is removed. The aircraft is then washed and rinsed, and any paint remained is removed mechanically. On the F-15 aircraft, this process yields approximately 6,800 pounds of hazardous waste, plus rinse water.

PLASTIC MEDIA BLASTING DEFINED

Plastic media blasting (PMB) has been authorized as an alternative to chemical paint removal operations. PMB is similar to sand blasting in operation. The aircraft is first washed, as with the chemical depaint process, causing a waste stream then particles of plastic (acrylic, melamine, or polyester) are propelled at the paint using compressed air. The process generates from 1,500 to 3,000 pounds of dry hazardous waste per F-15 aircraft. This waste is made up of plastic media which has been abraded to particles too small to be recycled and the paint removed from the aircraft. It is the lead, cadmium, and chromium from the paint chips which cause the end product to be classified as a hazardous waste.

Plastic media also tends to ingress the aircraft through seals and seams in the aircraft skins. Man-hour intensive masking operations are used to reduce this ingress to a minimum. After the PMB process, media removal may also be required.

CARBON DIOXIDE PELLET BLASTING - A POSSIBLE SOLUTION

As a possible solution to these problems, Warner Robins Air Logistics Center (WR-ALC) has been investigating the use of CDPB as a potential new aircraft coatings removal process. Carbon dioxide pellet blasting is similar to PMB and sand blasting, in that small particles are accelerated toward a painted surface using compressed air. However, the first two processes rely primarily on the abrasive action of the particles to remove the old paint; the new pellet blasting process uses carbon dioxide in its solid form--namely, dry ice. Carbon dioxide pellets can be blasted at the coating at subsonic, sonic, or supersonic speeds, depending on the particular application. In stripping paint, these particles provide not only an abrasive action, but also a thermal shock from the -109 degree Fahrenheit temperature of the particles. After the carbon dioxide pellets strike the paint, they simply sublime (evaporate directly from a solid to a vapor without a liquid phase), leaving only the removed paint as residue. Because the stripping action is not degraded by surface grime and the media is not recycled, there is no requirement to wash the aircraft prior to using this process.

A demonstration of this process in April, 1990 illustrated that this process had matured to the point that it could safely be used to strip coatings on certain parts of the aircraft. It was also established that this process could reduce the hazardous waste produced from each F-15 aircraft to 240 pounds, as compared to the other stripping methods listed above. This hazardous waste consists only of the paint removed from the aircraft.

Because the carbon dioxide pellets sublime, there is no intrusion problem. Any media which intrudes the aircraft simply sublimes and does not have to be removed. Masking of the aircraft is reduced to only what is required to protect certain delicate materials such as aircraft canopies.

CARBON DIOXIDE TEST PROGRAMS AND PRELIMINARY RESULTS

There are several WR-ALC sponsored test programs under way to evaluate the CDPB process. The major aircraft stripping effort is being undertaken by the F-15 Product Directorate under an engineering services contract to Mercer Engineering Research Center (MERC) and has, so far, prototype stripped one F-15 aircraft. This first aircraft demonstrated that the process could be applied, with limited success, to the F-15 aircraft. However, there were several areas of concern identified that require resolution prior to production implementation.

The first concern is the slow strip rate of the process. Mercer Engineering Research Center has indicated, in its preliminary Phase I Test Results (11 Jan 91), that instantaneous strip rates varied from 1.0 square-foot-per-minute to 0.1 square-feet-per-minute, depending on the substrate that was stripped. The net average strip rate on the F-15 aircraft was reported to be approximately .189 square feet per minute of nozzle on-time (.13 square feet per minute with worker effectiveness factored in). This net average strip rate is marginal at best. The slowest strip rates were experienced on Alclad surfaces. While only 20 percent of the F-15 has Alclad surfaces, other Air Force aircraft have up to 80 percent Alclad surfaces. Strip rates of aircraft with a larger percentage of Alclad will be extremely slow.

The second problem concerns the surface condition of Alclad and other aircraft surfaces. The MERC Phase I Test Results Report indicates that the current process does not remove all of the paint from the Alclad surfaces. The surface that is left is not readily repaintable. The residue paint must be removed by other processes in order to provide an adequate surface for repainting. This is obviously a potential problem for any aircraft with Alc'ad skins.

The third problem concerns the ergonomics involved with using the process. The weight and thrust of the blast nozzles and hoses has been identified from the very first test as a human factor fatigue problem. The hose and nozzle weigh approximately 20 pounds when held at chest level. If blasting underneath an aircraft surface, the blast process thrust adds approximately 10 pounds to this weight. MERC has indicated in the preliminary Phase I Test Results that the ergonomic difficulties "reduce stripping effectiveness (speed) by 150 percent" for the F-15 aircraft. Workers in the F-15 demonstration manual CO2 booth (Bldg 137) currently switch the blasting duty every fifteen minutes. This requires two personnel for each blast nozzle. MERC recommends the use of some type of assistive device to reduce operator fatigue and thus increase stripping speed and quality.

The fourth concern is that the process, as currently defined, is too aggressive for use on thin skin aluminum alloys which are unsupported. Aluminum .032 inches thick and thinner are susceptible to peening-type damage at the pressures required for effective paint stripping. This is only a minor problem for the F-15 aircraft, but could prevent the use of CDPB on up to 20 percent of cargo aircraft skins.

In order to resolve the above-listed concerns using CDPB, several projects are currently underway. The problems with the ergonomics will be resolved by a robotic system which has been installed and should be operational by Aug 91. This robotic system will, at least, double the current strip rate. The precision of the robot may provide an acceptable surface condition (especially on Alclad) as compared to what is seen with the manual CDPB process. Other assistive devices, such as manipulator arms, could also be utilized instead of robotics.

Also under development is combined CDPB and flashlamp stripping head, which may boost stripping speed to three square-feet-per-minute. The combined system will also significantly reduce the aggressiveness of the process, allowing even thin skin materials to be stripped without damage. This project will be completed by 15 May 92, with the demonstration of the system on an F-15 aircraft at WR-ALC.

A WR-ALC and MERC test effort will look at the positive and negative attributes of using paint softeners prior to using the CDPB system. Paint softeners will allow the use of much less aggressive parameters allowing the use of CDPB on even thin-skin materials. Stripping speed should also increase to well over one square-foot-per-minute. The possible drawbacks of using chemical softeners include an increase in disposable waste, additional aircraft processing steps, and the potential effects of the softeners on aircraft materials.

In order for CDPB to be widely accepted as an aircraft coatings removal process, it must be capable of adequately removing the entire coating system from aircraft materials without damaging substrate materials. In order for CDPB to be cost effective as a coatings removal tool, as compared to other paint removal processes, it must remove coatings a rate of .25 to .70 square-feet-per-minute (depending on the aircraft).

CONCLUSIONS

The intent is for the F-15 aircraft to achieve an acceptable strip rate and surface condition Increases in using the robotic system currently undergoing installation at WR-ALC. Alclad surfaces may still require some alternative paint removal prior to repaint. Advances in strip rate and Alclad surface condition through the use of combined flashlamp and CDPB systems, or the use of paint softeners, will make this an even more cost-effective system. strip rate, the ability to strip thin skins without damage, and complete paint removal are all required prior to the acceptance of this process on the WR-ALC managed cargo aircraft.

ALTERNATIVE TECHNOLOGIES FOR ENVIRONMENTAL COMPLIANCE

J. Michael Locklin
Douglas Aircraft Company
Long Beach, California

ABSTRACT

Many Americans have hailed the 1990's as the decade of the environment. While some consider this recognition long overdue, others see it as a narrow, self-serving scheme to create jobs for environmentalists that will ultimately degrade the worldwide manufacturing competitiveness of the United States. For many years now, increasing public awareness has led to more and more specific environmental regulations throughout the country, but especially in Southern California. These regulations have had, and will have, a serious impact on productivity in American industry. Whether this impact will be advantageous will depend upon the efforts of environmental professionals who research and implement alternative technologies. The direction we take with these new processes will contribute greatly to the long-term manufacturing capability of the United States.

Over the past couple of decades we have seen several foreign nations displace the United States as world leaders in a variety of product lines. For example, Japan's impact on the automobile industry is well-documented, as are their efforts in the computer and electronics fields. Korea is rapidly becoming a world leader in the manufacture of steel. And the soon to be declared European Economic Community is already well represented in the transport aircraft community by the Airbus consortium; not to mention other European aerospace leaders such as British Aerospace (Great Britain), Dassault and Arianespace (French), Deutsche Aerospace, MBB and Dornier (German), Aeritalia and Aerospatiale (Italy), CASA (Spain), and Fokker (Netherlands).

The competition is serious. However, all is not lost. A prompt and thoughtful resolution of today's environmental concerns may provide the opportunity needed to re-establish the world manufacturing leadership that the United States has enjoyed in the past. While foreign nations have been directing efforts toward increased manufacturing capability, environmental concerns often have taken a back seat to the lure of immediate profits and a higher consumer standard of living. But there is a strong movement in the United States to clean up the environment. We now have the opportunity to establish the U.S. as a leader in both environmental and manufacturing technologies.

DOUGLAS AIRCRAFT COMPANY PROGRAMS

At the Douglas Aircraft Company (DAC) in Long Beach, California, a division of the McDonnell Douglas Corporation, we are working to continuously improve our manufacturing technologies while simultaneously reducing the impact of these technologies on the local and global environment. The following alternative technologies are being investigated/implemented at the Douglas Aircraft Company:

1. **High solids topcoats** - in lieu of conventional topcoats and/or exempt solvent topcoats.
2. **Chromium elimination** - in paints, primers, sealants, process chemicals, etc.

3. **Alkaline/aqueous** degreasing technologies in lieu of solvent **vapor degreasing**.
4. Alternative **Handwipe Solvents**/cleaners.
5. **CFC Elimination**.
6. **Resource Recovery/Waste Minimization**.

Each of these projects contributes to the ultimate goal of eliminating the negative impact of manufacturing processes upon the environment. Perhaps the future will provide methods that are even beneficial to the environment. Meanwhile, achieving this goal will require considerable training and "retooling" of american industry. At DAC we are working diligently with our suppliers and subcontractors to develop, test and implement new alternative technologies.

High solids topcoats is an alternative technology to painting aircraft with conventional topcoats which contain high levels of solvents for sprayability and drying. While it was recently acceptable to simply substitute exempt solvents, such as 1,1,1 trichloroethane, in order to comply with regulations, it is now apparent that exempt solvents also have damaging environmental properties and will eventually be banned by the governing agencies. The Montreal Protocol set guidelines for the elimination of many solvents including exempt solvents. Locally, the Southern California Air Quality Management District Rule 1124 requires topcoat paints to contain fewer than 420 g/l of VOC as of July 1, 1991.

The term "high solids" implies less solvent/more solids, i.e., resins and heavy metals. The application of the new high solids paints will require some training and familiarity since they will have different characteristics. Tests have indicated that they generally require approximately 10% more dry time than conventional topcoats, although experimentation with accelerators continues. The pot life of the new paints is shorter by one half. High solids primers are displaying approximately a 200% longer dry time than their conventional counterparts.

As of April, 1991 we have painted three (3) MD-11 customer aircraft with high-solids topcoats. Since these coatings are typically more glossy than the old topcoats, the need for a clear coat is eliminated. Both the customer and - The Air Quality Management District (AQMD) are pleased.

Chromium elimination covers a variety of processes, such as painting, sealing, plating, and chemical processing. Chromium has long been a main ingredient of many airframe processes because of it's excellent corrosion and wear resistant properties. But, since having been identified as a human carcinogen, efforts are now underway to reduce its' usage to the point of eventual complete elimination.

In 1990 DAC expanded the use of a thin film sulfuric acid anodize process to include some commercial work, thus reducing the use of the popular chromic acid anodize. Some justification for this change was provided by military contracts which, in fact, specify the sulfuric process. A dilute chromic acid seal is currently being used with this process. We are testing other seals to eventually replace even this usage of chromium but, as yet, have not approved an acceptable alternate. Our best results to date have been with organic seals.

Chrome-free aircraft sealants are also being investigated by DAC personnel. Working with our sealant suppliers, who are also aware of evolving environmental regulations, we are testing newly developed chrome-free and lead-free formulations for compliance and efficiency. Although none has yet been approved, several show promise, including a manganese cure system which still requires some work.

Non-chromated deoxidizers are also being researched. Airframe manufacturers commonly use deoxidizers for cleaning, brightening, and removing corrosion from aluminum prior to subsequent processing. For this reason, chromium is a common ingredient of aluminum deoxidizers. Finding an acceptable substitute that will provide the same protection and a safe, durable airframe but will not negatively impact the worker or the environment is no easy task. Again, our suppliers are working diligently to develop alternate deoxidizers while our laboratories are testing the new formulations to aircraft specifications. We hope to have an acceptable non-chromated deoxidizer approved and in place by the end of this year (1991).

Alternative Plating Technologies are also being investigated under the auspices of chromium elimination since chrome plating has always been one of the industry's favorite metal plating treatments due to its' aforementioned wear, and corrosion resistant properties. Three potential alternatives now available are:

1. Detonation Gun Coatings. This thermal plating technology, while proprietary to Union Carbide Corporation, provides excellent wear and corrosion resistant properties but can only be accomplished by Union Carbide. Since it adheres to titanium better than chrome plating (which tends to flake off), this process had already been approved for use on certain titanium airframe components prior to the concerns about chromium. It is being considered for additional airframe applications as a replacement for chromium processes.

2. Plasma Spray. This non-proprietary thermal plating process is approved for use in certain applications. Since it is a chrome-free process, it is also being considered for additional applications.

3. Electroless Nickel. This process uses different percentages of nickel and phosphorus to achieve the desired hardness and corrosion resistance. It has been approved at DAC for use on certain areas of the landing gear. We are still evaluating even harder formulations that include boron and thalium which may provide additional applications in the future.

Vapor Degreasing is a cleaning process that uses solvent vapors alone to effectively remove a variety of contaminants from the workpiece. It is a relatively simple, one step process that provides a clean, dry part ready for subsequent processing. However, this popular process has now been identified as a major contributor to ozone depletion. DAC presently uses 1,1,1 trichloroethane (TCA) as the solvent of choice for vapor degreasing. Some areas of the country use trichloroethylene (TCE), but TCE has been prohibited in California. Tests are presently being conducted on various immersion type cleaners to replace solvent vapor degreasing. Some of the candidates are aqueous cleaners, terpene based cleaners, and the use of ultrasonic technology with immersion cleaners.

Aqueous cleaners are typically alkaline in nature, their pH being in the range of 9-11. Many chemical suppliers already provide alkaline cleaners on the open market. In fact, alkaline cleaning is presently approved for certain applications at Douglas Aircraft. Alkaline/aqueous cleaning is presently the leading contender to replace vapor degreasing, but the implementation of this will require some change in process and equipment which will, in turn, require some operator training and/or familiarity.

As mentioned above, the vapor degreasing process provides a clean and dry workpiece directly from the vapor degreaser equipment. This will not be the case with the new immersion cleaners. Since the new cleaners will not evaporate as quickly as the solvents with which we are familiar, the workpiece will come out of the cleaning tank wet. And, since these new cleaners may also leave a residue that could impact subsequent processing or the workpiece itself, a rinse cycle is likely to become necessary with the new process. Of course, the part will come out of the rinse cycle in a wet condition also. Therefore, it may be necessary to add a dry cycle to the process in order to prevent corrosive action.

Lab testing is presently in process at DAC on a variety of candidate cleaners, most of them alkaline in nature. Several companies provide a terpene-based metal cleaner, which contains the active ingredient d-Limonene. Although these cleaners have a low vapor pressure, they are high in volatile organic compounds (VOCs). A major defense contractor has also reported that terpenes do not respond well to purification via ultrafiltration. The use of new immersion cleaners will likely require some method of removing the contaminants gathered in the cleaning solution or it will rapidly become less effective. There are several methods, and many commercial filtration units are available. The best one for a specific application will depend upon the choice of which cleaner you choose and equipment.

Handwipe solvents are used extensively for clean up and repair during the manufacture and assembly of transport aircraft. Not unlike the solvents used for vapor degreasing, these solvents may be ozone depletors and/or carcinogens. Again, we are working with our suppliers to develop cleaners that will work effectively at ambient temperatures to remove the common aircraft industry contaminants. One of the most difficult to remove contaminants is a semi-cured polysulfide based sealant used extensively throughout the aircraft. This work has typically required exhorbitant solvent usage, much to the detriment of the ozone layer! Preliminary testing at ambient temperatures has indicated that the aforementioned terpene based cleaners seem more effective against this contaminant than the aqueous cleaners. Additional testing is required.

CFC elimination. Chlorofluorocarbons (CFCs) are used as cleaners of small electronic parts such as printed circuit boards (PCBs), in wire assembly areas, and in many maintenance tasks. They are also used in air conditioners and machine tool chillers, and as propellants in aerosol can applications. One of our immediate substitution efforts at DAC is directed at replacing CFCs as aerosol propellants. Again, our suppliers are cooperating as they are aware of the need for change in their products. It is important to evaluate the impact of any changes to products that are governed by Mil Specs.

There are several lubricants and mold release compounds at DAC that are comprised of CFCs. These are not used in high volume, but do provide an opportunity for reduction of CFC emissions. As yet, we have not been able to concentrate on replacing these compounds, but have begun the process of substituting the propellants used in some of the aerosol lubricants (see above).

Resource recovery/waste minimization is another broad category that encompasses many technologies: chemical processing, waste disposal, recycling and housekeeping are just a few. There are many opportunities for improvement under this category as environmental technology continues to advance.

Waste minimization was highlighted during DAC's recent implementation of the Total Quality Management philosophy. This effort enlightened and encouraged every employee to consider his or her impact on the environment and the workplace. The impact of Foreign Object Damage (FOD) was emphasized during this introduction since it is a familiar topic to those in the aircraft industry. This type of damage can be attributed to NOT minimizing waste.

As an example of resource recovery, DAC's recently completed chemical processing facility includes a controlled room for applying the chemical milling maskant which contains perchloroethylene, a toxic air contaminant. Through state-of-the-art controls, we are able to capture some of the perchloroethylene and sell it back to our maskant supplier. Another case of recycling technology is found in the same new processing facility, where a Caustic Etchant Regeneration (CER) unit purifies spent chemical milling solution by hydromechanically removing aluminum in solution and collecting it as aluminum hydroxide. This, in turn, can be used as raw material for various manufacturing technologies.

An extremely cost effective example of waste minimization was recently accomplished by simply reducing the size of our vendor-provided wipe rags. An on-site survey conducted to evaluate the usage of wipe rags discovered that the three foot square rags were too large for convenient wipe operations. The supplier agreed to provide smaller rags at no additional contract cost to DAC, thus reducing both the volume and weight of rag-generated wastes. Because of the wide variety of uses for wipe rags, they are liable to become contaminated with many products including hazardous substances that require the disposal of these rags as hazardous waste. We anticipate reducing rag waste volume by 50%, thus realizing a cost savings in the hundreds of thousands of dollars.

DAC has contracted with an outside recycling company to continuously recycle on-site such things as machine tool coolant and hydraulic fluids. The effective life of these solutions is thus extended, further reducing the procurement and disposal costs of additional, potentially hazardous, fluids and chemicals.

The Douglas Aircraft Company is located in southern California and is surrounded by residential neighborhoods. For this reason, DAC has always tried to be careful to minimize its wastes. DAC has had waste-water treatment systems operational for many years. One system uses sulfur dioxide to first convert hexavalent chromium to trivalent chromium. It is then neutralized and precipitated out as metal hydroxides. After further clarification and filtration, it is eventually discharged to the public sewer system. Nevertheless, we continue to train employees to use proper operating procedures and encourage measures to further reduce the generation of wastes. Some of these measures have been mentioned above, such as elimination of chromated process solutions and recycling. Other examples that everyone is familiar with are: improvement of general housekeeping practices and turning off lights and computers when not in use.

SUMMARY

DAC continues to investigate and implement new environmental technologies with the coordinated efforts of the different programs and core groups. McDonnell Douglas employees, like so many other Americans today, are becoming aware of the many advantages of an aggressive environmental compliance/technology program. A prime example of the growth of environmental concerns is the fact that ten years ago disposal costs were seldom considered when justifying a new expenditure (whether for installing a new machine tool or purchasing process chemicals). But today, responsible planning for the handling of tomorrow's wastes can not only improve our environment, but can prove to be very cost effective in the long term. With all the attention that environmental concerns are getting today, the attending rules and regulations can sometimes be very confusing. While the Environmental Protection Agency oversees national policies, its major concerns may not address certain critical concerns at the local level. Balancing the requirements of both agencies may require a change to a governing specification. As new technologies are developed, specifications are being updated. But the industry is growing rapidly and different agencies have authority over different aspects of environmental concerns. Local, national and international agencies assume responsibility for environmental legislation. Therefore, any change in process that may effect the environment requires a thorough investigation of the governing rules and regulations.

In conclusion, alternative technologies are becoming increasingly necessary to meet the ever-tightening demands of an aware public when it comes to environmental legislation. It is noteworthy that environmental professionals tend to share technological developments and breakthroughs in spite of the fact that they may work for competing companies. This indicates the importance of making our planet a better place to live. Although it sometimes requires a delicate balancing act to satisfy the various applicable agencies, regulations, and specifications, we now have the opportunity to lead the countries of the world to a better, cleaner, healthier future.

HIGH PRESSURE SUPERCRITICAL CARBON DIOXIDE EFFICIENCY IN REMOVING HYDROCARBON MACHINE COOLANTS FROM METAL COUPONS AND COMPONENTS PARTS

Robert F. Salerno
Organic Material/Surface Modification
EG&G Mound Applied Technologies
Miamisburg, Ohio

High pressure, supercritical carbon dioxide efficiency in removing hydrocarbon machine coolants (production process contaminants) from metal coupons and component parts was evaluated. Solubility experiments were performed on Cimperial 1011, Cimperial 15, Gulfcut 11D machine coolants. Extraction experiments were conducted on machine coolant contaminated aluminum and 303 stainless steel coupons (1.5 in. x 0.25 in.), as well as detonator production components. The solubility/fractionation experiments were conducted in a screening supercritical carbon dioxide system. The solubilities of Cimperial 1011 and Cimperial 15 were measured at 50°C, 13.8 Mpa. The solubilities of Cimperial 1011 and Gulfcut 11D were also measured at 35°C, 13.8 Mpa. In addition, coolant fractions were collected for gas chromatography analysis. Extraction (cleaning) experiments were conducted in a supercritical carbon dioxide feasibility system, utilizing a 300 ml process vessel. Cleaning trials were conducted at 35°C, 13.8 Mpa with carbon dioxide contact times of 15-30 minutes. Residual machine coolant concentrations on coupons and components cleaned with high pressure, supercritical carbon dioxide were determined by a hexane rinse/capillary gas chromatography analysis procedure.

The studies revealed that the three machine coolants are all soluble in supercritical carbon dioxide with solubilities ranging from 1.06 weight percent (wt %) to 4.69 wt %. Also, cleaning experiments conducted showed variations in the amount of residual machine coolants on coupons and components regardless of carbon dioxide contact time. Residual contaminants ranged from 3.0 to 840 mg.

Under the stated experimental conditions this study has demonstrated that high pressure supercritical carbon dioxide shows potential as a cleaning media for removing hydrocarbon machine coolants from metal substrates. If optimized in production cleaning applications the use of such a process would reduce plant waste streams significantly.

INTRODUCTION

Until recently, production cleaning processes have relied on halogenated solvents as cleaning media for removal of production process contaminants. However, government regulations concerning the use and disposal of these products are becoming more and more restrictive. As a result, considerable interest has been generated in developing cleaning processes that use environmentally acceptable cleaning agents and that reduce or eliminate hazardous waste.

As part of the Department of Energy (DOE) waste minimization efforts, EG&G Mound Applied Technologies (MAT) has been working to develop a final cleaning process for production parts using supercritical carbon dioxide as a substitute for halogenated cleaning agents. The objective of this study was to evaluate the efficacy of high pressure, supercritical carbon dioxide in removing hydrocarbon machine coolants (production process contaminants) from stainless steel and aluminum coupons and component parts (Inconel-glass ceramic).

SCOPE OF EXPERIMENTATION

Four solubility trials and eleven extraction experiments were conducted. The solubility/fractionation experiments were carried out in a small screening system. Four trials on the three coolants were completed. Solubilities of Cimperial 1011 and Cimperial 15 were measured at 50°C, 13.8 Mpa. The solubilities of Cimperial 1011 and Gulfcut 11D were also measured at 35°C and 13.8 Mpa. In addition to determining solubilities, various fractions of the coolants were collected for subsequent analysis.

Cleaning experiments were conducted in a Feasibility System, utilizing its 300 ml process vessel. Cleaning trials were conducted at a temperature of 35°C, a pressure of 13.8 Mpa, and contact times of 15-30 minutes. The objective of the program was to demonstrate that supercritical carbon dioxide could remove contaminants to a residual level on the order of 1-10 mg/cm^2. Determination of the effects of operating pressure and temperature on the solubility of the three machine coolants was also sought. Residual oil concentrations on coupons and components cleaned with supercritical carbon dioxide were determined by a hexane rinse/capillary gas chromatography analysis procedure.

PROCESS EQUIPMENT

Screening Unit

The solubility/fractionation experiments were conducted in a Screening Unit, which consists of a 300 mL extractor, one 70 mL separator, a solvent pump, heat exchangers, and a back pressure regulator. A diagram of the screening unit is shown in Figure 1.

A screening unit operates as follows: Liquid carbon dioxide from a dip-tube storage cylinder is subcooled using a glycol-cooled heat exchanger to prevent vaporization in the solvent pump. For carbon dioxide, cylinder pressure is the saturation pressure at ambient temperature, about 6.2 Mpa. The subcooled solvent is compressed from cylinder pressure to the extraction pressure, 13.8 Mpa for this series, using a reciprocating, packed-plunger pump.

An electric heat exchanger raises the high pressure solvent to the extraction temperature, which was maintained in the range of 35°C - 50°C for this series. The solvent, now at supercritical conditions, flows continuously upward through a sample of machine coolant in the extractor vessel. Electric band heaters under temperature control maintain the extractor at the desired temperature.

Supercritical solvent, containing a soluble fraction of the contaminants or test material from the feed, leaves the extractor and flows through a back pressure regulator, which is electrically heated. Flow through this valve reduces the stream's pressure to atmospheric so that the fluid entering the separation vessel is now a gas and no longer has good solvent properties. Electric heaters on the separator vessel are used to control the separation temperature. The material previously dissolved in the solvent precipitates in the separator as a solid or liquid that can be easily removed from the system.

For the solubility/fractionation experiments, the solvent flow was stopped, the separator removed, and the soluble fraction collected after each approximately 600 g of carbon dioxide passed through the system. In this way, multiple fractions (3-4), which could be averaged to determine an approximate solubility, were collected. In addition, the fractions could be physically examined and analyzed to observe their different physical characteristics.

Feasibility Unit

The cleaning experiments were conducted in a feasibility unit. A diagram of the unit is shown in Figure 2. The feasibility unit consists of a 300 mL extractor, one or two 70 mL separators, and associated pumps, heat exchangers, and pressure control valves. It is used to conduct batch extractions of small samples of solids or liquids to determine feasibility of desired separations. The unit includes a co-solvent pump for adding a small amount of liquid co-solvent to the supercritical fluid to modify the solubility characteristics of the solvent system. In addition, the unit has a recirculation pump, which moves supercritical fluid through the extraction vessel at a high velocity. Also, the unit has been designed to allow flammable supercritical solvents, such as hydrocarbons, to be safely employed.

The feasibility unit operates as follows: Liquid carbon dioxide from a dip-tube storage cylinder is subcooled using a glycol-cooled heat exchanger to prevent vaporization in the main pump. For carbon dioxide, cylinder pressure is the saturation pressure at ambient temperature, about 6.2 Mpa. The subcooled solvent is compressed from cylinder pressure to the extraction pressure, 13.8 Mpa for this series, using a reciprocating, packed-plunger pump. An electric heat exchanger raises the high pressure solvent to the extraction temperature, which was maintained at 35°C for this series. The solvent, now at supercritical conditions, flows continuously upward through a batch of materials (feed) in the extractor vessel. Electric band heaters under temperature control maintain the extractor at the desired temperature.

Supercritical solvent, containing soluble components from the feed, leaves the extractor and is split into two streams. The majority of the solvent is recirculated back into the extraction vessel via the recirculation pump. A small flow enters a separation area contained in a temperature controlled oven. Within the oven, the solvent flows through a pressure control valve, which is electrically heated. Flow through this valve reduces the stream's pressure so that the fluid entering the separation vessel is more "gas-like" and no longer has good solvent properties. Control of the oven temperature and of the electric heaters on the separator vessel itself determines the separation temperature. The material previously dissolved in the solvent precipitates in the separator as a solid or liquid that can be subsequently removed from the system.

The solvent leaving the first separator can flow through an additional control valve into a second separator, where material may be collected under different pressure and temperature conditions. Solvent leaving the second separator can be directed through a cold trap, cooled by dry ice, to collect very volatile components. From the cold trap, the solvent stream passes through a dry test meter to measure flow rate and is then vented to the atmosphere.

TEST COUPONS AND COMPONENTS

To determine the efficiency of supercritical carbon dioxide in the removal of machine coolants, thirty test coupon disks, 1.5 in. diam x 0.25 in. thick, were used. Twenty of the disks were 303 stainless steel and ten were aluminum. An additional twenty small Inconel glass ceramic components, having an overall cylindrical shape and approximately 0.5 in. diam x 0.625 in. long, were tested.

The coupons were cleaned two at a time. They were positioned on edge and suspended one above the other in the center of the cleaning vessel. Small pieces of stainless steel screen were used to keep the coupons in a vertical position and away from the vessel walls. Carbon dioxide entered through a port and distributor at the bottom of the cleaning vessel, passed up over the surfaces of the two coupons, and exited the cleaning vessel through the cover. This assembly provided high solvent velocity and good contact across the coupon surfaces.

MATERIALS

Carbon Dioxide

Commercial grade liquid carbon dioxide from Liquid Carbonic Corporation was used for all experiments in this program. The standards for this grade of carbon dioxide are shown in Table 1.

Machine Coolants

The three hydrocarbon machine coolants studied were Cimperial 1011, Gulfcut 11D, and Cimperial 15.

EXPERIMENTAL PROCEDURES

Cleaning

The coupons were coated with oil in the following fashion: First, a clean, dry coupon was weighed. Approximately 50 mg of machine coolant was then poured onto a non-linting, absorbent towel. The oil was wiped onto the surface of the coupon with care to evenly coat the entire surface. Finally, the coupon with machine coolant was weighed again; the weight gain was recorded. This procedure resulted in an average of 5.5 mg of oil being deposited onto the coupon.

After completing the coating procedure, the coupons were loaded, two per experiment, into the 300 mL cleaning vessel. The vessel was closed, and the system was purged with low pressure carbon dioxide for about 5 minutes. After purging, the vessel was pressurized via the solvent pump. Control of system pressure was maintained by adjusting the pressure control valve on the outlet of the extractor.

Solvent flow was established through the system and controlled by varying the stroke rate of the solvent pump. Simultaneous with the pressurization of the extraction vessel, temperatures were brought to the desired levels and maintained via electric heaters. Once operating conditions were reached, the recirculation pump was turned on, and a high flow of solvent was maintained up through the cleaning vessel. A steady flow of supercritical solvent was maintained so that the solvent-to-feed ratio increased with time.

When the desired cleaning time had been reached, solvent flow was stopped by turning off the solvent pump and the recirculation pump. The system was depressurized by slowly venting the solvent through the separator and out the vent system. Once the extraction vessel had returned to atmospheric pressure, the vessel was opened and the coupons were carefully lifted out for analysis.

ANALYTICAL

Hexane Rinse

The cleaned coupons and components were carefully removed from the extraction vessel and rinsed with two, 4.0 mL aliquots of high purity hexane. The two hexane aliquots were combined and evaporated down to about 1.0 mL in volume. This sample was then transferred to a reaction vial and further evaporated to 500 mL.

Capillary GC Analysis

All samples were analyzed on a Perkin Elmer #8320 Capillary Gas Chromatography equipped with a flame ionization detector. A 15-m, RTX-1 non-polar column (0.53 mm i.d., 0.5 micron film) was used. The details of the chromatographic method are provided in Table 2. Hexane concentrate injections of 1.0 mL were made with the GC in a split injection mode. Figures 3 and 4 are chromatograms of the hexane rinses of the coupons and components after cleaning with CO_2.

RESULTS

The three machine coolants studied, Cimperial 1011, Gulfcut 11D, and Cimperial 15, are all soluble in supercritical carbon dioxide. A table of the solubility measurements is as follows:

Oil	Temp.	Pressure	Wt. % Solubility
Cimperial 1011	35°C	13.8 Mpa	2.17
Gulfcut 11D	35°C	13.8 Mpa	4.69
Cimperial 1011	50°C	13.8 Mpa	1.33
Cimperial 15	50°C	13.8 Mpa	1.06

Although the table of solubility measurements is incomplete, it appears that Gulfcut 11D is more soluble than Cimperial 1011, which is more soluble than Cimperial 15.

Supercritical carbon dioxide can clean Cimperial 1011, Gulfcut 11D, and Cimperial 15 from coupons and components at the operating conditions studied. Average residual contamination on the order of 0.65% of the initial loading was found at the conclusion of nearly every cleaning experiment, independent of experimental conditions studied to date.

CONCLUSIONS

1. Supercritical carbon dioxide is very effective in removing Cimperial 1011, Cimperial 15, and Gulfcut 11D from the surface of aluminum and stainless steel coupons at operating conditions of 35°C and 2000 psig. Residual oil concentrations equivalent to less than or equal to 0.15 mg/cm^2 (99.94% removal efficiency) were achieved.

2. Supercritical carbon dioxide is very effective in removing Gulfcut 11D from the surface of Inconel-glass ceramic components at operating conditions of 35°C and 13.8 Mpa. Residual oil concentrations equivalent to less than or equal to 8.20 mg per component (99.97% removal efficiency) were achieved.

3. At the operating conditions used for cleaning (50°C, 13.8 Mpa), 15 min of contact with supercritical carbon dioxide is sufficient to clean test coupons and Inconel-glass ceramic components. A longer contact time of 30 min did not result in significantly cleaner parts.

4. The solubility of the three machine coolants studied varies as a function of their composition. Although the table of solubility measurements is incomplete, it appears that Gulfcut 11D is more soluble than Cimperial 1011, which is more soluble than Cimperial 15.

5. For the one material measured at both 35°C and 50°C, Cimperial 1011, solubility increased by more than 50% at the lower temperature.

DISCUSSION

Under the stated experimental conditions this study has demonstrated that high pressure, supercritical carbon dioxide shows potential as a cleaning media for removing hydrocarbon machine coolants from metal substrates. It appears that the machine coolant's solubility in supercritical carbon dioxide was greatly dependent on CO_2 density. This was apparent when the solubility of one of the machine coolants increased dramatically as the temperature was dropped from 50°C to 35°C with the pressure maintained at 13.8 Mpa. This drop in temperature, while maintaining a constant flow rate, increased the density of CO_2 from 0.675 g/cc to 0.825 g/cc and the solubility of Cimperial 1011 from 1.33 wt % to 2.17 wt %. This change in solubility as a function of CO_2 density under controlled conditions clearly indicates the potential utility of supercritical CO_2 as a versatile cleaning media for production cleaning operations.

ACKNOWLEDGEMENTS

EG&G Mound Applied Technologies appreciates the efforts that Supercritical

Processing Inc. put forth in completing the requested experiments for this study.

Charpentier, B. A. and M. R. Sevenants, ACS Symposium Series 366 (1987).

Davidson, P., R. D. Gray, Jr., M. E. Paulaitis, and J. M. L. Penninger, Ann Arbor Science Publishers (1983).

Johnston, K. P. and J. M. L. Penninger, ACS Symposium Series 406 (1988).

Krukonis, V. J., M. A. McHugh, J. M. L. Penninger, and M. Radosz, Volume 3 (1985).

Table 1 - LIQUID CARBON DIOXIDE MANUFACTURING SPECIFICATION

The standard for liquid carbon dioxide produced in a Liquid Carbonic plant is as follows. It is not applicable to " as-delivered" product.

Component maxima are in parts per million by volume (ppm v/v) unless otherwise noted.

Carbon dioxide, % v/v	99.95
Water	8.00
Hydrogen	20.00
Oxygen	8.00
Nitrogen	60.00
Carbon monoxide	1.00
Methane	20.00
Other volatile hydrocars	1.00
Sulfur dioxide	0.00
Hydrogen sulfide	0.10
Phosphine	0.30
Carbonyl sulfide	0.50
Non-volatile residues, w/w	5.00
Odor	None
United States Pharmacop	Passes

Table 2 - CAPILLARY GC CONDITIONS

Column - RTX - 1 (non-polar)
- 15 m length
- 0.53 mm i.d.
- 0.5 micron film thickness

GC Parameters

Initial Oven Temperature - 120°C
Hold Time - 2 min
Ramp 1 - 30°C/min

Second Oven Temperature - 268°C
Hold Time - 5 min
Ramp 2 - 20°C/min

Final Oven Temperature - 310°C
Hold Time - 1 min

Injector Temperature - 350°C
Detector Temperature - 350°C
Pressure - 19.0 psig

GC ran in split injection mode with a vent flow of about 6 mL/min.

Sample Preparation:

- The sample was rinsed two times with 4 mL of high purity hexane.
- The two rinses were combined in a sample jar.
- The hexane rinse was evaporated to about 1 mL volume and transferred to a 1 mL reaction vial where it was evaporated to a volume of 500 mL.
- A 1 mL sample was injected onto GC column with syringe.

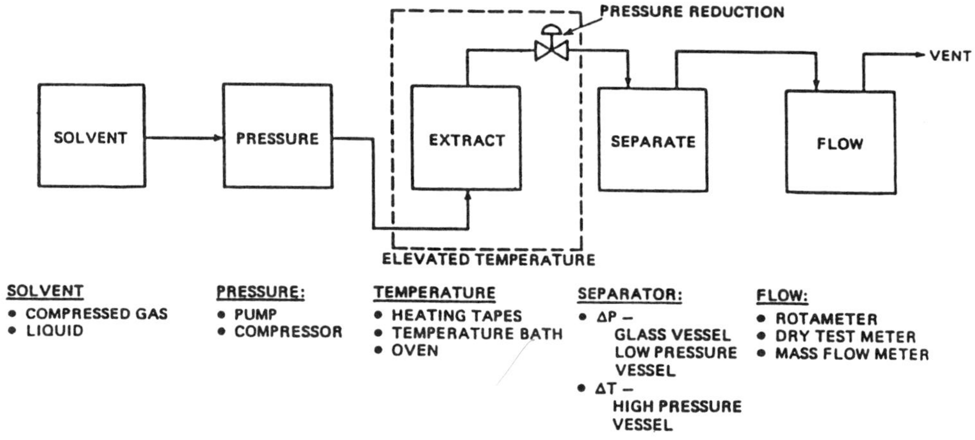

Figure 1

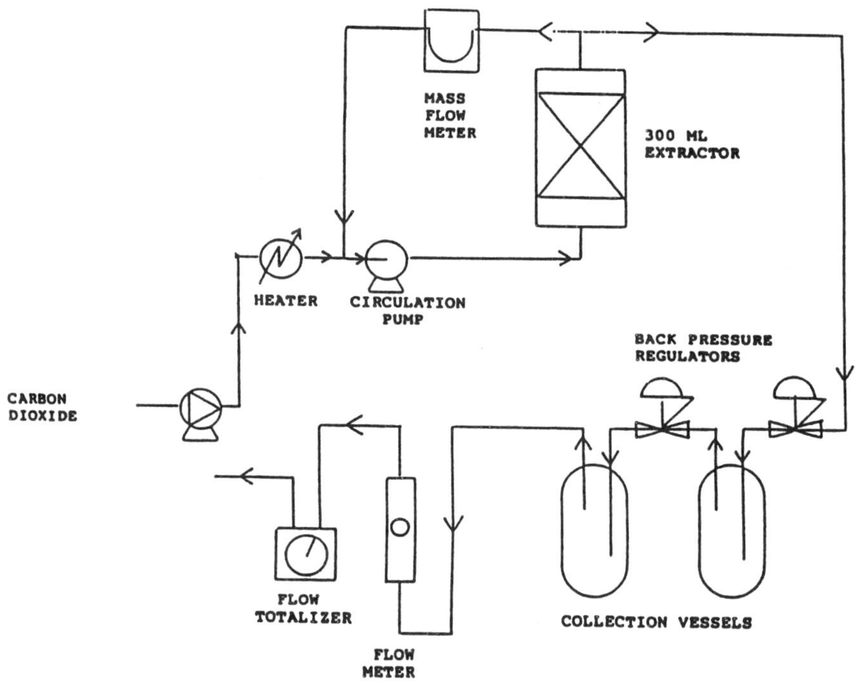

Figure 2

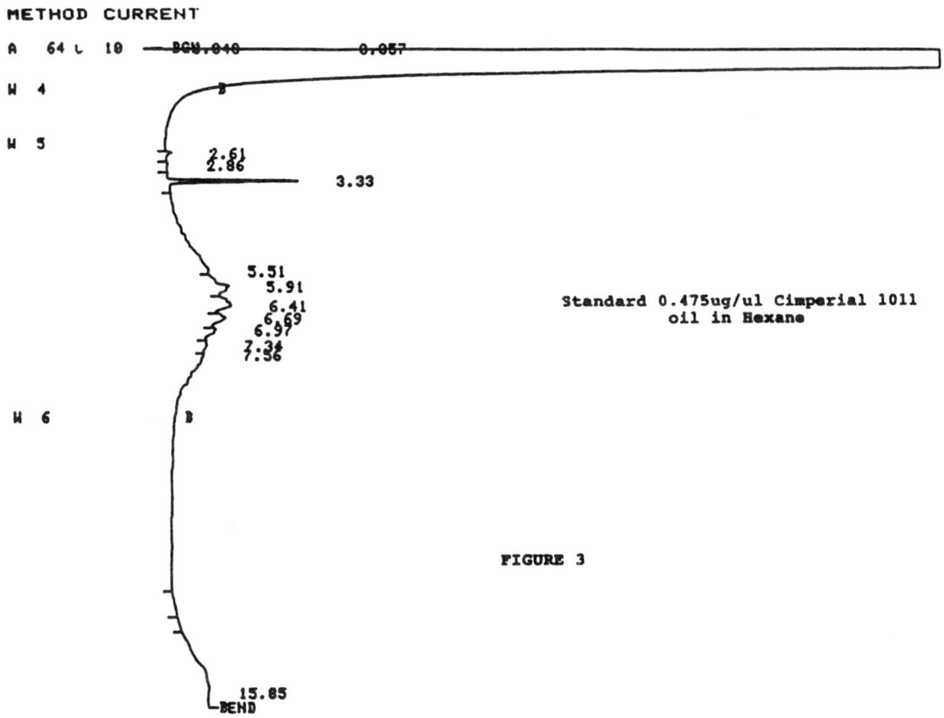

Standard 0.475ug/ul Cimperial 1011 oil in Hexane

FIGURE 3

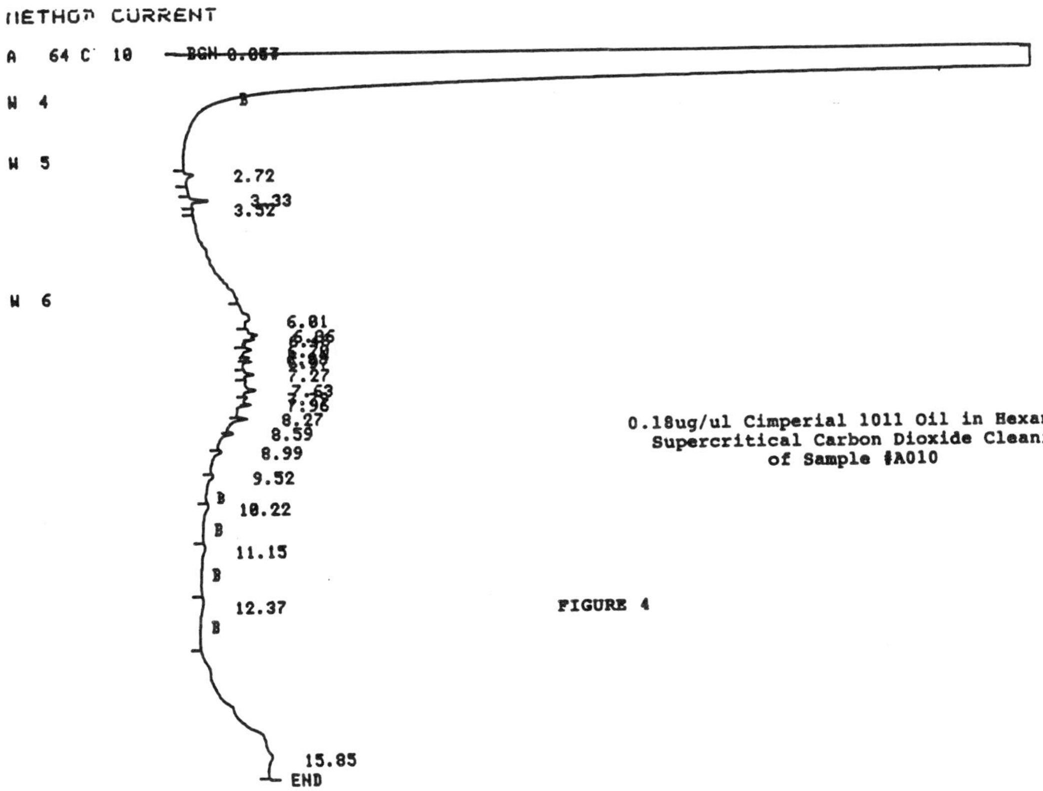

0.18ug/ul Cimperial 1011 Oil in Hexane After Supercritical Carbon Dioxide Cleaning of Sample #A010

FIGURE 4

CLOSED LOOP ALTERNATIVE TO THE USE OF HAZARDOUS CHEMICALS IN INDUSTRY

Dan Suciu
TEC/NIQUES International Ltd.
Benton Harbor, Michigan

ABSTRACT

With the advent of new technology in connection with alternatives to CFCs and the lowering of VOCs worldwide, the search for alternatives that are the most practical, productive, and cost-effective is a monstrous, time consuming, and expensive undertaking.

This paper deals with the research and development of a "Closed Loop" system which will assist industry in making those decisions. The system is a time-saving, money-saving, effective and efficient method of replacing old technology, equipment, and chemical waste treatment with new, viable and state-of-the-art alternatives.

This "Closed Loop" system has been developed by a coalition of experts in the following three fields: aqueous and semi-aqueous chemicals; suitable types of equipment for application of those chemicals; and various methods of waste treatment for aqueous and semi-aqueous spent and contaminated effluent. The end result is a system that works cost effectively and efficiently. It eliminates the need of investing massive amounts of time and money in many individual interviews in the search for proper equipment and materials. Decision making will be greatly simplified with less confusion and with minimal disruption in the changeover to a viable and totally functional system.

INTRODUCTION

The "Closed Loop" system is formed through a coalition of three companies representing aqueous and semi-aqueous chemicals, aqueous and semi-aqueous chemical cleaning equipment, and effluent waste treatment and waste minimization. The members of the coalition, TEC/NIQUES International, are: Environmental Research and Development, Inc. (ERAD); The Ransohoff Company; and 3D, Inc.; representing treatment, equipment, and chemicals for solvent substitution. This coalition eliminates the need for plant environmental and operations personnel to become experts in the available substitute chemicals, chemical cleaning equipment, and waste treatment technologies, while minimizing the time required in researching prospective vendors and reviewing various literature and papers describing all the "best" technologies. At the same time the substitution can be custom designed to incorporate existing and expected needs of the company. In this custom fit, the equipment, technologies, or chemical cleaners may not all be manufactured by members of the coalition, but the procurement and implementation will be coordinated through the coalition, thereby providing a complete operating product. Each aspect of the coalition is discussed separately below.

CLEANING CHEMICALS

Many industries have replaced perchloroethylene with 1,1,1-trichloroethane. This replacement only involved removal of the perchloroethylene from the vapor degreaser and addition of the 1,1,1-trichloroethane. No significant change in operation occurred. The operators only observed a name change. Substitution of chlorinated hazardous solvents with non-hazardous cleaners cannot be accomplished by this "easy" one-step operation.

Cleaning with the hazardous solvents offers good oil removal and cleaning capabilities. The solvent evaporated quickly and provided spot-free parts normally requiring no auxiliary corrosion protection. There is little treatment of the solvent required. The sludge (dirt, oil, grit) is pumped from the tank (vapor degreaser) to be drummed for disposal. The solvent is replenished regularly by the addition of new solvent. The solvent is lost due to drag-out and evaporation and most often requires replenishing at a significant rate.

Extremely efficient substitute aqueous and semi-aqueous cleaners are available to replace the hazardous solvents. The cleaner chosen for a specific job is dependent on many variables, such as the type of soil to be removed, the metal to be cleaned, and the process to follow the cleaning. A number of variations of these cleaners are available depending on the application equipment. For example, when used in a high pressure cabinet washer or when high pressure spray is applied to the surface being cleaned and foaming is undesirable, low foam or de-foamed versions are effective alternatives: Passivators can be used to prevent flash rusting on certain ferrous metals, and to prevent corrosion on non-ferrous metals during cleaning cycles or in the rinse cycle. All variables must be reviewed in order to choose the proper aqueous or semi-aqueous cleaning chemical that will bring optimal results.

Efficient use of the substitute aqueous or semi-aqueous cleaners cannot be achieved in the existing vapor degreaser. In many cases rinsing and drying of the part is required. However, equipment is available which makes the entire process a continuous, totally integrated operation.

Substitute aqueous and semi-aqueous cleaners will require treatment within the cleaning process to remove the grits and oils and to prevent them from adhering to the part and treatment of the spent substitute cleaners prior to disposal. Although the substitute cleaners are non-hazardous, the cleaning process may produce a hazardous waste by adding soils, oils and greases, and metals to the spent cleaning solution. In order to minimize the hazardous waste disposal cost and the long-term liabilities associated with the waste, the spent cleaning solution should be treated prior to discharge.

CLEANING EQUIPMENT

In order to successfully and effectively replace an immersion, spray or vapor-type solvent degreasing system with an aqueous-based process, the following considerations should be examined.

PARTS HANDLING

Parts handling is an important feature of an aqueous cleaning system. It must be designed to address the often complex configurations of parts being processed. If parts have intricate internal passages, pockets or crevices that tend to cup solution or if there are areas difficult to reach with a cleaning solution spray, part handling becomes critical.

The handling system must assure that all surfaces of the parts are positioned for maximum spray impingement. Parts that are not free-draining must be rotated or tilted to prevent cross contamination or carry-out of process solutions. The part transfer design must prevent damage to delicate, machined parts and, most importantly, the handling system must be designed to integrate with existing methods of parts loading and unloading.

CLEANING PROCESS

Selecting the proper cleaning process will assure that part cleanliness and final appearance will conform to application requirements. Spray, agitating immersion, ultrasonic immersion or a combination of these can be selected to address specific applications. Although single-stage processes are often adequate for many applications, it is sometimes necessary to utilize multiple-stage processes when part cleanliness and appearance are important.

An effective rinse is necessary to assure that cleaner residue is completely and thoroughly removed. If the plant water is of poor quality and water spotting is a major concern, it may be necessary to include a final deionized water rinse. A controlled part temperature and a well designed, high-velocity air knife dry-off will also help prevent part spotting or streaking.

If a corrosion-preventive treatment is called for, it can easily be applied during the rinse cycle, or just prior to dry-off.

The quality of part rinsing and part dry-off in the aqueous cleaning/treating process is important if the process is to equal the performance attained with the hazardous solvent cleaning system.

BATH CONTAMINATION

The type and amount of contamination being removed from the parts must be considered when designing an aqueous process. Historically, very oily parts would dictate solvent cleaning because distillation could separate solvent from the accumulated oil for reuse.

Recent advances in oil-removal methods for aqueous cleaning solutions have made it possible to deal with high volumes of non-soluble oil entering the cleaning bath on a continuous basis. The oil removal system should be a decant style, designed as an integral part of the solution holding tank. It must prevent the oil that is removed from being recirculated through the process pump. This prevents the centrifugal pump from blending or emulsifying the oil, making separation more difficult. This integral oil trap should then skim off collected oil into a decanting chamber where a more complete separation can be achieved. Also note that the cleaning chemical utilized should be one that readily releases or separates the collected oil.

In applications where a high volume of water-soluble oils that do not easily separate from the cleaning solution are being removed it is recommended that a continuous overflow approach be used with direct cleaner injection into the overflow supply line. Non-soluble oils are much easier to deal with in an aqueous cleaning process.

Various methods of solution filtration also can be applied to the cleaning system. Filtration will remove fine, suspended particulates that tend to build up in the cleaning bath and can ultimately redeposit on the parts. The selection of the proper filtration system coupled with an effective method of oil removal will assure optimum life of the cleaning bath.

CONSTRUCTION MATERIALS

The aqueous cleaning system should be constructed from materials that are compatible with the process cleaning solutions. Mild steel construction usually is acceptable for an alkaline cleaner stage as the alkaline detergent offers adequate protection against corrosion. With mildly acidic washes, fresh water or deionized water rinses, it is recommended that a type-304 grade stainless steel be used. Mild steel construction should also be acceptable for the dry-off section.

There are various types and styles of aqueous cleaning systems available. They range from small, single-station dip tanks to complex automated-transfer, multi-stage systems. Several aspects need to be considered in determining the proper machine type for the application. These include the basic machine considerations relative to continuous part flow or batch part flow along with part considerations, such as: whether the parts are delicate, susceptible to marring or scratching; part sizes; part segregation; part orientation and volume of parts to be processed. Other considerations are the area of the plant in which the equipment will be installed; adequate sizes of clearances; floor load limitations; pad and pad curbing; floor drains and containments; and location of the waste treatment facility.

WASTE TREATMENT

Determination of the applicable treatment of the cleaning waste effluent is dependent on the type of cleaner used, the availability of existing treatment processes, the type of soil being removed, and the availability of space for installation of the treatment system. Point source treatment of the cleaner can, in some cases, be implemented into the design of the cleaning equipment. Point source treatment can be complete, treating the waste to a level required for discharge into surface water or a sewage treatment plant or it can treat the waste to a level required for discharge into an existing industrial waste treatment plant. In some applications, the treated water can be recycled into rinse tanks or other process lines, such as electroplating, requiring make-up water. In some of the applications, the treated water is a better quality than the present plant water.

Effluent waste treatment processes include chemical precipitation, chemical oxidation, biological oxidation, absorption on granular activated carbon, ion exchange separation, filtration, distillation, and separation of the oils and greases. Filtration removes the dirt from the cleaning solution and permits reuse without these dirt particles adhering to the cleaned part. However, in many cases, the cleaner is degraded with use through break down of the active ingredients or dilution. In some cases, the cleaner life can be extended by pulling off a side stream for treatment and adding new cleaner or active ingredients of the cleaner. In this manner, only the side stream of cleaning solution and any oils and greases separated in the cleaning need to be treated. Presently, the oils and greases are not treated in the plant, but are shipped at a nominal cost for processing. They are either burned or processed for reuse.

Chemical precipitation processes are used to remove metals and organics from the waste stream. The organics are co-precipitated through the addition of alumina or ferric iron and the appropriate flocculents. Metal precipitation to discharge limits can be achieved with the sodium sulfide/ferrous sulfate metal treatment process while minimizing the production of sludge. The sludge produced in these processes must be disposed of as a hazardous waste.

Chemical precipitation processes require the selection of the proper clarification system to enhance the coagulation and settling of the particulates. The clarifier design is dependent on the process flow, type of particulates, size of particulate, and the available space for process implementation. The alternative cleaners may impact the clarification process, requiring the use of additional iron or alumina to achieve clarification.

CONCLUSIONS

This technological approach for the replacement of toxic or hazardous solvents with environmentally safe cleaners offers significant advantages for users. The foremost advantage is that one organization, consisting of a coalition of selected experts in the fields of treatment, equipment and chemicals works together as a team on each problem ensure that every aspect in the solvent substitution process operates as a cohesive, fully integrated "CLOSED LOOP" system.

Section II

ALTERNATIVE SOLVENTS

BIODEGRADABLE SOLVENT SUBSTITUTION

Anne E. Copeland
Directorate of Environmental Management
Tinker Air Force Base, Oklahoma

INTRODUCTION

The Air Force Logistics Command (AFLC) overhauls and repairs jet engines, aircraft and various components. The cleaning processes, which must precede all inspection and repair, use a variety of solvents including perchloroethylene, methyl chloroform, trichlorotrifluoroethane, and PD-680 (stoddard solvent). Because of the environmental and health hazards associated with these solvents, AFLC has recognized that these solvents must be replaced with safer products.

In 1987 the Air Force Engineering Services Center (AFESC), at the request of AFLC, began the research and development project titled "Biodegradable Solvent Substitution." The objective of the program was to find safe substitutes for the solvents used for metal cleaning at AFLC installations. The work, contracted to EG&G Idaho, Inc., was performed at Tinker Air Force Base. This paper presents the user's perspective of the program.

APPROACH

The program was structured in three phases, each lasting one year. EG&G performed a majority of the work at a pilot plant facility, located at Tinker's industrial wastewater treatment plant (IWTP). This pilot plant is a small scale replica of the IWTP.

Phase I - Phase I consisted of data collection, establishment of criteria for substitute cleaners, a market search of available products and screening tests.

EG&G surveyed AFLC to determine their current cleaning processes, what solvents were used and in what applications. The survey revealed the primary solvent use was in the metal cleaning using perchloroethylene, methyl chloroform, trichlorotrifluoroethane, and PD-680 (stoddard solvent) to remove oils, greases, carbon, and masking wax used for selective plating.

The criteria AFLC established for new substitute cleaners included: (1) efficiency - substitutes must be at least as efficient as current solvents; (2) flashpoint - flash point must be greater than or equal to 200 degrees Fahrenheit; (3) biodegradability - a product must biologically degrade, as measured by its chemical oxygen demand (COD), in six hours (actual retention time) in Tinker's IWTP to the NPDES permit limit of 150 mg/l; (4) corrosiveness - products must not cause corrosion rates to exceed 0.3 mil/yr on specified metals (see Table 1) as measured by ASTM Methods F483-77, "Total Immersion Corrosion for Aircraft Maintenance Chemicals," and F519-77 for hydrogen embrittlement.

After searching the market for available products, EG&G contacted 215 companies and selected 175 samples to screen. The products were screened for biodegradability, soil solubility, cleaning efficiency, and corrosiveness. Of the products that passed the screening tests, six were chosen for continued evaluation in the program. It should be noted that all of the products corrode magnesium at a rate greater than 0.3 mil/yr. Two of the six products were later dropped from the program, one for low flashpoint and the other for a toxic component. During Phase III a new product, already tested for performancein one of Tinker's overhaul shops, was incorporated in the program. Of these five final products, two are aqueous and three are organic, or not water dilutable.

Phase II - In Phase II the chosen products were subjected to extended performance tests. These tests included process enhancements (temperature, agitation, ultrasonics), cleaning capacity, rinsing requirements and the impact on Tinker's IWTP. Results showed that a process temperature of 140 degrees Fahrenheit coupled with pressurized spraying or vigorous agitation and rinsing gave the best results. Ultrasonic enhancement, due to its numerous variables, was not pursued further in this program. It was, instead, placed under its own separate program.

To determine the effects of these products on Tinker's IWTP, the candidate substitutes were processed through the pilot plant. The used products were sequentially fed through the pilot plant and their effects on each unit process were monitored. The results were that while all were biodegradable in the laboratory jar tests, only one product was successfully treated in the pilot plant. One problem encountered with the other products was that they floated the metal sludge that had been intentionally removed from solution and precipitated. While with some of the cleaners this effect could be counteracted by adding ferric chloride, the IWTP personnel were not in favor of adding another chemical to the process or the associated costs of this addition. Another problem, which occurred with only one of the products, was that when the cleaner was mixed with the rest of the waste stream, the bacteria would not acclimate to it. In other words, the bacteria preferred the other "food" available and would not consume our cleaner. It therefore passed through the plant untreated.

Phase III - In Phase III the cleaners were tested in full scale production. The process conditions shown to be optimum in Phase II were demonstrated in an agitated immersion tank and a cabinet spray washer, similar to a large dishwasher. Agitation was achieved by recirculating the cleaner through a pump located outside the tank and reinjecting in the tank through submerged jet spray nozzles. The pump rate turned over the volume of the tank once every two minutes. Only the aqueous cleaners were tested in the spray washer due to the explosion hazards associated with heating and atomizing the organic products. The tests were conducted in two production shops at Tinker using actual engine and aircraft parts. The parts were soiled with plating wax, oil, grease, light carbon deposits, and heavy, baked-on carbon from the hot sections of the jet engines. These parts, normally scheduled for chemical cleaning, vapor degreasing or cold solvent cleaning, were rerouted to our test process. The acceptance criteria for product performance levied by our process engineers were that the parts had to be clean enough (1) to undergo fluorescent penetrant inspection (a method of nondestructive inspection) and (2) to accept paint. Only four of the five products were tested in production since two of the products are chemically very similar.

RESULTS AND CONCLUSIONS

Overall the aqueous cleaners, 3D Supreme and Fremont 776, performed far better than the organic products. They successfully removed oil and grease from 100% of the parts and light carbon deposits from approximately 80% of the parts subjected to the tests in 5-15 minutes process time. The shop operators reported the aqueous products cleaned better than vapor degreasing. These products did not, however, completely remove the masking wax or the heavy baked-on carbon, even after 90 minutes in the immersion tank.

Of the two organic products, Orange-Sol's De-Solv-It removed the masking wax moderately well, with a process time of 30-45 minutes. It was, however, restricted by part configuration. Tinker's "worst case" part was very intricate and so all wax was never removed. Other less intricate parts were successfully cleaned. Exxon's Exxate 1000 also removed the wax, though not as well as the De-Solv-It. One disadvantage of the organic cleaners is odor. De-Solv-It, a terpene based cleaner, has a heavy citrus odor and the Exxate 1000, an acetate ester, is very pungent. Exxate 1000 did cause headaches when not vented.

The baked-on carbon deposits from the hot sections (combustion, turbine, exhaust, and afterburner sections) was never successfully removed within acceptable process times. Though in some cases the aqueous cleaners succeeded with much time and effort, they did not remove heat scale or corrosion. Since the solutions currently used to remove scale and corrosion (acids, bases, and oxidizers) also remove the heavy carbon very quickly, they will not be replaced with any of the cleaners from this program.

As a result of this program, Tinker's overhaul and maintenance shops are beginning to replace their vapor degreasers and cold solvent tanks with both of the aqueous products, 3D Supreme and Fremont 776. Even though the 3D Supreme won't be treated through the base IWTP, it cleans better than Fremont 776 in some applications and has numerous advantages over halogenated solvents. lternate treatment and possible recycling methods for 3D Supreme will be pursued nder another program, an expansion of this one, sponsored by the Department of Energy (DOE). All test data from both the AFLC and DOE programs will be available through G&G's Solvent Handbook

Table 1. Metal Samples Used for Corrosion Testing

Copper, CDA110 ETP
Nickel 200
Aluminum, AL2024
Steel, C4340
Aluminum, AL7075
Aluminum, AL1100
Stainless, 410
Admiralty Brass, CDA443
Carbon Steel, C4340, C1020
Stainless, 310S
Inconel 750
Monel MK-500
RMI Titanium
Waspaloy Alloy
Magnesium AZ31B

REFERENCES

1. Wickoff, P.M., Schober, R.K., Harris, T.L., Suciu, D.F., McAtee, R.E., Carpenter, G.S., Pryfogle, P.A., Beller, J.M., Substiution of Wax and Grease Cleaners with Biodegradable Solvents: Phase I Report, ESL-TR-89-04, Air Force Engineering Services Center, Tyndall AFB, Florida, September 1989.

2. Hulet, G.A., Lee, B.D., Espinosa, J.M., Larsen, D.J., Gilbert, H.K., Schober, R.K., Substitution of Cleaners with Biodegradable Solvents: Phase III, Full-Scale Performance Testing, Draft Final Report, Idaho National Engineering Laboratory, Idaho Falls, Idaho, November 1990.

3. Chavez, Angela, Principle Investigator, "Solvent Utilization Handbook" (project underway), EG&G Idaho, Inc., Idaho Falls, Idaho, (208)526-7834.

4. Poor, Kevin, Principle Investigator, "Solvent Recycle/Recovery" (subtask) "Chlorinated Solvent Substitution Program" (project underway), EG&G Idaho, Inc., Idaho Falls, Idaho, (208) 526-7841.

DOE/DOD SOLVENT UTILIZATION HANDBOOK

A. A. Chavez and M. D. Herd
Idaho National Engineering Laboratory
Idaho Falls, Idaho

Chlorinated hydrocarbon solvents and chlorofluorocarbons are used extensively in cleaning operations throughout the Department of Energy (DOE) defense program, the nuclear weapons complex, the Department of Defense (DOD) weapons refurbishment facilities, and in industry. The objective of the Solvent Utilization Handbook Task Force is to provide guidelines for the selection of nontoxic environmentally safe substitute solvents for these operations. The information contained in this handbook will include cleaning performance, corrosion testing, treatability operations, recycle/recovery techniques, volatile organic compound emissions and control techniques as well as other information. The handbook will be updated on an annual basis with information on new solvent substitutes that appear in the marketplace. Toxicological information, handling and disposal, and economics of solvent usage will also be included in the handbook when available.

The near term objectives of the handbook program will be to screen replacement solvents for cleaning/degreasing operations. The operations to be conducted in FY-91 will be cleaning performance testing, corrosion testing and handbook data base revisions. Longer range objectives will be to perform extended performance testing which will include extended corrosion testing, hydrogen embrittlement testing, continued screening of new solvents, and full scale demonstrations. These longer range objectives will be performed in FY-92 and beyond.

All samples received for testing in solvent utilization undergo an initial screening process to determine their suitability for testing. MSDS information is examined to determine if the solvent is low toxicity. Currently the flash point must be $\geq$200 °F in order for the solvent to meet combustibility criteria. If the solvent meets the criteria of low toxicity and flash point, the solvent is entered into the cleaning performance testing program. The criteria for cleaning performance is determined by weight loss of the soiled test sample. The solvent must remove 95% of the soil on the test sample, currently a SS pipe nipple. Upon passing the cleaning performance test, the solvent is then entered into corrosion testing. The corrosion tests are done in accordance with ASTM F483-87, using 1,1,1-trichloroethane as a control. If the solvent causes corrosion $\geq$ than found with 1,1,1-trichloroethane for any metallurgy, it is then is entered into further testing for VOC emissions, recycle/recovery and treatability.

Approximately 300 solvents have been tested with soils specific to DOD and have been evaluated for their cleaning performance and their corrosion characteristics. An additional 100 solvents are currently being tested for their cleaning performance on soils specific to DOE and DOD facilities. The 16 soils currently being used to evaluate the solvents are: vaseline, WD-40, hydraulic fluid, Sani-Tuff handcream, dioctylphthalate, Trim Sol machining fluid, Krylon Acrylic Spray overcoat, Versamid 140 epoxy curing agent, Epon 828 epoxy resin, Kester 197RMA solder flux, carnauba wax, Amber B^2 wax, Ram 225 mold release, MS122 mold release, lanolin, and molybdenum sulfide grease. The alternate solvents are also being tested for their corrosion characteristics on metals and alloys that are specific to DOE/DOD facilities. The corrosion testing will be according to ASTM method F483-87 and will be conducted with the following 25 metallurgies: Alloy 25, Alloy 52, Al 1100, Al 2024, Al 5456, Al 6061, Al 7075, Al A356, carbon steel 1020, carbon steel 4340, copper alloy CDA 110, copper alloy CDA443, copper oxygen-free, Inconel X-750, Inconel 625, Kovar, Mg AZ31B, Monel K500, Ni 200, SS 304L, SS 310, SS 410, SS 17-4 PH, Ti grade 2, and Waspaloy.

The handbook database is currently under revision. Modifications to the database include utilizing a PC-based system with a graphical user interface operating under a Unix operating system and possibilities for future network capabilities. All data collected for the solvent utilization program will be included into the handbook database, as will data obtained from collaborations with Boeing Aerospace, Hughes, and others.

Phase II of the solvent utilization program will be initiated in FY-92. Phase II will involve toxicity testing (ATP), solvent loading and lifetime studies, rinsing requirements, extended corrosion testing, and hydrogen embrittlement studies.

Phase III will be full scale demonstrations of the alternative solvents selected and will be performed at Tinker AFB. Phase III will be done in later years.

THE ELIMINATION OF CHLORINATED, CHLOROFLUOROCARBON, AND OTHER RCRA HAZARDOUS SOLVENTS FROM THE Y-12 PLANT'S ENRICHED URANIUM OPERATIONS

D.H. Johnson, R.L. Patton and L.M. Thompson
Oak Ridge Y-12 Plant
Martin Marietta Energy Sytems, Inc.
Oak Ridge, Tennesse

INTRODUCTION

The Oak Ridge Y-12 Plant* is one of several plants which make up the Department of Energy's manufacturing facilities for production of nuclear weapons components and subassemblies. The plant is a comprehensive manufacturing facility with operations encompassing material manufacture, component fabrication and subassembly generation. All associated inspection and certification functions are performed in the plant. The plant's two primary products are uranium and lithium materials.

A major driving force in waste minimization within the plant is the reduction of mixed radioactive wastes associated with operations on highly enriched uranium. High enriched uranium has a high concentration of theuranium-235 isotope (up to 97.5% enrichment) and is radioactive, giving off alpha and low level gamma radiation. The material is fissionable with as little as two pounds dissolved in water being capable of producing a spontaneous chain reaction. For these reasons the material is processed in small batches or small geometries. Additionally, the material is completely recycled because of its strategic and monetary value.

Since the early eighties, the plant has had an active waste minimization program which has concentrated on substitution of less hazardous solvents wherever possible. The following paper summarizes efforts in two areas--development of a water-based machining coolant to replace perchloroethylene and substituion of an aliphatic solvent to replace solvents producing hazardous wastes as defined by the Resource, Conservation, and Recovery Act (RCRA). A summary of the plant's overall solvent substitution and reduction program can be found elsewhere[1].

*Managed by Martin Marietta Energy Systems, Inc., for the U.S. Department of Energy under contract DE-AC05-84OR21400

A WATER-BASED MACHINING COOLANT FOR USE WITH ENRICHED URANIUM

A 50% mixture of perchloroethylene (tetrachloroethylene) and mineral oil had been used in Y-12 for machining enriched uranium for nearly twenty years, but changing regulatory conditiions made its use very difficult. Both the Clean Water Act and the Toxic Substance Control Act listed perk as a hazardous substance and RCRA declared waste sludge of perchloroethylene to be hazardous. For these reasons, a new coolant was developed.

Perchloroethylene has several properties which make it ideal for use as a machining coolant for uranium. It is non-reactive with uranium and all known machine tool materials, it can extinguish small uranium chip fires, it enhances nuclear criticality safety due to the presence of the chlorine-35 isotope which is a neutron poison, and it facilitates recycle of chips due to the ease with which it evaporates. Any new coolant has to maintain these characteristics. Additionally it has to be safe for humans and generate no RCRA hazardous wastes with the over-riding issue being nuclear criticality safety at the expense of any other desirable characteristics.

A secondary issue was an operating philosophy based on best management practices which encourages the use of generic rather than proprietary chemicals. Examination of available literature shows a number of commercial, water-based coolants are available which can be used with uranium after modification to insure nuclear criticality safety. However, formulations are not specified, are subject to change, and vary from lot-to-lot; all of which conditions are unacceptable conditions in nuclear operations and support a decision to utilize a specified formulation.

The coolant formulation selected consists of a 50/50, by volume, mixture of water and propylene glycol to which is added 90 g/L sodium borate, 1000 ppm sodium nitrate, and a few drops of Azure Blue dye. The sodium borate is a neutron poison and provides the necessary criticality safety margins for the coolant. The sodium nitrate is a corrosion inhibitor and the dye is a coloring agent added to facilitate quick visual verification that the coolant in use is nuclear safe.

The new coolant was implemented in January, 1985. Perchloroethylene usage in Y-12 dropped from 1,200,000 pounds in 1984 to less than 130,000 pounds in 1986; however, all chlorocarbons or chlorofluorocarbons were not eliminated since degreasing agents and water removal chemicals were still required. Any residue of sodium borate left on the machining chips must be removed prior to chip recycle in order to maintain the required nuclear characteristics of the material stream. The chips are washed in distilled water to clean off the borate residue which leaves absorbed water on the chips. This water is displaced by dipping the chips in Freon-113 which is immiscible with water. The displaced water then floats on the Freon-113 and is skimmed off for recovery of any residual uranium or disposal. These operations led to an increase in Freon-113 usage in-plant of about 60,000 pounds, yielding a net reduction in controlled substance usage of approximately 1.1 million pounds.

ELIMINATION OF THE GENERATION OF RCRA HAZARDOUS WASTES FROM SHOP FLOOR OPERATIONS

Background

On December 20, 1989, Region IV of the United States Environmental Protection Agency (EPA) issued a regulatory interpretation memo[2] concerning solvent wipers which said that solvent wipes and rags "used in cleaning and degreasing operations with any solvent or mixture of solvents identified under the RCRA hazardous waste codes, F001-F005, at 40 CFR $261.31" are "considered to be a listed hazardous waste (i.e., a spent solvent)."

On April 12, 1990, in the U.S. District Court for the District of Colorado, Judge Lewis T. Babcock issued a Memorandum Opinion and Order[3] in a civil suit between the Sierra Club, Plaintiff, versus the U.S. Department of Energy and Rockwell International Corporation which stated that "Atomic Energy Act" process residues are regulated under RCRA as mixed radioactive waste until the radioactive components are separated from the RCRA waste components.

These two rulings required an immediate change in the way Y-12 was doing business. Firstly, all shop floor cleaning operations which used Freons, methyl chloroform, or any volatile organic compound (VOC) which produced wastes classified as RCRA characteristically hazardous had to be treated as RCRA hazardous wastes and were now subject to manifesting and associated control requirements prior to disposal. Since these wastes were incinerated to recover any uranium residues, treatment facilities for such wastes now had to be permitted as hazardous waste treatment facilities, almost an impossibility for such "land-banned" wastes. Therefore, an immediate program was undertaken to eliminate generation of all RCRA wastes from enriched uranium operations.

SELECTION OF A NEW SOLVENT SYSTEM CALLED WATERCHASER 140 FOR SHOP FLOOR USE

Several criteria were considered in selecting a substitute solvent system. They included requirements that the new solvent should clean as well as the solvent it was replacing, yield non-hazardous wastes as defined by RCRA, have low or minimal toxicity, be a non-air pollutant, be compatible with all weapons materials, and require minimal changes in the plant's operational areas in order to comply with applicable safety and fire codes. Additional considerations were the universality of the solvent and potential costs.

Examination of these requirements drives one to the conclusion that the solvent should have a flash point greater than 139°F to meet Occupational Health and Safety Act (OSHA) requirements and be a Class III liquid as defined by the National Fire Protection Association (NFPA) in order to allow open shop usage. A study of Y-12's production operations showed that the primary solvent in use for part and component cleaning was Freon-113 with lesser amounts of methyl chloroform, naphtha and various low flash point alcohols. Comparison of the characteristics of these solvents with available materials using Hansen Solubility theory[4,5] led to the conclusion that a medium weight aliphatic hydrocarbon mixture with a small amount of a polar co-solvent additive would meet the cleaning requirements as well as code requirements. Figure 1 shows a comparison of cleaning characteristics of a number of solvents and methods using electron spectroscopy for chemical analysis (ESCA) methodology to determine cleanliness. As can be seen Water Chaser 140 fulfills the stated needs.

A survey of available commercial solvents shows that a number of blends are available which meet these general needs. Usually they are called hydrocarbon blends or "varsols," and are almost always refinery fractions. As such, they are subject to the variability inherent in these operations. To prevent regulatory liability due to the presence of uncontrolled or unknown (primarily aromatic) chemicals in the solvent due to lot-to-lot variability, generic specifications requiring certification of contents were generated and used for procurement.

The solvent selected for general shop usage consists of a mixture of aliphatic hydrocarbons with 5% dipropylene glycol monomethyl ether (DPM) which Y-12 calls Water Chaser 140. The DPM has alcoholic functional groups as part of its structure which lends some polar character to the solvent and causes water to bead up on uranium surfaces. Because of the strong hydrophilic nature of the surface oxide film which forms when uranium is exposed to air, water forms a tightly bonded film on uranium. The DPM acts as a surfactant which breaks the bonds, causing the water to bead so that it can be easily wiped from the part or component. Table 1 gives a summary of the procurement specifications for the solvent and Table 2 gives a summary of the pertinent characteristics of the material.

Based upon the data shown, Water Chaser 140 meets the majority of the characteristics desired when the study started; however, some obvious problems arise. The flash point causes the material to be classed as combustible, leading to increased potential fire loads in shops where either chlorocarbons or chlorofluorocarbons were previously used. Secondly, the material has a toxicity rating of 2 based upon the TLV-TWA. For these reasons, best management practices would indicate that the solvent wipes and rags should be stored in closed, nuclear-safe, approved containers. The low vapor pressure increases drying time, but also prevents exceeding the TLV if any ventilation at all is present. The increased drying time is not a problem in practice if operators wipe parts to remove excess liquid and, in fact, leads to cleaner parts because wiping is a much better cleaning method than air drying which is the general practice with highly volatile solvents.

Procurement was accomplished via standard bid processes with bids received from three chemical companies. In-house certification of the material indicated all specifications had been met by the low bidder and the material was accepted for use as a weapons-approved cleaning agent.

Implementation occurred during June, 1990. No major problems surfaced. To-date, no use related health incidents have occurred; i.e., no dermatitis, allergic reactions, or skin irritations have been reported. Industrial hygiene monitoring has been on-going and to-date no incidence in which the TLV was exceeded has occurred. On one occasion, a worst case scenario for exposure was mocked-up in which a two-gallon bucket of solvent was poured onto a machine; the maximum measured air concentration was 25ppm. With wiping, part cleaning times are comparable to previous experience.

CONCLUSIONS

Two major generators of RCRA wastes in the Y-12 Plant have been eliminated. A water-based machining coolant has been implemented to replace a perchloroethylene based coolant and an aliphatic hydrocarbon based solvent has been implemented to replace previous solvents which produced RCRA hazardous wastes when used in shop floor degreasing and cleaning operations.

REFERENCES

1. Thompson, L.M.; Simandl, R.F.; and Richards, H.L.; Y-xxxx, "Chlorinated Solvent Substitution Program at the Y-12 Plant"; January, 1991.

2. Memo: James H. Scarbrough, Chief, RCRA Branch, United States Environmental Protection Agency Region IV Office, Atlanta, Georgia; to RCRA Branch Personnel; Subject: Regulatory Status of Solvent Wipers; December 20, 1989.

3. Memorandum Opinion and Order in the United States District Court for the District of Colorado; Lewis T. Babcock, Judge; Civil Action No. 89-B-181; Sierra Club, Plaintiff vs. United States Department of Energy, and Rockwell International Corporation, a Delaware corporation; Filed April 12, 1990.

4. Hansen, C.M.; "The Universality of the Solubility Parameter," IE & C Product Research and Development; ACS Publication Volume 8, Number 1; 1969.

5. Barton, Allan F.M.; CRC Handbook of Solubility Parameters and Other Cohesion Parameters; CRC Press, Inc.; Boca Raton, Florida; 1983; 594 pp.

TABLE 1
SUMMARY OF PROCUREMENT SPECIFICATIONS FOR WATER CHASER 140

Flash Point	141°F Minimum per ASTM-D-56 TCC Method
Specific Gravity	0.777 - 0.827 @ 60°F
Evaporation Residue	<200 micrograms per gram
Acidity	Neutral per ASTM-D-1093
Volume % Aromatics	5% maximum per NMR Methodology
Doctor Test (for Sulfur)	Negative per ASTM-D-235
DPM	5 +/- 1% per G C Mass Spec or NMR
Aliphatic Hydrocarbons	> 90% per G C Mass Spec or NMR

TABLE 2
APPLICATION CHARACTERISTICS OF IMPORTANCE FOR WATER CHASER 140

Flashpoint	142°C
NFPA Class III Liquid	Allows open usage without necessity to store in a flammable storage cabinet during off shift
TLV-TWA	100 ppm

* Can cause irritation to eyes
* Prolonged exposure to skin can cause dermatitis
* Excessive inhalation can cause irritation, headaches, or asphyxiation

Vapor Pressure	0.50 mm Hg

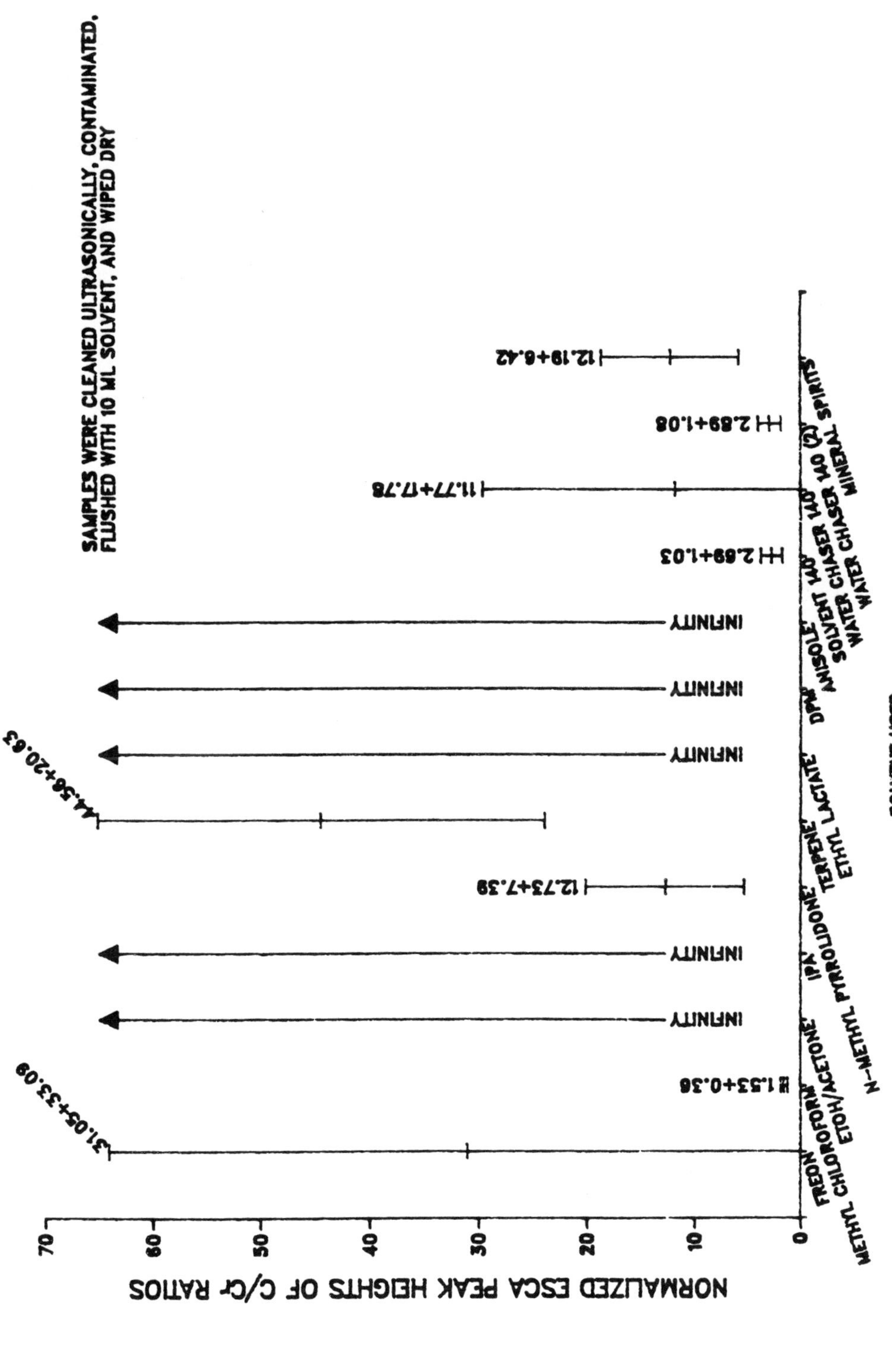

FIGURE 1 — ABILITY OF SOLVENTS TO REMOVE LIGHT OILS FROM 304L SS

PRINTED CIRCUIT BOARD DEFLUXING: ALTERNATIVES TO OZONE DEPLETING SUBSTANCES

Katy Wolf
Institute for Research and Technical Assistance
Los Angeles, California

INTRODUCTION

The two solvents most widely used for removing the flux from printed circuit (PC) boards after components have been soldered to them are 1,1,1-trichloroethane (TCA) and 1,1,2-trichloro-1,2,2-trifluoroethane (CFC-113). These solvents will be banned worldwide over the next decade or so because they contribute to stratospheric ozone depletion. Over the short term, users must begin to adopt methods of reducing the use of these substances and over the long term, users must identify and implement alternatives. This paper discusses the methods for reducing or eliminating the use of the two solvents in PC board defluxing.

OZONE DEPLETION AND THE REGULATORY REGIME

In 1974, two professors first put forth the theory of ozone depletion. There were substances called chlorofluorocarbons or CFCs that were very stable. They were so stable that they survived in the atmosphere without decomposing for some 100 years. During that time, they made their way from the troposphere or lower atmosphere to the stratosphere or upper atmosphere. Once there, ultraviolet light decomposed them and the chlorine they contained was released. This chlorine was then available to catalytically react with ozone, depleting the protective ozone layer that shields us from harmful radiation in the so-called B range. Each chlorine atom was capable of reacting with 100,000 times its own mass in ozone.

In 1978, the U.S. unilaterally banned the use of CFCs as propellants in nonessential aerosol applications. For some years, this action was assumed to be adequate because such applications represented about one-third of world CFC use. In 1985, the Antarctic ozone hole was discovered and the world became aware that ozone depletion was a serious problem. In September of 1987, most major world nations signed the Montreal Protocol, the agreement to limit the use of ozone depleting substances. The Protocol called for a phasedown to half the 1986 production level of the CFCs by 1998. In 1988, the Ozone Trends Panel reported that ozone depletion was more serious than had been previously thought. In June of 1990, at a meeting in London, the Montreal Protocol was strengthened. The London Amendments called for a complete phase out of the CFCs by the year 2000; TCA was added to the protocol and would be phased out somewhat later--in the year 2005. The Clean Air Act, recently reauthorized in the U.S., calls for a domestic CFC ban in 2000 and a ban on TCA in 2002. As an additional incentive to discourage the use of zone depleting substances, Congress placed a tax of $1.10 per pound on CFC-113 in January of 1990 and a tax of $0.137 per pound on TCA in January of 1991.

PRINTED CIRCUIT BOARD DEFLUXING

In the PC board assembly process, flux is first added to the boards to facilitate solder flow, to effect heat transfer and to prevent oxide formation. Flux is generally composed of a vehicle, usually alcohol, and activators which are solids. The components are then soldered to the boards. In the case of through hole boards, the components are soldered through holes that have been drilled in the boards. In the case of surface mount boards, the components are soldered to the surface of the boards. There is a trend toward surface mount technology because a higher density of components can be obtained. CFC-113 or TCA combined with alcohol are the solvents most commonly used for removing the flux and other contaminants from the boards.

Military specifications (mil specs) currently govern the use of solvents in the cleaning process for those with military contracts. The mil specs have become the worldwide de facto standard. Mil-Std-2000 calls out the solvents that can be employed in the cleaning process. It allows the use of CFC-113, TCA, various alcohols or combinations of the solvents with alcohols. Water can be used only with special permission. The standard also requires the use of a particular flux--rosin flux. Although the use of recycled solvent is not expressly forbidden, the standard refers to solvent purity specs that cannot be met with recycled solvent. The moisture level requirement is very low and it is doubtful that most virgin solvent could meet the standard either.

In recognition of the fact that CFC-113 and TCA will be banned, the Department of Defense (DOD) is aware that Mil-Std-2000 must be changed to allow the use of other cleaning methods. Although two different groups are involved in the changes, there is no indication that the standard will be changed in the near future.

METHODS OF REDUCING OR ELIMINATING THE USE OF ZONE DEPLETING SUBSTANCES

In the light of the ban on CFC-113 and TCA, users must employ measures for reducing the use of the substances over the short term and eliminating their use altogether over the long term. Short term measures include conversion from CFC-113 to TCA; adopting improved equipment; use of recycled solvent; and vapor recovery. Long term alternatives include substitution of flammable solvents, combustible solvents, HCFC or HFC solvents, aqueous based cleaning, no clean flux, and controlled atmosphere soldering.

TCA is currently four times less costly than CFC-113. Its ozone depletion potential is also lower--at only one-eighth that of CFC-113. Users can reduce their costs and improve the environment by converting to TCA immediately. TCA is a more aggressive solvent than CFC-113 and compatibility testing must be conducted to assess its suitability for a particular board. Polycarbonate materials, for instance, are incompatible with TCA and some PC boards contain it.

Many defluxing units allow high solvent emissions. Adding refrigerated freeboard chillers and extending the freeboard can better contain the solvent vapors and reduce emissions. Automated hoists which remove parts at a standard 11 feet per minute can reduce dragout losses on boards. Hot vapor recycle which relies on superheating the vapors can also reduce dragout.

Better solvent equipment with close to zero emissions is being offered by European and Japanese manufacturers. Tiyoda, the manufacturer of one new unit, guarantees that losses will be no grater than 10 kilograms per month. Such tight equipment is generally very expensive--between $125,000 and $200,000.

For users who will convert to alternative non-solvent processes, the capital investment would be significant for a short period of time. For some users who plan to continue using CFC-113 or TCA for several more years, this equipment could be cost-justified.

The Congressional tax applies only to virgin, not recycled solvent. Users can purchase and use stills on-site and use the reclaimed solvent in place of virgin solvent. Recyclers also offer recycled solvent that can be used in place of virgin solvent at a reduced cost. At this stage, the demand for recycled solvent is very high and it is not always easily available. Users subject to the mil spec probably cannot employ recycled solvent until Mil-Std-2000 is changed.

Vapor recovery methods can be used to reduce solvent vapor losses. Traditional carbon adsorption with steam desorption has been used for many years with CFC-113 based solvents. TCA is extremely unstable to hydrolysis and steam is not a good choice for desorption. Carbon can be used with inert gas regeneration, however. Another method, the Brayton Cycle Heat Pump, is being demonstrated in various solvent applications. It employs carbon and inert gas for regeneration and is said to reduce energy use.

Flammable solvents like alcohols and mineral spirits were used more frequently in the past and they are an option to replace CFC-113 and TCA. These low molecular weight hydrocarbons pose a workplace danger because they are flammable and, because they are photochemically reactive, many local air districts will not grant a permit for their use. EPA has issued a final test rule requiring toxicity testing for isopropyl alcohol, one of the potential alternatives.

So-called combustible solvents are higher molecular weight hydrocarbons with flash points in the combustible range. These include solvents like terpenes, Dibasic esters (DBE), N-methyl pyrollidone (NMP) and alkyl acetates. These solvents are not volatile and they require a water rinse. The semi-aqueous process involves spraying with solvent, rinsing in an emulsion bath, rinsing in a water bath and drying. Equipment has been designed to use these solvents. Although they are photochemically reactive and require a permit, emissions are likely to be relatively low. D-limonene, a major ingredient of terpene formulations, has given a positive carcinogenicity test in male rates. Most of the other solvents have not been tested for chronic toxicity or adequately scrutinized for their environmental effects.

DuPont and Allied are marketing mixtures of HCFC-123 and HCFC-141b with alcohol for PC board defluxing. The HCFCs have lower ozone depletion potential then CFCs because they contain hydrogen and therefore break down more readily in the lower atmosphere. The two HCFCs are more aggressive than CFC-113 but less aggressive than TCA. They are not compatible with polycarbonate. Depending on the specific blend, the boiling points of the solvents are in the 80 to 90 degree F range. Solvent losses will be very high unless the chemicals are used in extremely conservative equipment that has been designed to minimize losses. HCFC-123 and HCFC-141b are currently being tested by a consortium of international CFC producers for chronic toxicity. The results of the tests should be available in the 1993/1994 time frame.

Asahi Glass is examining another HCFC, HCFC-225. It is an extremely gentle solvent with properties that are very close to those of CFC-113. It has just gone into animal testing and the results are likely to be available after 1995. The solvent is composed of two isomers, one of which appears to be very toxic. It is not clear whether a manufacturing process that selectively produces the non-toxic isomer can be developed.

One HFC is being marketed for PC board defluxing. Pentafluoropropanol has not been tested for chronic toxicity and is not likely to be so tested in the future. It has a very low workplace exposure level, reflecting its high acute toxicity.

Aqueous cleaning formulations can be used in two ways. First, water can be employed with rosin flux and a surfactant. Second, water can be used with organic acid or water dispersible flux; no surfactant is necessary because the flux is water soluble. Closed loop water recycling cannot be used with surfactant because it builds up in the system; on-going research is attempting to solve this problem. Water is a suitable cleaner for through hole boards. Some users claim that water cannot be used for surface mount boards where the spacing between the board and the components is less than 10 mls. They argue that water, because of its contact angle, cannot penetrate under the components and carry away the contaminants. The effluent from boards that have been cleaned is likely to require treatment to remove the lead from the solder. This can be done using an ion exchange. Water based cleaning requires more energy for drying and generally more floor space.

Low solids flux contains only about 3 to 5 percent solids compared to traditional flux which contains 25 to 30 percent solids. Because of the low solids content, the flux does not have to be removed from the board after soldering. A problem with low solids flux is that much more stringent process control is required to deposit the solids on the board in a uniform manner. Not all users will be sophisticated enough to employ the technique.

Controlled atmosphere soldering using a reducing atmosphere like hydrogen or an inert atmosphere like nitrogen is another option. The equipment is expensive and more stringent process control is again required.

CONCLUSIONS

CFC-113 and TCA the major solvents used today for defluxing PC boards will be phased out over the next decade. Over the short term, users can adopt measure that will minimize the use of the solvents. Over the long term, users can test and implement alternative processes that will eliminate the use of the ozone depleting substances. Virtually all of the alternatives have limitations and many of them have unknown and unexplored health and environmental effects. The most suitable alternative will have to be selected on a case-by-case basis.

ELECTRONIC ASSEMBLY SOLVENT SUBSTITUTES

Alex Sapre, Ph.D.
Environmental Technology
Hughes Aircraft Company
Los Angeles, California

SUMMARY

In the worldwide movement to protect the environment, perhaps the most significant event so far has been the signing of the Montreal Protocol in 1987, where over 40 countries agreed to phase out the production of chlorofluorocarbons and other ozone-depleting chemicals (ODCs). In June 1990, the London Amendment to the Montreal Protocol substantially accelerated the phase-out schedule and also covered additional ODCs, methyl chloroform and carbon tetrachloride.

The defense electronics industry is significantly affected by these decisions because ODCs are used in the manufacture and assembly of high quality electronics hardware. Moreover, in most instances such use is dictated by military specifications (milspecs) or industry-wide standards.

Many technology alternatives are being developed to replace ODCs and other solvents. However, no technology has completed all the necessary requirements for wide-spread implementation at this time, particularly in defense electronics applications.

The cheapest and the quickest way to reduce the consumption of ODCs is to improve operating and, wherever practical, engineering practices. There is a long list of recommended operating practices such as preventing drafts around degreasers, covering degreasers, optimizing production schedules and solvent replenishment, repairing leaks, increasing operator awareness, and training employees.

Engineering controls can also help reduce consumption of ODCs. Some examples are increasing freeboard height, retrofitting automatic hoists and programming them for proper entry and exit speeds, installing automatic covers, installing extra cooling coils etc.

Organic solvents such as ketones, aromatics and alcohols can remove solder fluxes and many polar contaminants. However, they are typically volatile, flammable, and their emissions are considered VOCs, which in combination with nitrogen oxides cause smog. Accordingly, they are not desirable alternatives.

Chlorinated compounds such as trichloroethylene, perchloroethylene, and methylene chloride are effective cleaners, but they all are considered carcinogenic. Hence, they also will not be viable alternatives.

Water is an excellent solvent for removing ionic contaminates and water soluble fluxes. Water in combination with a saponifier can remove water insoluble substances such as oil and rosin fluxes. The saponifier chemical does not dissolve rosin but reacts with it to form an alkaline/amine salt which is water soluble. In most instances, aqueous saponifier cleaning is claimed to be as effective as CFC blends in removing ionic and nonionic rosin flux residues. Aqueous saponifier cleaning may also be used as an interim step to long term conversion to organic water soluble fluxes.

Saponifier cleaning does have several drawbacks. Aqueous saponifier cleaning requires more process control than CFC solvent cleaning to achieve consistent cleaning efficiency. The active concentration of saponifier in the wash tank is reduced over time as a result of reaction with rosin and by the addition of water to the tank to replace losses from drag-out. Saponifier is an alkaline chemical which is corrosive to solder joints if not totally removed. This can put limitations on geometry of designs. Also, saponifiers can be incompatible with certain materials.

Regulations placing limitations on waste water contaminants which can be discharged into a public sewer without pretreatment are also an important consideration when converting to aqueous cleaning.

A newer technology for aqueous removal of rosin flux residue is the use of terpene solvent cleaning. The terpene is mixed with a large amount of a surfactant which acts as an emulsifier so that the terpene and rosin can be flushed off the circuit boards with water. The terpene actually dissolves the rosin rather than reacting with it as in the case with saponifier cleaners. The solvent is non-corrosive and a much lower amount of lead is dissolved into the rinse water.

There are some drawbacks to this cleaning process. Conventional aqueous spray cleaning equipment cannot be used because the emulsifier generates excessive foam. The mixture represents a potential safety hazard because it is combustible, with a flash point of approximately 120°F and is used at elevated temperatures. Terpenes are considered VOCs. In addition, water discharge regulations need to be taken into account even in this case.

In the arena of new solvents, Hughes is developing two unique technologies. One, known as the reacting aqueous defluxing systems (RADS), involves aqueous cleaning, while the other takes advantage of unique solvent properties of some substances under supercritical conditions.

RADS is a four-part process consisting of a reaction phase, neutralization rinse, final rinse and drying. In addition, a separate loop for regenerating the RADS solvent is added to minimize the effluent from the process. The reaction phase operates at about 140°F with a pH of 10.5-11. The neutralization rinse uses deionized water at 120°F, while the final rinse consists of deionized water at 100°F. The drying sequence can consist of any viable water removal method.

Unlike other aqueous systems, RADS is an excellent defluxing technology with no VOC emissions and with desirable environmental characteristics. Its material compatibility behavior also appears very acceptable. Substantial work in optimizing the process and in further minimizing its environmental impact is currently underway for this technology.

Some chemicals develop good solvent properties under supercritical conditions. Currently, carbon dioxide is being evaluated for cleaning flex cables, cryogenic detector components, bearings, hybrid micro-circuits and navigational gyros. Initial results on solder flux removal from a printed circuit board, designed to function as a power control unit, are very promising.

The use of organic water soluble fluxes and low solids "leave on" ("no clean") fluxes offers another way tc eliminate the use of solvents in defluxing.

Organic, water soluble fluxes have been used for wave soldering of circuit boards for many years. Nearly half of electronics companies in the U.S. use water soluble flux. It has been used in the computer telecommunications, industrial, automotive and consumer electronics market segments. Water soluble fluxes generally possess higher activity than rosin fluxes so their soldering performance is better than rosin fluxes, particularly when boards or component leads are excessively oxidized. Improvements in the technology over the last several years have resulted in fluxes which result in better ionic cleanliness and surface insulation resistance characteristics than previously attained.

Most water soluble fluxes possess a chloride content of 2-3 percent. Most are acidic, highly ionic and potentially corrosive so circuit board assemblies must be designed so that flux entrapment is avoided.

Surface insulation resistance after cleaning is usually reduced when using an organic, water soluble flux. Reliability problems with high impedance circuitry, are also experienced with these fluxes.

Very few halide-free water soluble fluxes are currently available for wave soldering of circuit board assemblies. Their thermal stability is not as good as chloride-containing water soluble fluxes in that they occasionally tend to form insoluble residues.

Water soluble fluxes offer the same water discharge considerations as aqueous cleaners. Because of their acidic nature, water soluble fluxes can leach lead from the circuit boards. The large amount of organic material in the wash tank discharge can make total recycling of the wash water for reuse on-site impractical.

One way to eliminate both the use of ozone depleting solvents and also the potential water discharge problems is to use low solids "no clean" fluxes. Also, they offer additional benefits. For example, manufacturing floor space required for a given line can be reduced by elimination of cleaning equipment. Components, which were hand soldered onto an assembly after wave soldering because they could not withstand the cleaning operation, can be inserted onto the circuit board prior to wave soldering.

The "no-clean" fluxes are halogen-free with a solids content of 2-5 percent. Carboxylic-type acids are the main activation system for the flux. Most assemblies will appear cosmetically clean after soldering. A small amount of residue is left behind after soldering so the flux must not be capable of degrading reliability of the assembly to be suitable for use.

The range of process parameters at which optimum soldering performance can be obtained is narrower for low solid fluxes than that for the traditional flux types. Therefore, it would be more difficult to implement these fluxes. Significant effort is necessary to develop "no-clean" fluxes for high reliability applications.

So far, the Department of Defense has not viewed favorably the use of water soluble and, particularly, low solid, "no-clean" fluxes.

Fluxless-soldering, inert atmosphere soldering and organic solders (conductive adhesives) offer another set of alternatives to reduce/eliminate solvent usage in cleaning of electronics hardware.

A project is underway at Hughes to develop a series of lead/tin base solder alloys that will not require fluxes for joining of aerospace electronics. Small amounts of a highly reactive material such as lithium are added to chemically reduce the oxides of lead and tin when the solder is fused. The products of the solder oxide reduction reactions are inert oxides of the additives. Several eutectic lead/tin solder alloys containing lithium or indium sulfide particles in the range of 0.1 to 0.5 atomic percent additions have been prepared using powder mixing techniques and melting characteristics of alloys. Improved techniques for incorporating lithium particles into the solder are under evaluation.

In the area of inert atmosphere soldering, nitrogen inerted wave soldering machines are attracting considerable attention. In addition to nitrogen, formic acid is sometime injected near the solder wave in order to scavenge residual oxygen in the nitrogen, as well as the oxides on the board. In order to promote hole filling and to promote wetting of the clipped leads, substances such as adipic acid are applied to the circuit assembly using a spray fluxer.

Potential organic solder substitutes are presently marketed as conductive adhesives, typically metal-filled epoxies, and are used for many hermetic microelectronics applications. Currently a project is underway at Hughes to develop a detailed specification describing the required physical, mechanical and electrical properties of the potential substitute materials. Of particular importance is preservation of these properties under effects of temperature and/or humidity cycling and vibration. Approximately 30 organic solder samples would be evaluated in the project.

Table 1

Estimates of Ozone Depletion Potentials (ODP) and Global Warming potentials (GWP) of Substances Included in the Montreal Protocol (Normalized with respect to CFC-11)

Chemical	ODP	GWP
CFC-11	1.0	1.0*
CFC-12	0.95	3.1
CFC-113	0.85	1.35
CFC-114	0.70	3.90
CFC-115	0.40	7.50
HCFC-22	0.05	0.35
HCFC-123	0.017	0.018
HCFC-124	0.020	0.096
HFC-125	0	0.58
HFC-134a	0	0.26
HCFC-141b	0.09	0.090
HCFC-142b	0.05	0.36
HFC-143a	0	0.74
HFC-152a	0	0.03
CC14	1.1	0.34
CH3CC13	0.13	0.024
Halon 1301	10.0	NA
Halon 1211	2.6	NA
Halon 2402	5.6	NA

*Global warming potential of CFC-11 is estimated to be 4,000 times the potential of carbon dioxide (CO_2) on a mass basis.

CHLORINATED SOLVENT SUBSTITUTION PROGRAM AT THE OAK RIDGE Y-12 PLANT*

L. M. Thompson, R. F. Simandl and H.L. Richards
Martin Marietta Energy Systems, Inc.
Oak Ridge, Tennessee

INTRODUCTION

In recent years, several regulations regarding chlorinated solvents have been established. The Montreal Protocol, which has been ratified by the United States, calls for a ban on production of chemicals such as methyl chloroform and trichlorotrifluoroethane due to their association with the depletion of the ozone layer. Other chlorinated solvents such as perchloroethylene and methylene chloride have been identified as suspect carcinogens. All of these solvents mentioned above are listed wastes under the Resource Conservation and Recovery Act (RCRA) which strictly restrains handling and disposal. These regulations make substitution of these solvents very appealing. The Oak Ridge Y-12 Plant has been actively seeking substitutions for these solvents for the past 7 years.

CONSIDERATIONS FOR SUBSTITUTION

The first step in our substitution program was to determine the uses of the chlorinated solvents. This step was done by conducting usage surveys in the plant and by compiling information of purchases from the Y-12 stores and other purchasing systems. The main uses of these solvents were determined to be for cleaning purposes. The uses included cleaning parts after machining and prior to inspection, cleaning chips in the chip cleaning facility, cleaning urethane foam guns, cleaning meter mix machines, and dissolving adhesives. In looking for substitutes for chlorinated solvents, there are several pitfalls one must avoid.

Large problems can result from overlooking small details. Factors which we take into consideration include compatibility, toxicity, flammability, means of disposal, effects on production, and ability.

Compatibility issues concerning the material to be cleaned are usually addressed by conducting submersion tests and looking for signs of corrosion. When addressing compatibility issues, one must look not only at the material to be cleaned but at the handling materials such as gloves, wipes, and dispensers. Degradation of these materials can transfer unwanted contamination to the part as well as risk personnel exposure. This issue is usually addressed by conducting submersion tests with the materials in the potential solvent substitute or conducting surface analysis studies on a sample which has been cleaned while using these materials.

Toxicity issues are addressed by searching for health properties of the solvent in sources such as *Sax's Dangerous Properties of Industrial Chemicals*, *Registry of Toxic Effects of Chemicals*, and the *Hazardous Substance Data Bank*. Our Industrial hygiene Department also evaluates the solvent and/or cleaning operation for health concerns and determines if monitoring is needed or the personal protective equipment to be used.

When it is necessary to replace a halogenated solvent with another organic solvent, one must begin looking at flammable solvents. Due to stringent requirements by the Occupational Safety and Health Administration (OSHA)

*Managed by Martin Marietta Energy Systems, Inc., for the U.S. Department of Energy under Contract DE-AC05-84OR21400.

regarding highly flammable solvents, our approach has been to use solvents with a flash point of 140°F or higher. This also enable you to be above the 140°F limit given as a characteristic RCRA waste. Other concerns which must be addressed concerning flammable solvents deal with the proper use of these solvents which includes storage, use, and dispensing. The ability of the solvent to form peroxides is examined so that the solvent could be handled safetly.

The means of disposal must be examined to prevent generation of a waste which cannot be handled. Generally, a biodegradable solvent is desirable although incineration is acceptable. Bacteria from our biodegradation pond are placed in a given amount of solvent. This mixture is then tested to determine if and when the solvent is degraded.

Effects on production include several items depending upon the next operation involved. When using a solvent with a high flash point, the evaporation rate is much slower than a chlorinated solvent. This results in changes in production whether it be simply waiting longer before doing the next step or drying with a paper towel. Other effects include effects on bond strength, effects on welding, or effects on handling.

In testing the ability of a substitute solvent or cleaning method, comparative studies between possible substitutes and the current cleaning method are first conducted on small samples. Surface analysis is conducted using X-ray Photoelectron Spectroscopy (XPS/ESCA). Using this technique, a surface is bombarded with X rays and the energy of the electrons emitted from the surface is measured. The electrons from different elements or elements in specific bonding states have different binding energies. Thus, one can determine the specific elements or combination of elements on a surface from the measure of these different energy levels. XPS/ESCA is capable of examining microlayers of a surface. Data are recorded for a contaminated surface to get a feel for what elements are present due to the contamination. Peak height ratios of the main element associated with the contamination to the base metal are calculated from the XPS/ESCA data. These ratios are compared to determine the effectiveness of the solvent and/or cleaning operation. The lower this ratio, the cleaner the surface. The ability to remove other elements associated with the contamination is also examined. After conducting these studies, the solvent and/or cleaning operation is tested on a larger scale to determine if there are any problems associated with its use.

ULTRASONIC CLEANING

Substitution efforts at Y-12 have been divided into two main efforts. The first effort was the replacement of large vapor degreasers utilizing chlorinated solvents with ultrasonic cleaners using aqueous detergent and water. Ultrasonic cleaning works by cavitating a liquid and forming small micro bubbles which burst on the surface to be cleaned. This provides mechanical as well as chemical cleaning action. Three variables can influence the effectiveness of ultrasonic cleaning: 1) the frequency of the ultrasonic cleaner, 2) the liquid medium, and 3) the coupling between the cleaner and the liquid. In order to cavitate the liquid, a frequency of at least 18 kHz is required.

Ultrasonic cleaning has been shown to perform as well as if not better than cleaning with vapor degreasers as shown in Fig. 1. In this study, samples of uranium/6% niobium were initially cleaned ultrasonically for 8 h in a detergent, isopropanol, and demineralized water solution at 50°C, rinsed in demineralized water, fingerprinted, and handled thoroughly. One sample was retained as a control specimen while the remaining specimens were dipped in a rust preventative oil and allowed to dry. The specimens were cleaned using the techniques described. All of the specimens with the exception of the specimens degreased in isopropanol had comparable cleanliness levels. The sample which was soaked in detergent had slightly higher levels of contamination. However, this level may be acceptable for some operations. In the majority of our operations, the power of ultrasonics was required to enable the liquid to clean in small cracks and crevices which would otherwise not be cleaned.

There are some drawbacks with ultrasonic cleaning. The equipment requires an initial capital investment, a rinse step must be included in the process and a drying step is also necessary.

SOLVENT SUBSTITUTES

The second phase of the substitution program has been the replacement of squirt bottle or specialty type operations with other organic solvents. The main application was the cleaning of parts after machining or prior to inspection by wiping. The contaminants being cleaned from the surface were the usual machine shop contaminants such as machining coolant, rust preventative oil, lapping oil, lubricants, and fingerprints. Initially, possible solvent substitutes were chosen using Hansen Solubility Parameter Theory.[2] Using this theory, solvents which have similar parameters have similar solvent properties. A wide range of solvent types have been tested for the different contaminants which are present. The experimental procedure used was to initially clean samples of type 304L stainless steel ultrasonically in aqueous detergent and water in order to establish a baseline level of cleanliness. A sample was retained as a control sample. The remaining specimens were coated with the contaminant and allowed to dry overnight. One contaminated sample was also retained to determine what elements are present due to contamination. Each sample was then squirted with a given amount of solvent being tested and wiped dry. The samples were submitted to XPS/ESCA for analysis. The main element present due to the contamination was carbon. Therefore, a peak height ratio of carbon to chromium (which represents the base metal) was calculated and comparisons were made of this ratio.

Figure 2 shows the results of the study comparing solvents for the cleaning of rust preventative oil. Solvents such as dipropylene glycol methyl ether (DPM), ethyl lactate, anisole, propylene glycol methyl ether acetate (pm acetate), ethanol denatured with acetone (EtOH/acetone), and isopropanol did not remove the rust preventative oil sufficiently enough to enable the ESCA to see the metal surface. A terpene-based cleaner and N-methyl pyrrolidone (NMP) worked as well as CFC-113. Solvent 140, which is a high flash mineral spirit, worked as well as the methyl chloroform and better than CFC-113. Ultrasonic cleaning with aqueous detergent yielded the cleanest surfaces.

Figure 3 shows the results of a study comparing the ability of solvents to remove lapping oil. DPM, ethyl lactate, and a terpene-based solvent yielded the dirtiest surfaces followed by anisole, pm acetate, NMP, and Solvent 140. A solvent blend, which consists of 95% Solvent 140 with 5% DPM, yielded the best results of the possible solvent substitutes compared to methyl chloroform and CFC-113. Ultrasonic cleaning with aqueous detergent again yielded the cleanest surfaces overall.

Figure 4 shows the results of a study comparing the ability of solvents to remove a water-based machining coolant known as Trim Sol. Solvents such as anisole, pm acetate, isopropanol, Water Chaser 140, and Solvent 140 yielded surfaces comparable to those cleaned with CFC-113. Surfaces cleaned with DPM, ethyl lactate, and a terpene-based cleaner were somewhat dirtier and the ethanol/acetone blend yielded the dirtiest surface overall. The cleanest surface was found by cleaning with ultrasonic cleaning with aqueous detergent.

Figure 5 shows the results of a study comparing the ability of solvents to remove fingerprints. The ethanol/acetone solvent mixture and the ultrasonic cleaning yielded the best results overall. Methyl chloroform, isopropanol, and Solvent 140 gave the next best results followed by Water Chaser 140. CFC-113 yielded the worst results and did not appear to remove the fingerprint oils. All of the organic solvents left behind inorganic contamination from the fingerprints such as sodium, nitrogen, sulfur, potassium, chlorine, and calcium.

Due to these results, the Y-12 Plant is currently in the process of changing to a strategy using Solvent 140 and Water Chaser 140. Since Solvent 140 is totally immiscible with water, a blend of Solvent 140 with DPM was developed. Adding the DPM enables the solvent to be slightly miscible with water which aids the solvent in its ability to "chase" water-based machining coolants. This also adds the power of a polar solvent to that of a nonpolar solvent. These solvents have low toxicity, are non-RCRA, are easily handled under fire code considerations, and are compatible with materials used at Y-12. Solvent 140 will be used in moisture-sensitive areas of the plant while Water Chaser 140 will be used in the remainder of the plant. The drawbacks of these solvents are that they are flammable and are slow evaporators which require a change in production operations.

REMOVAL OF URETHANES AND EPOXIES

Other uses of chlorinated solvents, namely methylene chloride, at Y-12 have been for the cleaning of urethane foam spray guns, dissolving of urethane adhesives, and removal of epoxies.

Three solvents have been found to be effective in the cleaning of urethane foam spray guns and dissolving of urethane adhesives. These solvents are anisole, dibasic esters, and N-methylpyrrolidone. The anisole works well but is a characteristic RCRA waste because its flash point is below 140°F. Dibasic esters work but are slow evaporators and one must be careful regarding compatibility with O-rings using this solvent. The N-methylpyrrolidone also works well but is also a slow evaporator.

Several experiments have been conducted regarding the swelling of epoxy bonds. Methylene chloride is a good solvent for this application because it is a small molecule and can penetrate into the epoxy structure easily and swell the epoxy. Substitutes such as anisole and N-methylpyrrolidone are large molecules and cannot penetrate the structure as easily. Thus, they act more slowly at swelling the epoxy. However, adding a solvent with a smaller molecular structure to these solvents appears to penetrate the structure so that the larger molecules can penetrate and work on the epoxy. A 10% solution of acetone in N-methylpyrrolidone or a 10% solution of methanol in anisole appears to cut back on the time required to swell epoxy compared to N-methylpyrollidone or anisole by themselves.

CONCLUSIONS

Chlorinated solvents are widely used throughout industry for cleaning purposes. However, because of health and environmental problems associated with their use, substitution of these materials has become desirable. The Y-12 Plant has successfully substituted ultrasonic cleaning with aqueous detergent as a substitute for large vapor degreasers and Solvent 140 and Water Chaser 140 for chlorinated solvents used in squirt bottle type operations. There are drawbacks associated with the use of these substitutes but these drawbacks can be overcome.

REFERENCES

1. Sax, N. Irving, *Dangerous Properties of Industrial Materials*, Van Nostrand Reinhold Company, New York, New York.

2. Hansen, C. M., *The Universality of the Solubility Parameter*, IE&C Product Research and Development, ACS Publication, Vol. 8, 1969.

FIG. 1 – COMPARISON OF ULTRASONIC CLEANING WITH OTHER CLEANING TECHNIQUES

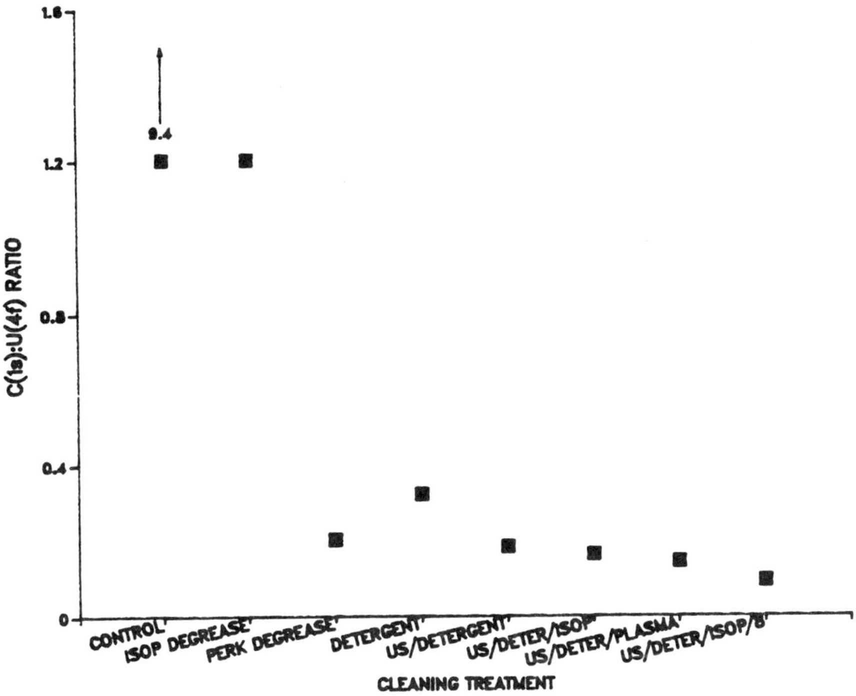

FIG. 2 – ABILITY OF SOLVENTS TO REMOVE LIGHT OILS FROM 304L SS

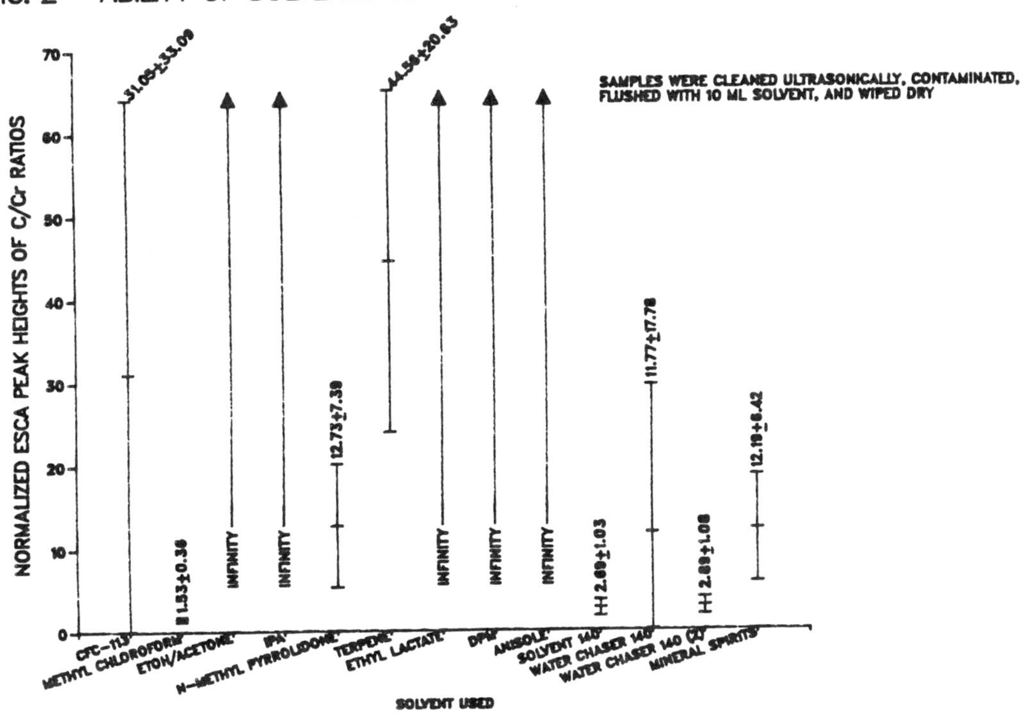

FIG. 3 – ABILITY OF SOLVENTS TO REMOVE LAPPING OIL FROM 304L SS

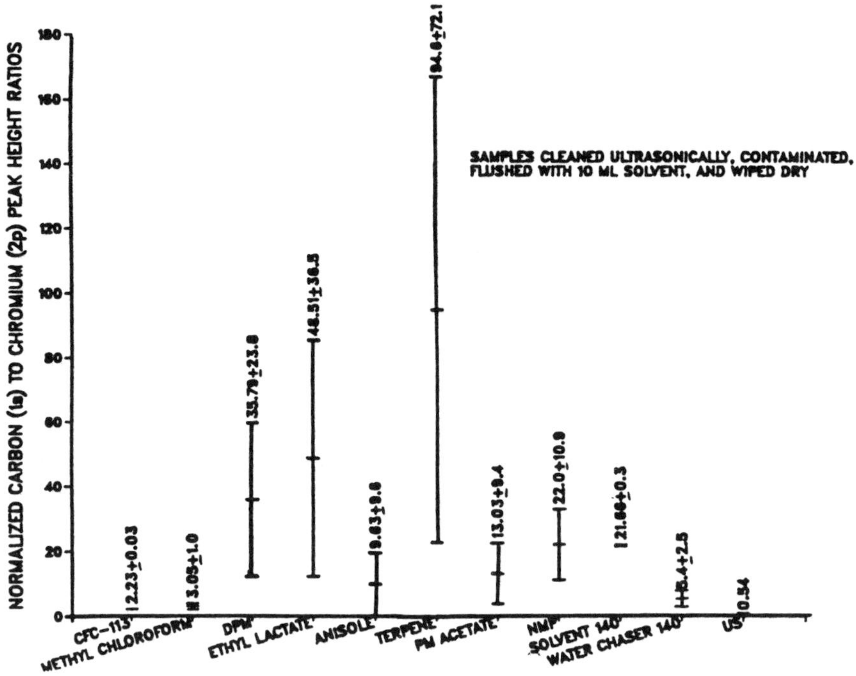

FIG. 4 – ABILITY OF SOLVENTS TO REMOVE TRIM SOL FROM 304L SS

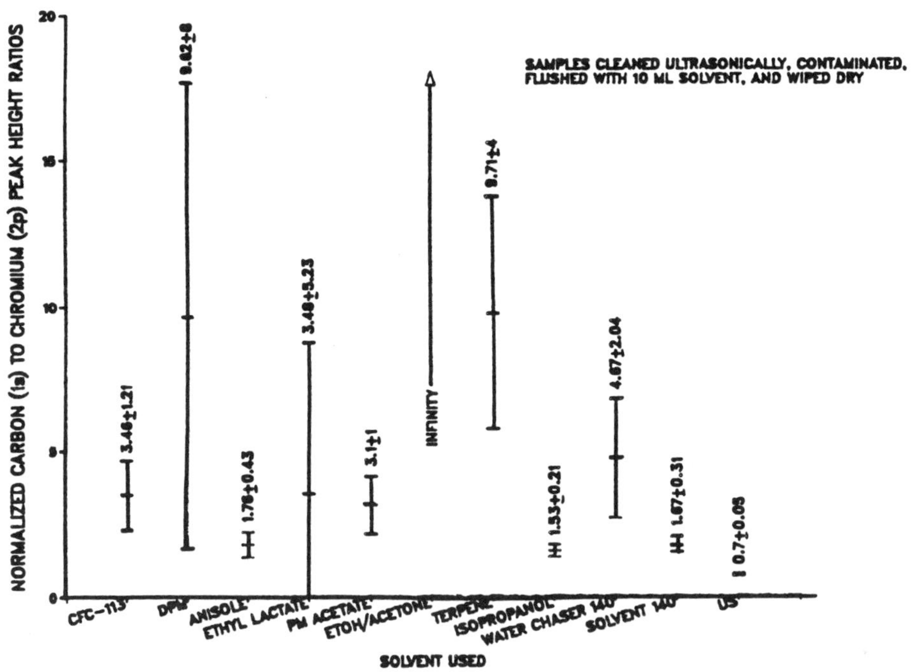

FIG. 5 – ABILITY OF SOLVENTS TO REMOVE FINGERPRINTS FROM TYPE 304L SS

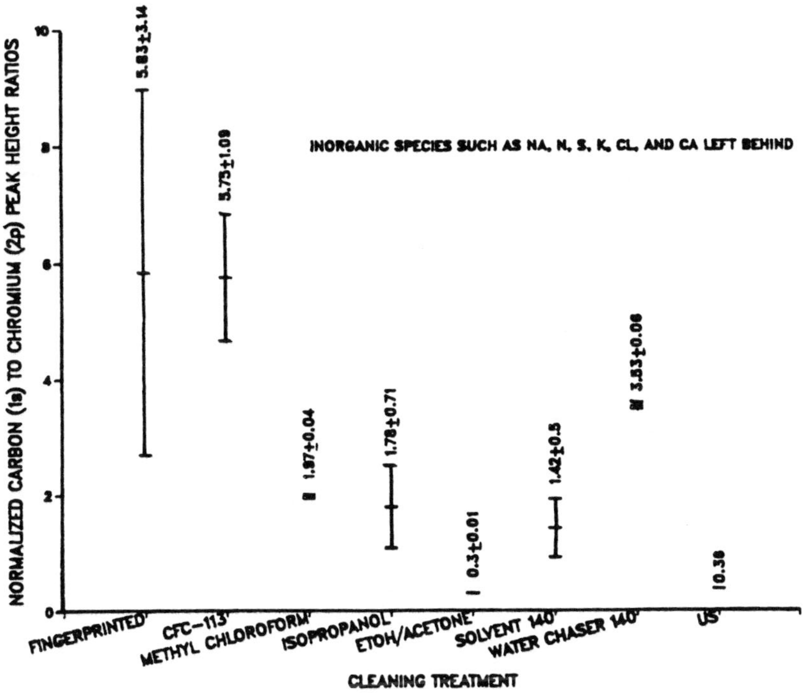

SOLVENT SUBSTITUTION FOR ELECTRONIC ASSEMBLY CLEANING

M. C. Oborny, E. P. Lopez, D. E. Peebles and N. R. Sorensen
Sandia National Laboratories
Albuquerque, New Mexico

ABSTRACT

The Department of Energy (DOE) is striving to eliminate the use of chlorofluorocarbon and chlorinated hydrocarbon solvents from weapons production. A major use of these materials is in the cleaning of electronic assemblies during production. In seeking to eliminate these materials, a screening study has been completed to identify alternate materials/processes. This screening study involved parallel investigations into: 1) the use of alternate cleaners for removing the rosin-based flux and mold release material currently being used, and 2) the use of water soluble fluxes in place of the rosin-based flux. The evaluation criteria used in this screening study were: the environmental, safety and health impact upon production operations, cleaning efficacy, corrosion potential, and, the bondability and high voltage breakdown resistance of cleaned surfaces. Upon completion of the screening evaluation, oxo-decyl acetate and a terpene cleaner have been selected for further study.

INTRODUCTION

Like industry in general, the DoE weapons complex utilizes chlorofluorocarbon and chlorinated hydrocarbon solvents for most cleaning and degreasing operations. However, environmental, safety, and health (ES&H) concerns and regulations, now dictate that the use of these materials must be minimized and eventually eliminated. Since electronic assembly cleaning processes account for a large percentage of total halogenated solvent usage within the weapons production complex, Sandia National Laboratories (SNL) and Allied Signal/Kansas City Division (AS/KCD) have established a joint program to identify, qualify and implement alternative materials and processes that would eliminate halogenated solvents from electronic assembly cleaning processes at AS/KCD. This program is presently focussed on a major electronic system. The fabrication of this system, which contains eight major electronics assembly modules and numerous secondary components, requires 49 separate solvent cleaning steps using trichloroethylene (TCE). These cleaning steps are primarily for the removal of rosin solder flux residues after soldering, cleaning of completed subassembly and assembly modules prior to foam encapsulation, and post-encapsulation cleaning of silicone mold release from encapsulated units.

HALOGENATED SOLVENT ELIMINATION OPTIONS

In seeking to eliminate halogenated solvents from electronic assembly cleaning processes, two separate options were investigated. The first of these options is the use of alternate cleaning materials in lieu of the halogenated solvents now being used for the removal of the rosin mildly activated (RMA) solder flux and silicone mold release residues. The second option involves alternative processing using water soluble soldering fluxes rather than the RMA solder flux now being used. Water soluble fluxes have the advantage that they can be cleaned with water, thus eliminating the need to use halogenated solvents for flux residue removal. Since mold release cleaning studies had indicated that the silicone mold release agent can be removed with isopropyl alcohol (IPA), it was hypothesized that a two step cleaning process, hot deionized (DI) water followed by IPA, would effectively remove residues from both the water soluble solder flux and the silicone mold release agent. Additionally, using IPA as the second cleaning step would remove any water remaining from the first step and rapidly evaporate to leave dry assemblies.

ELECTRONIC ASSEMBLY CLEANING PROGRAM

A three phase program was designed to carry out the studies necessary to identify, qualify, and implement alternative cleaning processes for electronic assembly cleaning. Phase 1 is an initial screening study, involving a limited number of alternative cleaning agents and water soluble solder fluxes, to identify a single cleaner or flux for further evaluation. Phase 2 involves more extensive testing and evaluation of the alternative cleaner or water soluble flux that was identified in Phase 1. Finally, given a successful outcome of the Phase 2 testing and evaluation, Phase 3 will be the implementation of the alternative cleaner or water soluble flux into production operations.

At this time, the Phase 1 screening studies have been completed. In these studies five substitute cleaning materials were evaluated as potential replacements for halogenated solvents: a terpene, oxo-decyl acetate, two proprietary aqueous cleaners, and isopropyl alcohol. Additionally, two water soluble soldering fluxes, Kester 2120 and Kester 2224, were evaluated as possible substitutes for the Kester 197 RMA solder flux now in use. A two step cleaning process, hot DI water followed by IPA, was used in the water soluble flux evaluation studies. Evaluation criteria that were used in the Phase 1 screening studies were: ES&H impact at AS/KCD, cleaning efficacy, cleaner corrosivity, bondability of cleaned surfaces, and high voltage breakdown resistance of high voltage assemblies after cleaning. The bondability and high voltage breakdown resistance testing were necessary due to specific issues related to the production and function of this particular system.

ES&H IMPACT ASSESSMENT

An ES&H impact assessment was done by environmental, fire safety, and industrial hygiene personnel at AS/KCD. As a result of this assessment it was determined that, properly used, none of the alternative cleaners or water soluble fluxes would present a significant ES&H problem with either current or anticipated future regulations.

CLEANING EFFICACY STUDIES

Cleaning studies were carried out at both SNL and AS/KCD. The SNL studies were primarily focussed on the removal of solder flux and silicone mold release agent from four substrate materials common to the electronics system under study. These materials were bare copper, bare copper which had been fluxed and Sn/Pb solder dipped, 17-4 PH stainless steel, and E-glass/polyimide printed wiring board material. The bare copper and solder dipped copper substrates were used to simulate electronics materials before and after soldering. The 17-4 PH stainless steel is used as the structural housing material for the electronics assemblies and the E-glass/polyimide printed wiring board material is used in all of the electronics assemblies in the system.

For the alternate cleaner studies, coupons of the four substrate materials were contaminated with each of the two contaminants: the Kester 197 RMA solder flux and the silicone mold release agent. These coupons were then cleaned using TCE or one of the five substitute cleaning materials. The TCE-cleaned samples were necessary to provide a basis for comparison between the current cleaner and the alternative cleaners. After cleaning, the coupons were analyzed to determine cleaner efficacy for each combination of substrate, contaminant, and cleaner.

In the water soluble flux studies, coupons of the four substrate materials were contaminated with each of the two water soluble fluxes and then cleaned using the two-step cleaning process. As in the alternate cleaner studies, the cleaned coupons were then analyzed to determine cleanliness levels for each combination of substrate material and water soluble flux.

The solder-flux contaminated coupons from both the alternate cleaner and water soluble flux studies were analyzed visually and with either Auger electron spectroscopy (AES) or x-ray photoelectron spectroscopy (XPS) to identify and quantify surface contaminants. An additional set of solder dipped copper coupons was analyzed using an Omegameter to measure residual ionic contamination levels.

All silicone mold release-contaminated coupons from the alternate cleaner studies were analyzed visually and also with either AES or XPS. Silicone mold release cleaning efficacy for the alternate cleaners was also determined using contact angle goniometer measurements. For these measurements, copper coupons were contaminated with silicone mold release agent and then cleaned with TCE or one of the five alternate cleaners. After cleaning, water drop contact angle measurements were made to determine relative surface cleanliness levels.

In addition to the above SNL cleaning study, cleaning studies were also carried out at both AS/KCD and SNL to determine the effectiveness of TCE, the five alternative cleaners, and the two step DI water/IPA cleaning process in removing a number of general contaminants that are present in the production area. These studies were necessary because experience has shown that these materials occasionally end up on production units and must be removed. These studies were performed on bare copper, bare aluminum, and solder dipped copper substrates using various oils, greases, mold releases and body oils. Cleaning efficacy in these studies was determined visually, by weight loss measurements, MESERAN (Measurement and Evaluation of Surfaces by Evaporative Rate ANalysis) values, water drop contact angle measurements, and Grazing Angle Reflectance-Fourier Transform Infrared Spectroscopy.

Analysis of data from all the SNL and AS/KCD cleaning studies indicated that, of the five substitute cleaners, the oxo-decyl acetate and terpene cleaner were the most effective in removing the RMA solder flux, silicone mold release agent and general contaminants that were studied. These data also showed that the oxo-decyl acetate and terpene cleaner both cleaned as well as TCE in these studies. Data from the water soluble flux and silicone mold release cleaning studies indicated that the two water soluble fluxes and silicone mold release agent can be effectively removed using a two step DI water/IPA cleaning process. However, the general contaminant studies found that this cleaning process would be less effective for removing some of the production area contaminants.

CLEANER CORROSIVITY STUDIES

Two types of tests were conducted at SNL to evaluate the relative corrosivity of the substitute cleaners and TCE. The first of these tests was an immersion test to determine dissolution rates in each of the cleaning solutions. In this test, preweighed coupons of copper and Sn/Pb solder were immersed in each of the cleaners for one week, at ambient temperature. At the end of this time, the coupons were removed, rinsed, dried, and reweighed. These weight loss measurements provided an indication of the relative corrosiveness of each cleaner. As expected, the two aqueous cleaners exhibited greater dissolution rates than the organic cleaners (TCE, IPA, oxo-decyl acetate and the terpene).

The second part of the corrosion evaluation

testing was designed to assess potential long-term corrosion problems due to cleaner residues left from incomplete rinsing after cleaning. In this test, bare copper and Sn/Pb solder coupons were prepared under no rinse, partial rinse, and full rinse conditions for each of the cleaners. These coupons were then placed in an environmental chamber and aged for 30 days at 40°C, 70% relative humidity. At the end of this time, the coupons were removed and a visual assessment was made of the relative amounts of corrosion present. Results of the immersion and cleaner residue tests indicated that all five substitute materials were acceptable.

BONDABILITY TESTING

Post-cleaning bondability of materials was investigated at SNL. These tests were necessary due to the large number of bonding and encapsulation processes which occur during production of the system. In these studies coupons of 17-4 PH stainless steel, and E-glass/polyimide wiring board were contaminated with the RMA flux and the two water soluble fluxes. An additional set of Sn/Pb solder-dipped copper coupons was prepared by dip soldering bare copper coupons that had been fluxed with the RMA flux and the two water soluble fluxes. Coupons prepared using the RMA flux were cleaned using TCE or one of the five alternate cleaners while coupons prepared with the water soluble fluxes were cleaned using the two step DI water/IPA cleaning process. After cleaning, thin-wall steel cylinders were bonded to the cleaned substrates with an encapsulating resin and torqued to failure to determine the short-term adhesive shear strength between the resin and coupon. Within experimental scatter, all coupons yielded the same short-term adhesive shear strength.

HIGH VOLTAGE BREAKDOWN TESTING

The final evaluation testing in Phase 1 involved high voltage breakdown testing at AS/KCD. High voltages are present in several modules of the electronic system under study and previous experience had shown that high voltage breakdown problems are often the first indication of incomplete cleaning and/or contamination of these modules. High voltage breakdown testing was done using a specially designed high voltage test assembly. Three separate groups of test assemblies were fabricated using the RMA flux and the two water soluble fluxes. After fabrication, the RMA fluxed assemblies were cleaned with TCE or one of the five alternate cleaners while the water soluble fluxed assemblies were cleaned with the two step DI water/IPA cleaning process. After cleaning, each test assembly underwent high voltage stress testing to determine breakdown behavior. Results of this testing indicated no significant high voltage breakdown failures due to the use of any of the cleaners or water soluble fluxes.

CONCLUSIONS

At the completion of Phase 1 testing, both the oxo-decyl acetate and terpene cleaner were selected for further evaluation in Phase 2 of this program. This selection was based upon cleaner performance and systems-level concerns associated with the use of aqueous or semi-aqueous cleaning processes.

As has been previously discussed, the oxo-decyl acetate and terpene cleaner were found to be as effective as TCE in removing the RMA flux, silicone mold release agent, and general production area contaminants. Additionally, in contrast to the proprietary aqueous cleaners and water soluble solder fluxes, these cleaners neither contain nor require water for their use. Although the use of aqueous processing was not a concern when these investigations were started, during these studies a systems level decision was made that, if possible, no moisture should be introduced into the weapon system as a result of aqueous processing methods. This decision hindered any possible use of the proprietary aqueous cleaners and water soluble solder fluxes.

At this time, Phase 2 studies are underway. Among the issues being addressed in these studies are: materials compatibility, electronic functionality of cleaned assemblies, and the long-term reliability of cleaned assemblies.

ACKNOWLEDGEMENTS

The authors wish to acknowledge the efforts of numerous individuals at Sandia National Laboratories and Allied Signal/Kansas City Division who contributed to this program. The general contaminant cleaning studies at AS/KCD and SNL were done by M. G. Benkovich and P. J. Nigrey, respectively. J. E. Reich performed the contact angle goniometer studies. M. E. Smith provided technical assistance for the XPS and AES measurements, and G. A. Poulter assisted in the corrosion studies. Bondability testing was done by T. R. Guess, M. E. Stavig and D. L. Zamora. B. N. Harnden was responsible for the high voltage breakdown testing at AS/KCD. This work was supported by the U.S. Department of Energy.

ALTERNATIVE SOLVENTS/TECHNOLOGIES FOR PAINT STRIPPING

M. N. Tsang and M. D. Herd
Idaho National Engineering Laboratory
Idaho Falls, Idaho

Paint stripping is a necessary part of maintenance at U. S. Air Force Air Logistics Centers (ALC). The waste from Air Force paint stripping operations contains toxic chemicals that require special handling and must be disposed of as hazardous waste at considerable cost. Emissions from these solvents into the atmosphere as volatile organic compounds (VOC) are another source of pollution. These wastes are hazardous to the environment and to operating personnel. The paint stripping wastes are regulated by the U. S. Environmental Protection Agency (USEPA), which can impose fines on operations whose wastes exceed the established limits.

The purpose of this program is to identify and test alternative solvents and/or technologies to strip paint from aircraft parts and equipment efficiently with the overall objectives of minimizing hazardous wastes and volatile organic compound (VOC) emissions. Commercially available chemical paint strippers will be tested for paint stripping efficiency, biodegradability, and corrosion characteristics. An extensive literature search has been completed to obtain background information, abstracts, patents, military specifications, and ASTM testing standards for paint stripping. Several mechanical paint stripping methods have been discovered during the literature search and are being monitored for their applicability in Department of Defense operations. A joint program has been established between Boeing Aerospace, Pacific Northwest Laboratory, and the INEL with the goal to expand the collaboration effort with other industries.

Phase I of this program has been completed. Phase I involved gathering baseline information, conducting laboratory screening of potential alternative paint strippers for their ability to remove paint, determining the biodegradability of the solvent, and determining the corrosion characteristics of the solvent. Phase II is currently being implemented in FY-91 and will involve extended performance testing of the alternative paint strippers surviving Phase I testing. Phase III will be conducted in later years and will involve full scale demonstration of the paint strippers selected.

Paint stripping efficiency was evaluated by determining the ability of the stripper to remove various types of paint systems from metal coupons. The test methods were developed from military and federal specifications for paint stripping. A preliminary test was conducted on all samples to eliminate those that cannot remove paint under moderate conditions. The effects of each stripper on the paint system were determined by visual inspection of the coupon after paint stripping, since this is the standard procedure at the ALCs. For the preliminary test, Al 2024 and an epoxy paint system were selected as the representative paint system. Paint strippers passing the screening requirements of $\geq$50% paint removal were subjected to a more stringent test to provide accurate performance data. The second test used Al 2024 and carbon steel 1010 as the substrates and utilized six paint systems for the test.

The paint systems used were:

1. Epoxy polyamide primer with epoxy polyamide topcoat.

2. Elastomeric polysulfide primer and urethane topcoat.

3. Water-thinned epoxy primer and CARC urethane topcoat.

4. Zinc chromate primer and alkyd topcoat.

5. Epoxy polyamide primer, polysulfide sealant, epoxy polyamide primer, and a urethane topcoat.

6. Epoxy polyamide primer with epoxy polyamide topcoat that differed in formulation from #1.

The painted coupons were subjected to accelerated aging by immersion in 2% hydrogen peroxide for 18 hrs. This accelerates oxidation, which normally occurs with ultraviolet (UV) light and time. Coupons for the preliminary test were not aged before testing.

Corrosion testing will be performed in accordance with ASTM F483-87, Immersion Corrosion Testing. Those paint removal solvents passing the extended performance testing will be subjected to corrosion testing, hydrogen embrittlement testing by ASTM F519-77, and biodegradation studies.

Currently 60 paint removal formulations have been screened. Out of this screening test, 10 immersion paint strippers passed into extended testing. These ten paint removal solvents are: Chemical Methods CM-3707, Chemical Solvents SP-800, Fine Organics FO 606, Frederick Gumm Clepo Envirostrip 222, GAF M-Pyrol, McGean-Rohco Cee Bee A245, McGean-Rohco Cee Bee A477, Patclin 126 Hot Stripper, Rochester Midland PSS 600, and Turco T-5668. These solvents are currently undergoing extended testing, VOC studies, and recycle/recovery studies.

Several new process technologies for paint removal have also been identified and are being monitored by the INEL. These new technologies include: wheat starch blasting, CO_2 ice blasting, ice blasting, water jet blasting, flash lamp stripping, laser stripping, and sodium bicarbonate blasting.

A PROPOSED "MORE DEMANDING" PWB DESIGN AND TEST PLAN TO EVALUATE AQUEOUS AND SEMI-AQUEOUS CLEANING TECHNOLOGIES

K.K. Asada, K.S. Hill and M.D. Walley
Radar Systems Group
Hughes Aircraft Company
Los Angeles, California

ABSTRACT

A "more demanding" PWB design and a test plan were developed through a project jointly funded by Hughes Aircraft Company and California's South Coast Air Quality Management District (SCAQMD). This test plan follows the guidelines developed by the IPC/DOD/EPA committee for removal of flux residues using alternatives to CFC solvents. Hughes Aircraft Company also proposed two follow-on phases which will establish the performance of fluxes and solvents through screen testing and involve full qualification testing necessary to gain customer approvals.

The design and test plan were derived from an extensive literature and industry review of existing solvents, fluxes, and alternative testing programs. Based on the need for more complex test vehicles, two printed wiring boards were designed for more demanding cleaning requirement configurations for advanced complex electronics: one for surface mount and one for plated through-hole. The surface mount design consists of 15 different component types and sizes ranging from 2 to 408 leads with lead pitch from 0.050 to 0.020 inch, and stand-offs from 0.001 to 0.013 inch. The plated through-hole design consists of 7 component types and sizes ranging from 2 to 68 leads.

The comprehensive test plan involved the modification of the IPC process flows for each board type, as well as the creation of the necessary travelers and process instructions. The test plan is divided into three segments which describe in detail cleaning media screening, assembly screen testing and customer qualification testing. This is the first comprehensive test program that addresses the evaluation of alternative cleaning technologies for defluxing electronic assemblies of aerospace complexities.

INTRODUCTION

In 1987 the Montreal Protocol was developed and adopted to establish a production control plan for chlorofluorcarbon (CFC) emissions and other ozone-depleting chemicals to result in a total phaseout by the year 2000. Given the potential for environmental damage, along with the pending regulations and new taxes on CFC usage, it has become imperative for the aerospace electronics industry to eliminate CFC usage wherever possible. CFC-based solvents are extensively used by the aerospace industry for defluxing electronic assemblies because of their compatibility with other materials and hardware, as well as their ability to meet the stringent cleanliness requirements mandated by the aerospace customers, in particular, the Department of Defense (DOD). Because of the need for industry to comply with the new laws and regulations, a joint committee consisting of the Institute for Interconnecting and Packaging Electronic Circuits (IPC), the DOD, and the Environmental Protection Agency (EPA) was organized to evaluate *Alternatives to CFCs for Printed Board Assembly Cleaning.* This committee has established a three-phase CFC replacement test plan for defluxing electronic assemblies.

The first phase establishes a benchmark of "how clean was clean" using CFC 113/methanol/nitromethane stabilized solvent. The second phase evaluates alternative cleaning media for rosin flux. The third phase evaluates alternative technologies, such as water soluble and no-clean fluxes.

One major concern was identified when Hughes Aircraft Company reviewed the IPC/DOD/EPA Phase 1 Test Program. The results from this test program would not validate the reliability of many aerospace electronics industry designs. One key quote from the purpose section of the test plan was, "in addition, this testing program does not address where printed wiring board assemblies have more demanding cleaning requirements; for example, where the spacing between conductors is significantly less than provided on the test board." Hughes, as well as other members of the aerospace electronics industry, decided that more testing would be required.

A three-phase program was designed and to evaluate alternative cleaning technologies to gain customer acceptance for defluxing electronic assemblies. Phase I consists of the development of a comprehensive test plan and the design of two printed wiring boards typical of complex aerospace configurations. This phase was co-funded by Hughes and the California South Coast Air Quality Management District, the results of which are represented in this report. Phases II and III are proposed as follow-up efforts to perform the evaluations of alternative fluxes and solvents. Phase II will establish the performance of fluxes and solvents through screening tests established in Phase I. Phase III will involve full qualification testing necessary to gain customer approvals based on the results of Phase II.

LITERATURE AND INDUSTRY REVIEW

In order to ensure a complete review of the issues related to existing and developing electronic defluxing technologies, a multifaceted approach was taken. First, an advisory committee of Hughes personnelconsisting of members from the six different Hughes Aircraft Company Groups was established. Second, a group of industry advisors was also formed. In both cases, the group charter was to focus on defluxing efforts within Hughes and throughout industry. A second facet was to contact several military customers in order to determine their position on defluxing issues, as well as their requirements. A third facet was to conduct an extensive literature review, including over 200 articles. The three topics reviewed were solvents and solvent development, fluxes, and industry testing. The last facet was to contact other companies in the industry in order to determine their progress in CFC elimination for defluxing applications.

Design Review of Aerospace Electronics

The objective was to review design requirements in the aerospace industry (e.g. PWB configuration, materials and performance) and compile a single set of requirements that would best represent aerospace electronics. Internal and external reviews were conducted to ensure that the needs of aerospace companies would be met.

A review was conducted at Hughes to determine the various design, material and performance requirements. This information was vital to the understanding of the requirements and for determining the configuration of the PWBs that needed to be designed for future cleaning testing. It has also reaffirmed the need for a test vehicle with more demanding cleaning configuration.

Eleven aerospace companies or affiliates external to Hughes in Southern California were invited to participate in the review of the design, materials and performance requirements identified by Hughes. Only five companies participated: AeroJet Electrosystems, Aerospace Corporation, Northrop - Hawthorne, Rockwell Rocketdyne and TRW.

Customer and Technical Requirements

The philosophy of this program was discussed

with the major military customers: the Navy, the Air Force and the Army technical research facilities. All three agencies voiced similar opinions on the use of alternative fluxes and cleaning media and were actively involved in the joint IPC/DOD/EPA Ad Hoc Working Group. All three agencies were acting individually in regard to testing reviews because there was no centrally identified agency that would speak for all branches. They all stated that as a minimum, the IPC/DOD/EPA test plans would have to be performed in order to obtain material approval for alternative fluxes and solvents. Additionally, testing would have to be performed on hardware configurations representative of the products to be cleaned. Finally, the consensus was that each individual contract would have to be modified to allow alternative fluxes and solvents because blanket approvals through changes to the existing military specifications would not be forthcoming.

Solvents Literature Review

A literature search was conducted on cleaning solvents and their associated equipment. The alternatives to CFC cleaning were identified and divided into four categories - drop-in, near term, semi-aqueous, and aqueous. Each system is defined as follows:

Drop in - A solvent that can be used in existing equipment with only minor or no modification to the equipment. Such modification might entail adjusting the thermostatic controls on the refrigeration coils (i.e., dilute CFC 113 and stabilized 1,1,1-trichloroethane (TCA), and TCA alcohol blends).

Near Term - A solvent that will be available in the near future. It may be used in the current process but will require new equipment or extensive modification to existing equipment, such as the addition of colder condensation coils (i.e., hydrochlorofluorocarbons (HCFCs).

Semi-Aqueous - In general, organic solvents that are used to remove flux and then rinsed from the printed wiring assemblies (PWAs) with water (i.e., hydrocarbons, surfactants with water rinse and isopropyl alcohol (IPA) and water).

Aqueous - Water with saponifiers and/or surfactants, or just water (i.e., saponifier and water, and water and surfactant).

There is not one specific material category that stands out as "the" likely CFC replacement; however, a few solvents have passed the IPC/DOD/EPA Phase 2 testing, and several materials are undergoing IPC/DOD/EPA Phase 3 testing.

Flux Literature Review

A literature review was performed to evaluate the current research use and of water soluble and no-clean flux formulations by electronic manufacturers and vendors. Both types of fluxes are used; however, specific industries consistently use each type. The no-clean flux formulations have been limited to commercial industries, while water soluble flux formulations are being used in the telecommunications, commercial, industrial, military and aerospace industries. The literature reviewed suggested that the use of no-clean flux formulations is limited because of its recent appearance into the flux market. The articles reviewed also suggested that both types of fluxes needed to be further understood to predict long-term reliability.

During the literature review, it was found that a single data base was not available for comparing the fluxes. In order to evaluate the various fluxes and match them to their corresponding assembly application, a data base would be beneficial. Numerous vendors were contacted and requested to submit technical data sheets and material safety data sheets (MSDS) for their available products. A compilation of this data resulted in an extensive, detailed centralized flux data base containing over 100 materials available for soldering electronic hardware.

Alternative Testing Review

An industry literature search was performed to gather current and historical information on alternative cleaning methods and testing for rosin and non-rosin fluxes. This search was accomplished through the review of published articles, attending technical conferences, and reviewing past technical reports and technical conference proceedings. Very little data was available and that most people felt that implementation of alternative cleaning techniques or non-rosin fluxes would require specific testing depending on an individual company's hardware configuration, geographic location and financial ability.

In the past 30 years, there have been many testing programs that have evaluated either alternative cleaning media or alternative fluxes (non-rosin based) with their associated cleaning media. However, only three major military funded programs were identified that address alternative cleaning media or alternative fluxes. The first was in 1966, sponsored by the Army and performed by Radio Corporation of America (RCA). The second was in 1983, sponsored by the Navy and performed by the Navy's Electronic Manufacturing Productivity Facility (EMPF) at China Lake. The third program is the IPC/DOD/EPA Ad Hoc Working Group's Phase 1, 2 and 3 testing, currently being performed by various government facilities and manufacturing companies.

The Army and Navy evaluations arrived at conflicting conclusions. The Army evaluation stated that water soluble fluxes (WSF's) could be used for military applications if specific precautions were taken, whereas the Navy evaluation stated that WSF's could not be used because of their devastating effects on surface insulation resistance (SIR) and ionic cleanliness.

Hughes conducted a survey of aerospace companies to determine the industry usage of non-rosin fluxes. Table 1 summarizes the results of the survey by identifying the company, application, materials and processes used.

TEST PROGRAM - ALTERNATIVE SOLVENT AND ALTERNATIVE FLUX TESTING

The alternative solvent testing program (IPC/DoD/EPA Phase 2) for more demanding cleaning requirement configurations is the evaluation of replacements for CFC solvents and associated equipment. This is limited to electronic manufacturing cleaning processes for the removal of MIL-F-14256 rosin-based fluxes. The testing includes:

a. Cleaning media screen testing - performed on glass slides.

b. Assembly screen testing cleaning media - performed on the IPC-B-36 PWB and on more demanding cleaning configurations (both surface mount and through-hole) using modified IPC/DOD/EPA Ad Hoc Working Group Phase 2 process flows, which utilized soldering processes other than the vapor phase and wave soldering process. The quantity of hardware required is not as extensive as for customer qualification testing.

c. Customer qualification - the testing sequence is the same as that for the assembly screen testing with the exception of the statistically acceptable quantity of hardware.

The alternative flux testing (IPC/DOD/EPA Phase 3) is the evaluation of non-rosin based fluxes and their corresponding cleaning systems as replacements for CFC solvents. This is limited to electronic manufacturing cleaning processes. The testing includes:

a. Flux screen testing - characterization of the flux based on material and performance.

b. Flux/equipment screen testing - glass slide testing of the combinations of alternative fluxes and their associated cleaning systems.

c. Assembly screen testing of fluxes and equipment - performed on the IPC-B-36 PWB and on more demanding cleaning configurations (both surface mount and through-hole) using modified IPC/DOD/EPA Ad Hoc Working Group Phase 3 process flows which utilize soldering processes other than the vapor phase and wave soldering process. This testing will provide preliminary test data.

d. Customer qualification - the testing sequence is the same as that for the assembly screen testing with the exception of the statistically acceptable quantity of hardware.

All pertinent travelers and process instructions were developed and written for the difference process flows and will be available through the SCAQMD. Upon the completion of this testing all IPC/DOD EPA Ad Hoc Working Group Test requirements should be satisfied. This testing will also provide additional data for the more demanding cleaning requirement configurations typically found on aerospace hardware.

PRINTED WIRING BOARD DESIGNS

Surface Mounted PWB Configuration

The HAC-SMT-615 PWB (see Figure 1) was designed to be used as the test vehicle for more demanding cleaning difficulties typical of aerospace Surface Mount Technology (SMT) component selection and placements. There are two unique features of the component selection and placements. First, the same 68 I/O LCC component that is used on the IPC-B-36 PWB has been designed into this board to provide correlation to the IPC boards. Second, axial and chip capacitor components are placed around the perimeter of the 408 and 224 leaded components to simulate densely populated assemblies. Because of the tight component spacings on many designs and their potential for inhibiting the cleaning of a PWB, these components were necessary. A list of the components along with their corresponding descriptions can be found in Table 2.

There are a total of 36 SIR patterns; three simulate the IPC-B-36 SIR patterns (0.006-inch line widths and 0.006-inch spaces between lines) and the remaining have 0.008-inch widths and 0.010-inch line spaces. There are also eight daisy chain patterns, utilizing circuit board traces, which are for reference only. The boards shall be fabricated to the requirements of MIL-P-55110D and made from standard epoxy material conforming to MIL-P-13949 (Prepreg, MIL-P-13949/12, and Plastic Sheet, MIL-P-13949/4) Type GF. The board thickness is 0.070-inch nominal with four layers, utilizing one ounce copper. To ensure the quality of the PWB's, chemical characterization and Group A testing per MIL-P-55110D shall be performed.

Through-Hole Mounted PWB Configuration

The HAC-PTH-616 PWB (see Figure 2) is designed to be used as the test vehicle for the more demanding cleaning requirement plated through-hole (PTH) configurations and consists of many different component configurations. A list of the components along with their descriptions can be found in Table 3. The selection of this design and these components is intended to present a more demanding cleaning requirement configuration typical of aerospace plated through-hole designs. The SIR patterns, fabrication requirements, and materials are the same as that for the surface mount configuration. The same chemical characterization and Group A testing per MIL-P-55110D shall also be performed.

CONCLUSIONS

This program has provided a comprehensive tool to assist aerospace companies in their quest to eliminate CFCs in electronic defluxing processes. Conclusions that can be made from this program are:

- There is a need for more complex PWB design test vehicles for cleanliness validation.

- The IPC/DOD/EPA testing programs provide the prerequisite data required by the military customers. The Hughes/SCAQMD program, which covers more demanding cleaning requirement configurations, parallels the IPC/DOD/EPA testing.

- Blanket approvals of materials in existing military specifications seem unlikely.

- No one material (technology) stands out as "the" likely CFC replacement. Selection of the materials for a company to pursue as a CFC replacement is a very complex process. Most replacement materials are ecologically better than CFC-113 but still create environmental damage. In some cases, damage results from smog and, in others, from water shortages and waste water products. Each company will need to make educated decisions on what to pursue based upon their product complexity, geographic location, and financial ability.

- The detailed test program - including the artwork, travelers, and process instructions-will be available through the SCAQMD.

Table 1. Industry Survey of Non Rosin Flux Users

COMPANY	APPLICATION	MATERIAL	PROCESS
IBM	Commercial	WSP[1] Alpha 1208, Kester 577 WSF[2] Kester 450 London Chemicals 3355-11	IR repair wave
IBM	Military MIL-STD-2000	none	
Litton Systems Canada, LTD	Military – Boeing, McDonnell Douglas MIL-STD-2000	none	
Litton Systems Canada, LTD	Military – McDonnell Douglas	none currently previously, Lonco 3355-11	tinning
Intel	Commercial – AT&T, Reuters	WSF Alpha Metals 850-33 WSP Kester OAR0577T Alpha WS 601	wave
TRW / EPI	CECOM	none currently previously - WSF Kenco	wave
HAC - EDSG	Military	WSF Superior 85	tinning
Magnovox, Indiana	Military radios sonobuoys	 no flux – proprietary "wax" Alpha 871	 wave wave
Texas Instruments, Texas	Military – Air Force	WSF	
Texas Instruments, Texas	Military – Air Force	WSF modified, Alpha 871-25 Cobar 353	wave & hand
Group Technologies	NSA	WSF RA, from Vichen	modified wave
Tektronix	testing only	WSP Kester R577-TOA Alpha 601	none
Link Flight Simulation Corp	Military MIL-STD-2000 MIL-STD-454 Req.17	none	
Link Flight Simulation Corp	Military	none	
Allied Signal Aerospace		Organic Flux Alpha 250 Blackstone 2508	pretinning
Allied Signal Aerospace	DOE	also Alpha 260 HF polyethylene, polypropylene, and glycol	pretinning PWB manufacturing
Allen Bradley, PC Division	unknown	WSF	through-hole

1 WSP – Water Soluble Paste

2 WSF – Water Soluble Flux

Table 2. Surface Mount PWB Parts List

Surface Mount PWB (HAC-SMT-615)	
ITEM	QTY/PWB
408 Lead 0.025-inch Pitch Quadpack	1
224 Lead 0.025-inch Pitch Quadpack	1
208 Lead 0.020-inch Pitch Quadpack	1
180 Lead 0.030-inch Pitch Quadpack	1
84 I/O 0.025-inch Pitch LCC	1
68 I/O 0.050-inch Pitch LCC	1
64 I/O 0.025-inch Pitch LCC	1
44 I/O 0.050-inch Pitch LCC	1
32 I/O 0.050-inch Pitch LCC	1
28 Lead Surface Mount Dip	1
18 I/O 0.050-inch Pitch LCC	1
16 Lead 0.050-inch Pitch Component	1
10 Lead 0.100-inch Pitch Component	1
Axial Leaded Device	8
Chip Capacitors	11

Table 3. Through-Hole PWB Parts List

Through-Hole Mount PWB (HAC-PTH-616)	
ITEM	QTY/PWB
Pin Grid Array	1
20 Lead Dual In-Line Package	2
28 Lead Dual In-Line Package	1
24 Lead Dual In-Line Package	1
28 Lead Dual In-Line Package	1
40 Lead Dual In-Line Package	1
Axial Lead Component	2

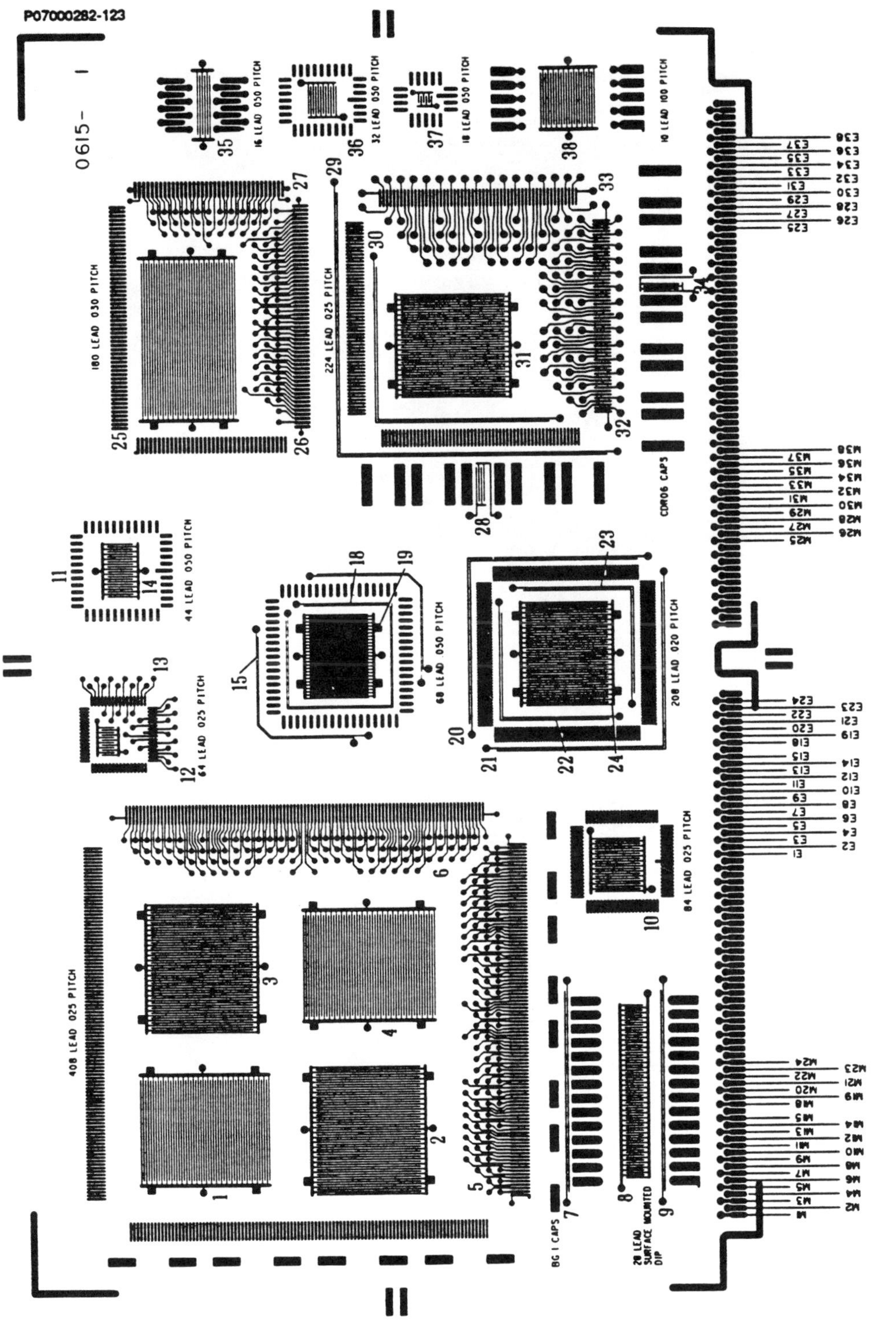

Figure 1. HAC-SMT-615 PWB

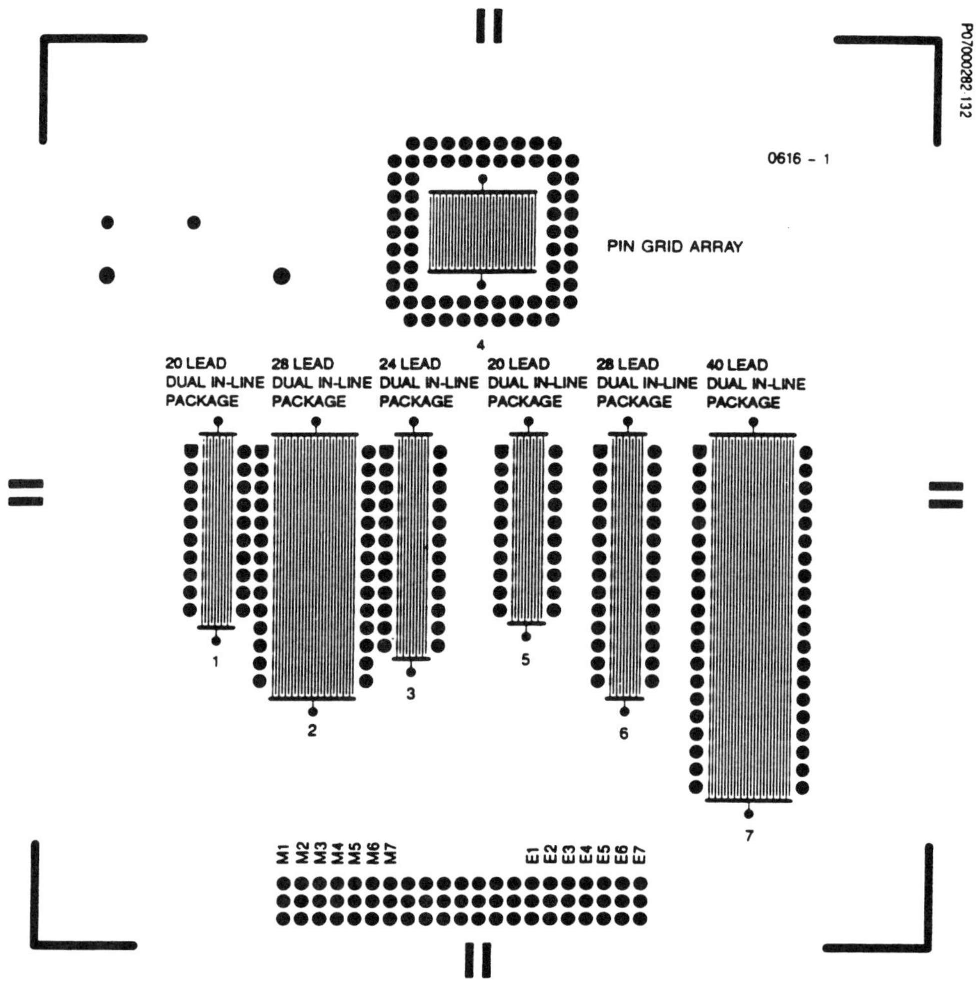

Figure 2. HAC-PTH-616 PWB

DEVELOPMENT OF A SOLVENT DATABASE SOFTWARE PROGRAM

Ralph D. Hermansen
Hughes Aircraft Company
El Sequndo, California

ABSTRACT

A novel computer database software program has been developed to assist materials scientists/engineers with decisions about solvent replacement or reduction in high VOC products, such as cleaners, coatings, primers, impregnants, etc., by presenting comparative data quickly and easily. The software development effort was jointly funded by the South Coast Air Quality Management District and Hughes Aircraft Company. The solvent database is unique in that it provides information not only about solvent physical properties, but also about environmental compliance status and safety characteristics as well. Altogether, 27 fields are filled for each solvent record. Over 150 solvent records have been entered. The solvent database is a stand-alone program, which runs on IBM PCs and compatibles. The program is menu driven and is designed to be user-friendly. Help screens explain functions of the database at a keystroke. Solvents can be added, modified, or deleted from the database. The search functions allow for combinations of requirements. The solvents found meeting the screening criteria can be visually examined or printed out. A manual was written to explain the functions of the database. Hughes Aircraft Company intends to make the solvent database available to interested parties in the near future.

BACKGROUND

The public is highly concerned about environmental problems and as a result, there have been stricter and more numerous regulations issued affecting solvents and solvent-containing products. Environmental compliance from a laboratory standpoint often involves reformulation of coatings, primers, impregnants, cleaning solvents, and other solvent-containing materials to eliminate, replace or reduce the proportions of offensive solvents. However, the selection of replacement solvents can be a complex task for the environmental material scientist/engineer because many other factors besides the specific environmental regulation must be addressed. One such group of factors to be addressed consists of those physical and chemical properties which caused the solvent to work properly in the original formulation. Examples of such properties are chemical family, evaporation rate, or solubility parameter. Another group of factors includes flammability, toxicity, and other areas of environmental compliance.

When environmental materials scientists/engineers attempt to reformulate noncompliant materials, budget and/or schedule restraints often prevent them from doing an in-depth search for replacement solvents or to evaluate the resultant new formulation for a wide spectrum of concerns. Thus, a software program is needed to display the necessary solvent information quickly and easily. The program should be user-friendly and should run on IBM-type microcomputers. The user should be able to add, modify and remove solvent data and should be able to conduct searches based upon individualized search strategies.

OBJECTIVE

The objective of the effort was to develop a solvent database software package, which would be readily available to environmental materials scientists/engineers to assist them in their reformulation assignments.

METHODOLOGY

The methodology followed in this effort consisted of the following phases:

Phase 1 - Survey of Environmental Engineers. In order to build a useful software tool, it was first necessary to interview the scientists and engineers who were working in the field to determine what problems needed to be solved and what information would be provided by the solvent database program.

Phase 2 - Literature Search of Solvent Properties. Mailings and telephone canvassing of solvent manufacturers was conducted in order to obtain technical data on the solvents. In addition, the Hughes Technical Library was used to obtain books and periodicals on the topic of solvents.

Phase 3 - Database Design and Programming. Many different properties could be included in the database, but an effort was made to prune the list down to those properties most needed and most useful. Two main groupings of fields were decided upon: physical and environmental. The solvent database software program was written in dBase III Plus using the programming capabilities of the utility to design a user-friendly, menu-driven program.

Phase 4 - Entry of Solvent Data. Solvent property values were taken from several sources including manufacturer's literature, Material Safety Data Sheets, technical handbooks, and databases. The source of each data item is included in the database and can be examined by the user.

Phase 5 - Completion of Prototype Software Package. A menu-driven program was designed. Detailed coding was written to have the program function in a coordinated manner. The dBase III Plus software from Ashton-Tate Company was used. It was compiled into a self-contained program using Clipper (Nantucket Software). A manual was written to describe the solvent database software.

Phase 6 - Field Testing of Prototype Software Package. The first version was distributed within Hughes for evaluation. The comments and suggestions from the reviewers were noted and plans made for upgrading the solvent database program.

Phase 7 - Incorporation of Improvements. The obvious "bugs" in the program were corrected. In addition, the program was made more user-friendly by adding help functions to explain the various functions as they are encountered. This version was shared with other agencies that are pursuing similar efforts. Specifically, two agencies were visited: Idaho National Engineering Laboratory (INEL) and New Mexico Engineering Research Institute (NMERI).

Phase 8 - Final Release of Software Package. The completed software package was assembled and fifty copies produced. Thirty copies were distributed to concerned individuals within the Hughes groups. Comments and suggestions from the users were recorded.

ACCOMPLISHMENTS

Self-Contained Software. The solvent database program was designed to be a self-contained unit. That is, the program does not require that another software package be resident in order for it to run. Although the solvent database program was written in dBase III Plus, the source code was compiled using Clipper. Henceforth, the software will run on any IBM PC or clone independent of dBase III. The package was designed so that the user could add new solvents to the database, modify existing ones and delete obsolete ones. The user can also run searches and print out information.

Description of Database Format. There are a multitude of attributes used to describe solvents. So, it was decided to restrict the list of attributes in the database to those commonly used in solvent comparisons. The format of the database consists of three groupings of fields as shown in Figure 1. Overall, there are 27 fields per record. The first grouping is a description of the solvent. Fields are: solvent name, alternate name, tradename, manufacturer, and type. All of these fields are string type fields (i.e., string of characters) except for tradename, which is a logical field (i.e., true/false). Type refers to chemical family or category for the solvent. The different categories are shown in Figure 2.

The second grouping of fields is called "physical characteristics." The fields are: specific gravity, flash point, freezing point, solubility parameter, evaporation rate, molecular weight, vapor pressure, low boiling point, high boiling point, and VOC. Although all of these fields appear to be numeric, they are actually string fields, which can be converted to numeric fields by the program as required. The function of these fields is self-evident by their name.

The third grouping of fields is called "environmental characteristics." The fields are: Rule 1401, AB2588 AI, AB2588 AII, VOC exempt, Prop 65, Rule 442, acute oral toxicity, acute dermal toxicity, CAS number, inhalation PEL, inhalation STEL, and ozone depletion factor. The first five of these environmental fields holds a Y or N (yes or no) character. The rest of the fields are string type.

User-Friendly Software Environment. The solvent database program is intended for use by hundreds of different engineers and scientists. Therefore, the software was written with "the user" in mind. First of all, the entire program is menu-driven for ease of use. Figure 3 shows the appearance of the main menu. From the main menu, one may get help, add new solvents, search through the solvents, print out a report of the solvents, or quit the solvent database program.

During the process of adding new solvents or searching through the solvents, one may forget the meaning of the fields as shown in Figure 1. Another user-friendly feature is available to facilitate its use. One merely places the cursor over the field of interest and presses the F1 key. A window appears and helpful messages about that field appear. During the search mode, helpful messages appear in windows to let the user know what the program is doing or has done. For example, a window appears for each numeric search field selected and asks the user whether the field value will be a maximum, a minimum, or a range. If it is to be a range, then the window message asks for the lowest and highest values of the range. During the search, a message announces that the program is searching. Upon completion of the search, a message is given telling how many solvents were found meeting the search criteria. The message also tells the user how to view the complying solvents or how to print out the data on them.

Listing of Data Sources. Some users will want to compare the data very rigorously. That process may involve a greater knowledge of how the data were generated and by whom. For each field value in the database, there is a source listed showing from where the value was obtained. In order to view these data sources, one goes to the search screen from the main menu, locates the solvent of interest, and presses "R" for reference. A window appears as shown in Figure 4. The window displays a listing of the data sources by field for that solvent.

Solvents in Database. The solvents in the database were selected so that fairly equal representation exists for the different solvent categories (see Figure 2). The philosophy used for selecting solvents was to include both "good" and "bad" solvents. Here, good solvents would be solvents which are currently regarded as safe and environmentally compliant. Bad solvents would have some obvious problem. The immediate temptation was to list only the"good" or "recommended" solvents. However, with further consideration, it was realized that these "good or bad" ratings change dynamically with time. Solvents which were favored not long ago (i.e., methylene chloride or 1,1,1-trichloroethane) are slated for elimination today. Therefore, it was decided to include any common solvent.

Manual in Software Package. The manual, which is part of the solvent database package, provides the user with several services. First, the manual explains how to install the software programs on your computer. Secondly, the manual describes the solvent database programs. For example, it explains the meaning of the fields in the database. It explains how to add new solvents, delete solvents or search for solvents which comply with your search strategy. Thirdly, the manual includes a listing of the solvent properties as the database would display them. This compilation of solvent properties is quite useful in itself.

Individualized Usage of Package. Each user of the software will want to adapt the software to his/her particular interest. This can be done primarily by adding those solvents of interest to the user and deleting solvents of no interest. The solvent data can be updated and the source of the data included in the reference database. Many hundreds more solvents can be added. The storage capacity of the computer is the limiting factor.

CONCLUSIONS

A useful software package has been developed to assist scientists and engineers with decisions concerning solvent replacements. The solvent database program is a user-friendly, menu-driven system to be used on IBM PC compatibles.

FUTURE WORK

A follow-on effort is planned to test and improve the solvent database. A special group will be selected to evaluate the software, wherein a diversity of company interests are represented. Modifications to the existing solvent database should be made based upon the prioritized comments and suggestions. To date, it has been suggested that the program be modified so that solvent blends can be handled by the database and that a version of the solvent database be written for MacIntosh users.

ACKNOWLEDGEMENTS

At Hughes Aircraft Company, Larry Lipp assisted in design of the database format. Patrick Shuss did the majority of dBase/Clipper programming and wrote the manual. Phyllis Kelleghan and Liza Julien researched and entered the solvent data into the database. In addition, Angela Chavez at INEL and Dr. Jonathon Nimitz of NMERI were very cooperative in sharing their findings on solvent properties and offering comments on the Hughes/SCAQMD solvent database.

FIGURE 1

SOLVENT NAME:

ALTERNATE NAME:
TRADENAME?: MANUFACTURER: TYPE:

PHYSICAL CHARACTERISTICS

SPECIFIC GRAVITY :	EVAP RATE BuAc=1 :	LOW BOILING PT (C) :
FLASH POINT (C) :	MOLECULAR WEIGHT :	HIGH BOILING PT (C) :
FREEZING PT (C) :	VAPOR PRESS (mmHg):	VOC (GM/L) :
SOLUB. PARAMETER :		

ENVIRONMENTAL CHARACTERISTICS

RULE 1401? :	PROP 65? :	CAS NUMBER :
AB2588 AI? :	RULE 442 (K1/2/3) :	INH PEL (ppm) :
AB2588 AII? :	ACUTE ORAL (LD 50) :	INH STEL (ppm) :
VOC EXEMPT? :	ACUTE DERM (LD 50) :	OZONE DEPLETION :

FIGURE 2

DATABASE TYPE	DESCRIPTION
ALCOHOL	Contains hydroxyl as primary functional group. (i.e., Methanol and Isopropanol)
ALIPH. HC	Aliphatic hydrocarbon. Contains only carbon and hydrogen. (i.e., Hexane)
AROM. HC	Aromatic hydrocarbon. Contains the six-carbon benzene ring structure. (i.e., Toluene)
ESTER	Organic Ester. Contains R-CO-O-R' structure. (i.e., Ethylene Glycol)
GLYCOL	Organic Glycol. Contains two adjacent hydroxyl groups. (i.e., Ethylene Glycol)
HALOCARBON	Halogenated Hydrocarbon. Contains either chlorine, fluorine, or both on a hydrocarbon backbone. (i.e., 1,1,2-Trichlorotrifluoroethane)
KETONE	Organic Ketone. Contains the structure R-CO-R'. (i.e., Methyl Ethyl Ketone)
NITROGEN	Organic Nitro Compound. Contains nitrogen. (i.e., Nitromethane)
MISCELLANY	Miscellaneous compound. Not covered in any other type.

FIGURE 3

SOLVENT DATABASE UTILITY

MAIN MENU

H GET HELP OR INSTRUCTIONS

A ADD NEW SOLVENTS

P PRINT A REPORT OF ALL SOLVENTS

Q QUIT THIS PROGRAM

Version 1.0 ©1990 Hughes Aircraft Company

FIGURE 4

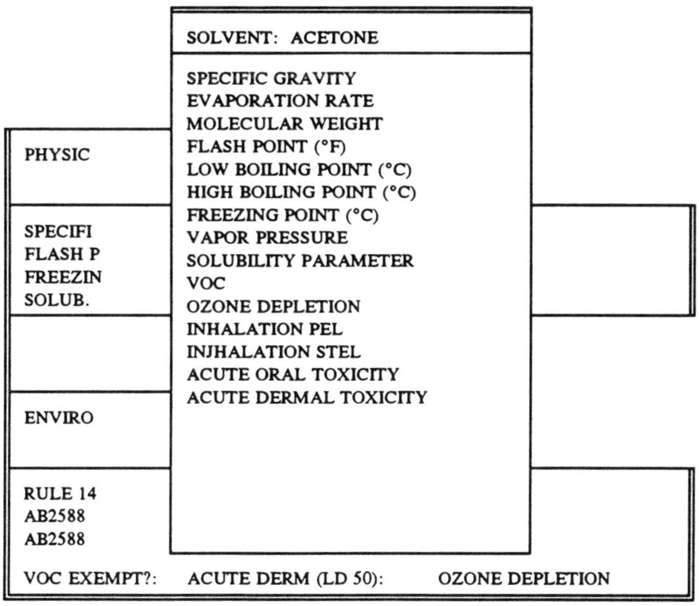

EVALUATION OF ALTERNATIVE CHEMICAL PAINT STRIPPERS[1]

Keturah Reinbold and Timothy Race
U.S. Army Construction Engineering Research Laboratory,
Champaign, Illinois

Ronald Jackson
U.S. Army Toxic and Hazardous Materials Agency,
Aberdeen Proving Ground, Maryland

Ronald Stevenson
Sacramento Army Depot, Sacramento, California

INTRODUCTION

Background

In 1984 the U. S. Environmental Protection Agency (U. S. EPA) placed a limit of 2.13 mg/L on the allowable concentration of Total Toxic Organics (TTO) which can be discharged from metal finishing operations. Some of the metal finishing operations find it difficult to comply with this regulation when using the popular cold chemical paint strippers. These strippers contain methylene chloride. Besides contributing to TTO, methylene chloride is a suspected carcinogen and is affected by restrictions on emissions of air toxics under amendments to the Clean Air Act. Many of the strippers are also classified as hazardous wastes after use. Disposal of the paint stripper wastes will become more difficult and costly as many of these wastes are banned from land disposal under the U. S. EPA schedule.

Objective

The objective of this study is to identify paint strippers which are operationally effective and environmentally acceptable replacements for methylene-chloride-based paint removers. The goal is to alleviate TTO compliance problems and to minimize environmental and health risks and disposal liabilities.

Approach

To test operational success, alternative stripper formulations were first evaluated in the laboratory. Materials meeting the criteria established for success in the laboratory were tested on a pilot scale. The environmental, health, and safety aspects of the selected strippers were also evaluated to ensure that the candidates are acceptable replacements. The final step in this process is a full-scale field test.

EVALUATION CRITERIA FOR AN ACCEPTABLE PAINT STRIPPER

Criteria for a successful paint stripper were developed in collaboration with Sacramento Army Depot (SAAD). The following criteria were selected: (1) acceptable stripping speed (SAAD upper limit of 2 hours), (2) effective for a broad spectrum of coatings, (3) not rapidly evaporated or depleted and easily replenished when it does become depleted, (4) no TTO contributing chemicals, (5) environmentally acceptable, (6) safe to use, (7) relatively easy to dispose, (8) commercially available, (9) easy to procure.

In addition, more specific criteria for environmental, health, and safety acceptability

----

[1] This research was supported by the U.S. Army Toxic and Hazardous Materials Agency (P147.01.16 Evaluation of Alternatives to Toxic Organic Paint Strippers).

of an alternative paint stripper were selected in cooperation with SAAD. These criteria included characteristics relating to toxicity (acute and chronic, human and environmental), environmental fate, safety (corrosivity, reactivity, and ignitability), and regulatory restrictions.

SELECTION OF COATINGS AND STRIPPERS FOR TESTING

Test coupons were prepared using three paint systems: zinc-chromate alkyd primer with an alkyd topcoat applied to aluminum, water-thinnable epoxy primer with a CARC urethane topcoat on aluminum, and epoxy polyamide primer with an epoxy polyamide topcoat on both aluminum and steel.

Candidate replacement strippers were solicited from industry and were previewed for probable success before inclusion in the test program. Organic strippers may contain any or all the following materials: (1) primary solvents, (2) cosolvents, (3) activators, (4) retarders, and (5) surfactants.

MS-111, which conforms to Mil-R-46116 (now cancelled), is a stripper which contains methylene chloride, phenol, and formic acid. It was included in this study as a control against which alternatives would be measured. Only one of the strippers evaluated contained methylene chloride or phenol both of which are TTO contributors. Common solvents in the other alternative strippers include 2-(2-butoxyethoxy) ethanol, n-methyl-2-pyrrolidone, monoethanolamine, and aromatic hydrocarbon solvents.

PROCEDURES AND METHODS

Stripper Performance

The laboratory evaluation of stripping performance was based on a laboratory-scale mockup of a typical stripping process. Steps in the test were immersion in stripper, caustic dip for acidic strippers, water rinse, and steam cleaning. Thirty-two strippers were evaluated for stripping efficiency under controlled conditions with the four coating/substrate combinations. Manufacturer's recommendations for optimal operational conditions were followed. Stripping conditions are listed in Table 1. Strippers were evaluated for percentage of coating removed for each paint system at specific time intervals.

Alternative strippers which performed adequately in the laboratory tests were evaluated in a 25-gallon pilot test using depot parts rather than test coupons. Stripping temperatures and dilution ratios with water were the same as in the laboratory tests.

Alternative strippers performing successfully in the pilot scale evaluation were considered for full scale production use. Full scale evaluation was performed during normal production operations at SAAD in a 1500 gallon tank. Qualitative results as well as periodic quantitative coupon analysis were reported. Quantitative coupon analysis was conducted employing prepared specimens similar to those used in laboratory evaluations.

Environmental, Health, and Safety Evaluation

After criteria for evaluation of environmental, health, and safety acceptability were selected, we developed a procedure to assign numerical ratings to permit a quantitative comparison of the hazards associated with each stripper. Each criterion was scored for each component of the stripper. A total weighted average score for the total stripper mixture was calculated by summing the result of the score for each component times the percent of that component in the stripper. A total score for each stripper was determined by summing characteristic scores. In addition, in order to evaluate the potential effect of a particularly hazardous component, a worst case hazard score was calculated for each stripper by summing the highest score for a component in each hazard category.

Several existing scoring procedures for individual criteria (toxicity, bioaccumulation, flammability, reactivity) were used (1,2,3,4,5). If a score were reported in the literature for a criterion for a stripper component, that score was used. In other cases, the same procedure was used to assign a score. If data for assigning a score were lacking, the values were calculated if possible.

RESULTS

Five of the alternative strippers removed all four coating systems in the laboratory tests. Eight other products removed three of the coating systems and partially removed the fourth in the proscribed 2-hour limit. The strippers evaluated in the pilot scale tests at SAAD were McGean-Rohco Cee Bee A-477, Fine Organics FO-606, Oakite ALM, Patclin 103B, Patclin 104C, and Turco 5668. It should be noted that laboratory, pilot, and production scale tests overlapped to a certain degree. The materials evaluated in later phases were judged relative to all the materials tested up to that time. Thus, in some cases, materials with good performance were not evaluated in pilot-scale and production scale tests while lesser products were.

All the strippers evaluated in pilot scale tests at SAAD except for Patclin 103B performed at a level consistent with laboratory test results. A GC-MS analysis indicated the presence of chloracetic acid in Patclin 104C. This Patclin product contains glycolic acid which is made by reacting NaOH with chloracetic acid. Incomplete conversion may have been the cause of the unacceptable chloracetic acid found in the stripper. Further consideration of Patclin 104C was withdrawn because of the chloracetic acid content. The remaining four strippers were considered to be candidates for full-scale production tests.

Environmental, Health, and Safety Evaluations

The results of the ratings of the most promising candidate strippers compared to MS-111 for environmental, health, and safety hazard are shown in Figure 1. The higher the total score, the greater the risks. The total weighted average scores show that any of the six candidate strippers are less hazardous than MS-111. Table 2 lists the order of relative hazard for the alternative strippers based on weighted average and worst case scores. Turco 5668 appears to be slightly preferable from an environmental and safety hazard point of view. Hazardous waste concerns and environmental regulations restricting discharges are also much more favorable for the candidate strippers compared to MS-111 (Figure 2).

FIELD TESTS

Oakite ALM was evaluated in a full-scale field test. Results were disappointing. Depletion of the active components, either through evaporation or absorption in the tank, led to an early decline in stripping power. The stripping time became excessive both for production parts and for uniform test coupons. Because of the nature of the production operations at the test site, the evaporation rate was above an acceptable level. The manufacturer analyzed the depleted stripper and provided a replenishment solution. The replenishment proved to be of marginal help. The full-scale evaluation of Oakite ALM was discontinued after approximately 9 months.

Besides the stripper performance evaluation, environmental monitoring was done during the field test with Oakite ALM. Results of analyses of stripper and rinse water revealed the presence of a complex mixture of aromatic hydrocarbon compounds besides the compounds reported by the manufacturer as components of the stripper. It is not surprising that the manufacturer did not know of the presence of these compounds in their stripper. A petroleum distillate such as was used in ALM is a complex blend of hydrocarbons which is characterized primarily by its boiling range. The identified compounds increase the hazard and decrease the environmental acceptability of the stripper compared to the estimates based on the composition listed by the manufacturer.

FO-606 was also evaluated in production-scale tests. Performance was initially acceptable with strip times of less than 2 hours noted for most production parts. The evaporation rate was quite high, and frequent additions of fresh stripper were necessary to maintain a sufficient depth in the tank. Stripper performance declined over the duration of the 12-month test. Significantly longer contact times were needed to remove the same coatings. Epoxy coatings were especially difficult to remove.

CONCLUSIONS AND RECOMMENDATIONS

Four candidate strippers demonstrated performance in laboratory tests consistent with the goals of the study. A total of six strippers were evaluated on a pilot scale. Four of the pilot-scale candidates tested were initially recommended for full-scale production use on an experimental basis. Oakite ALM was evaluated in full-scale production and did not meet the minimum requirements nor did it perform at a level consistent with laboratory and pilot-scale tests. In addition, chemical analyses of ALM stripper and rinse water samples revealed the presence of a complex mixture of aromatic hydrocarbon compounds which pose considerable health and environmental risk. FO-606 was also evaluated in production-scale tests. The user indicated initially acceptable performance which declined to marginal as evaporation of the stripper occurred. Material cost was unacceptably high, especially considering the relatively high evaporation rate. Cee Bee A-477 and Turco 5668 have also been recommended for use on a trial basis. One user has reported dissatisfaction in a trial with Turco 5668, but the authors have not conducted a comparable field test to compare that product with the two which have been fully tested.

ACKNOWLEDGEMENTS

The authors wish to acknowledge the contributions of S. Glascock, the technician who performed the laboratory stripping performance tests, and P. Hoglund and G. Barrett, research assistants who compiled the environmental, health, and safety data. Assistance from the Sacramento Army Depot during the project's planning, pilot testing, and field testing stages is also appreciated.

REFERENCES

1. Sax, N. I. Dangerous Properties of Industrial Materials, Sixth Edition, Van Nostrand Reinhold Company, New York. 1984.

2. Weiss, G., ed., Hazardous Chemicals Data Book, Noyes Data Corporation, New Jersey. 1980.

3. Comprehensive Environmental Response, Compensation, and Liability Act.

4. Vector Scoring System for the Prioritization of Environmental Contaminants. Prepared by CanTox Inc. and Senes Consultants Ltd. and Priority List Working Group, Ontario Ministry of the Environment. March 1988.

5. Reinbold K. A., and G. Barrett. Environmental Hazard Assessment of Chemical Paint Strippers. Draft Report. March 1988.

Table 1. Laboratory Testing Results

Stripper	Stripping Temperature	0.5 hr	1.0 hr	1.5 hr	2.0 hr
MS-111	Ambient	ABCD*			
FO-606	82 C	ABCD			
Patclin 104C (1:1 with water)	82 C	ABC	ABCD		
Turco 5668	82 C		AD	ABCD	
Cee Bee A-477	82 C	A	A	ABD	ABCD
Enthone S-26	Ambient	AC	AC	AC	ABCD
Enthone S-26 (1:4 with water)	Ambient	A	AD	AD	ABCD
Enthone S-26 (1:9 with water)	Ambient	A	AC	ACD	ACD B=95%
Patclin 126	88 C	CD	CD	ACD	ACD B=50%
Patclin 106Q	82 C	AD	AD	ACD	ACD B=20%
Intex 8798	82 C	A	ACD	ACD	ACD B=85%
Patclin 103B (1:3 with water)	82 C	AD	AD	AD	ACD B=65%
Ardrox 2302	71 C	AC	AC	AC	ACD B=50%
Patclin 125	82 C	D	AD	AD	ACD B=65%
Ardrox 5300-W	82 C	A	A	AD	ACD B=50%
Oakite ALM	82 C	A	A	A	ABD C=35%
Eldorado HT-2230	82 C	A	A	A	ACD
EZE 570-81	71 C	AD	AD	AD	AD C=30%
Omega R-824	71 C	A	A	AD	AD B=90% C=80%
Pentone R-3936	82 C	A	A	AD	AD C=5%
Pavco Decoater 3400	82 C	A	AD	AD	AD
Enthone S-26 (1:19 with water)	Ambient				AD B=40% C=20%
SafeStrip-66	54 C	A	A	A	AD B=10% C=15%
Chem-Lube X-177	93 C	D	D	AD	AD
Chemical Solvents SP-825	88 C		A	A	AD
Chem-Lube XH-36	93 C	D	D	D	D A=40% B=60% C=55%
Chemical Solvents SP-MNP	82 C	A	A	A	A D=90%
Key Chemical 570	79 C	A	A	A	A
NonMeth 120	49 C	A	A	A	A
NonMeth 140	49 C	A	A	A	A
Pavco Decoater 3321	82 C	A	A	A	A
Turco 5555-B	71 C	A	A	A	A
NonMeth 161	Ambient			A	A
Envirosolv L	Ambient				A=50%
Brulin Safety Strip 1000	Ambient				A=30%
Brulin Non-Chlorinated	82 C				

*A: zinc-chromate/alkyd (aluminum); B: epoxy polyamide/epoxy polyamide (aluminum); C: water thinnable epoxy/ CARC urethane (aluminum); D: epoxy polyamid/epoxy polyamide (steel).

Table 2. Total Stripper Hazard Scores from Worst to Best

Weighted Average	Worst Case
MS-111 -------- 16.3	MS-111, ALM ------ 22.0
Patclin 103 --- 12.9	
Cee Bee A-477 - 12.2	Patclin 103, 104 - 19.5
ALM ----------- 11.7	
Patclin 104 --- 11.5	
FO 606 -------- 11.1	Cee Bee, Turco,
Turco 5668 ---- 9.8	FO 606 --------- 16.0

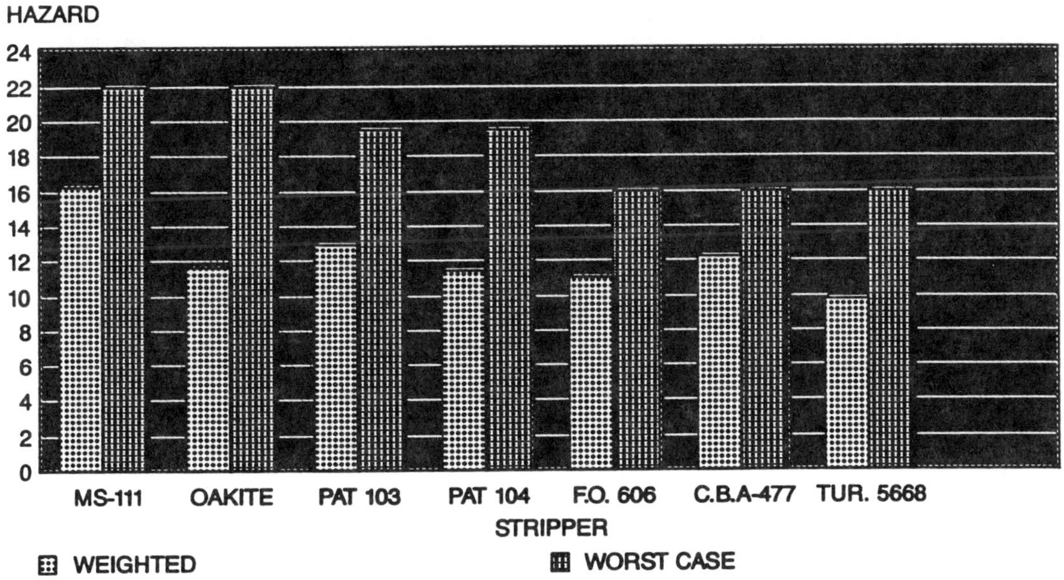

Figure 1. **STRIPPER HAZARD SCORES**

	HAZ. WASTE	WASTE WATER EFF.	TWA	TSCA	RCRA	CERCLA	CWA	FIFRA
MS-111	4	4	3	X	X	X	X	
OAKITE ALM	2	2	3	X	X		X	X
PATCLIN 103	1	1	1	X			X	
PATCLIN 104	4	1	N.E.	X				
F.O. 606	1	2	1	X	X			
C.B. A-477	0	0	2	X				
TURCO 5668	2	2	1	X		X	X	

Figure 2. REGULATORY RESTRICTIONS AFFECTING STRIPPERS

AQUEOUS DEGREASING: A VIABLE ALTERNATIVE TO VAPOR DEGREASING

J.T. Snyder
Martin Marietta Astronautics Group
Denver, Colorado

INTRODUCTION

The aerospace industry has used chlorinated hydrocarbons in vapor degreasers for over 35 years to remove dirt, greases and oils. TCE was used almost universally for a number of years as a degreaser solvent because it is a very aggressive cleaner with a high boiling point and will remove most oils, greases and waxes. Replacement of TCE by 1,1,1 TCA began in the late 60's and early 70's when the health hazards associated with the TCE usage were more clearly understood and publicized. Martin Marietta Astronautics Group (MMAG), a division of Martin Marietta Corporation, changed the solvent used in production degreasers from TCE to TCA around 1972. A number of companies continue to use TCE because changes are usually met with some kind of resistance and, for most vapor degreasing applications, TCE still performs best.

The Montreal Protocol of 1987 and recent updates of the Protocol are accelerating the phasing out of several CFCs primarily: R11, R12, 113, 114, 115. Two chlorinated hydrocarbons, TCA (trichloroethane) and carbon tetrachloride, have recently been added to the list. Martin Marietta's Environmental Management Task Force has directed the corporate divisions to develop a plan to meet the deadlines of the Montreal Protocol. A search for an available and an environmentally acceptable solvent substitute which could meet our requirements was unsuccessful. The performance, cost and possible environmental risks of substitute VOC's directed us toward the growing family of aqueous degreasers, which exhibit low safety and environmental risk potentials.

OBJECTIVES

The primary objective of this effort was to eliminate the usage of 1,1,1 TCA in our production vapor degreasers with strong consideration to be given to cost and performance of a substitute material. Vapor degreasers are utilized in the chemical processing area as a means of removing gross dirt in preparation for other chemical processing. With assistance from materials engineering and manufacturing, it was determined that vapor degreasing was not an indispensable processing step. Aqueous degreasing was a reasonable substitute. All detail parts and some assemblies processed in the factory require at least one of the following:

- Deoxidizing
- Pickling and passivation
- Etching for dye penetrant inspection
- Zinc phosphate coating
- Chromate conversion coating
- Acid etch for bonding prep
- Chem milling

All of the above processes were preceded by vapor degreasing and alkaline cleaning. The vapor degreasing step was used in every process to remove gross dirt before proceeding to alkaline cleaning. Because water is used in all processes following vapor degreasing, the aqueous solution doesn't pose a water problem on hardware.

An evaluation of the construction of our vapor degreasers revealed that conversion to aqueous immersion tanks was both practical and cost effective. Stainless steel was used exclusively in the fabrication of the degreasers.

TEST PROCEDURES

Six materials were selected as candidates for evaluation.

- •Bioact EC7
- •Simple Green
- •Daraclean 282
- •Quaker 624GD
- •Turco 3878
- •Coors Bio T

These materials were then screened using the following parameters:

- •health hazards
- •treatability in our industrial wastewater treatment plant
- •corrosion potential
- •performance

Material safety data sheets for each of the candidate materials were evaluated by our industrial hygiene department to determine health or safety concerns. Coors Bio T and Bioact EC7 were eliminated in this screening because of flammability concerns. Treatability studies of the remaining solutions eliminated Simple Green. Long term corrosion evaluation on 2014-T6 aluminum was initiated with the three remaining products. A two month immersion test in 10% solutions of the three materials revealed that Turco 3878 produced visible corrosion, while the Quaker and Daraclean products did not.

The remaining two candidate materials, Quaker 624GD and Daraclean 282, were tested in a 10% solution at 130°F with mild agitation for cleaning performance on 2014 samples coated with the following materials:

- •Fish oil (aluminum corrosion protection)
- •Mineral oil (ultrasonic thickness gaging couplant)
- •Glycerine (ultrasonic thickness gaging couplant)
- •Machine oil
- •Layout dye
- •Aluminum mill stamps

One cleaning material, Daraclean 282, remained after performance tests. This material was targeted for further testing required for approval and usage in a manufacturing process on production hardware. 2014 aluminum was selected for all corrosion and performance testing because of its extreme susceptibility to corrosion.

PROTOTYPE TESTING

A small vapor degreasing tank, measuring 4 ft. wide x 6.5 ft. deep x 12.5 ft. long, was selected for prototype testing in the production chemical processing area. Because the inside of the tank, boiling sump, and chiller coils were heavily coated with chloride scale, sand blasting was selected to remove the contamination. Sand blasting and saddle tank removal were the only requirements for converting the small vapor degreaser to an aqueous degreaser. The tank, constructed of 1/4-inch stainless steel, was then filled to a depth of 64 inches with a 10% solution of Daraclean 282 in tap water. An air agitation line was installed to enhance cleaning performance and the tank lid was removed to accelerate evaporation at 130°F. Parts were rinsed over the modified degreaser to eliminate dragout. Evaporation enhancement provided enough water loss to prevent overfilling the tank. Engineering requirements for inclusion of aqueous degreasing into a manufacturing process dictated that the new solution meet cleaning performance tests and the salt fog test for chromate conversion coated aluminum specimens specified in MIL-C-5541. Using vapor degreasing as a baseline cleaning performance tests indicated that the aqueous degreased specimens were visually cleaner and exhibited a more nearly water break-free surface than the vapor degreased specimens. Specimens processed through aqueous degreasing before chromate conversion coating exhibited superior salt fog corrosion resistance to those which were vapor degreased prior to chromate conversion coating.

RESULTS

Evaluation of test data and performance results by materials engineering and quality engineering found the new material to be suitable for use in production operations. A new manufacturing process entitled "Aqueous Degreasing" is now used as an alternative to vapor degreasing for processing production hardware. Following conversion of the large production vapor degreaser (7ft wide x 14 ft deep x 28 ft long) to aqueous completed in early 1991, 1,1,1 TCA usage at Martin Marietta Astronautics Group was reduced by over 95%. Conversion included adding shear spray rinse, filter, oil skimmer, and pump agitation. Transfer pumps, degreaser still and motorized cover were eliminated. The saddle tanks were sand blasted and cleaned for conversion to surge tanks. One of the saddle tanks also contains the oil skimmer. The aqueous degreasing process was later certified for usage on all titanium, stainless steel and carbon steel alloys.

CONCLUSIONS

- Performance of aqueous degreasing is superior to 1,1,1 TCA vapor degreasing.
- Manufacturing personnel readily accepted the change to aqueous degreasing because of improved performance and minimal safety concerns.
- No known environmental impact.
- Cheaper to operate and maintain than with TCA.
- Recyclable with simple filtration and oil separation.
- Cost of degreaser conversion to aqueous degreasing about 1/3 that of a new state-of-the-art vapor degreaser.

Figure 1

TOXIC CHEMICAL RELEASES

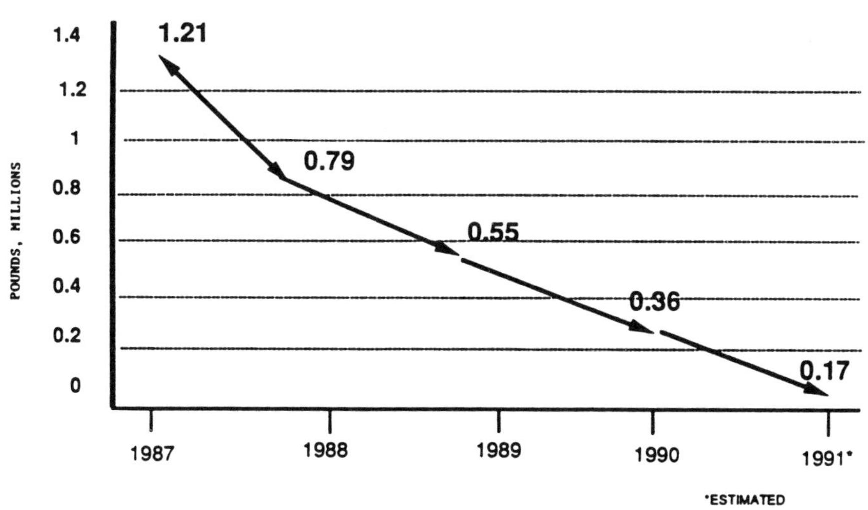

CHEMICAL SUBSTITUTION FOR 1,1,1-TRICHLOROETHANE AND METHANOL IN AN INDUSTRIAL CLEANING OPERATION

Lisa M. Brown and Johnny Springer
Risk Reduction Engineering Laboratory
U.S. Environmental Protection Agency
Cincinnati, Ohio

Matthew Bower
APS Materials, Inc.
Dayton, Ohio

INTRODUCTION

Passage of the 1984 Hazardous and Solid Waste Amendments (HSWA) to the Resource Conservation and Recovery Act (RCRA) of 1976 has redirected the U.S. environmental policy towards waste minimization to improve the quality of the environment. In its efforts to pursue the objectives set forth by Congress in the HSWAs to RCRA, the USEPA has established a national comprehensive pollution prevention program. Implementation of projects to achieve several of the pollution prevention program objectives is accomplished through research conducted by the Pollution Prevention Research Branch of the Risk Reduction Engineering Laboratory. This research addresses the intent of the Amendments to reduce the release and transport of hazardous, toxic, and non-hazardous materials through the air, water and solid media. The principal goal of the Pollution Prevention Research Branch is to encourage the identification, development, and demonstration of processes and techniques which result in less waste being generated, in order to promote a more rapid introduction of effective pollution prevention techniques into broad commercial practice.

1,1,1-trichloroethane (TCA) is used as a cold solvent degreasing agent in many industrial degreasing processes. In 1986 TCA was identified as a hazardous waste (F001) that must be managed under Subtitle "C" of the Resource Conservation and Recovery Act. As a result of this action, industries began looking for ways to avoid the use of TCA cleaning solvents. The EPA decided to target the metal finishing industry for participation in a joint research project to examine the possibility of substituting a terpene-based cleaner for TCA in degreasing operations. APS Materials, Inc., a facility in Dayton, Ohio participated in the research project. APS Materials, Inc. is a metal parts finishing company which generates TCA and methanol (hazardous waste F003) waste from cold solvent degreasing operations associated with their plasma spray deposition process.

PLASMA SPRAY DEPOSITION PROCESS

The plasma spray deposition process has emerged as a major means to apply a wide range of materials on diverse substrates. The deposition process is accomplished with the use of a plasma gun. The plasma, generated inside the gun, exits as a high velocity flame through the nozzle of the gun. A powdered feedstock is injected into the flame via a carrier gas (usually argon). The injected powder accelerates, melts, and is carried at sonic velocities to the substrate on which the particles impact and solidify rapidly, building a well-adhered protective coating.(1)

While APS Materials, Inc. employs the fundamental plasma spray deposition process, a few changes were made to better accommodate the plasma spray work performed by their company. First, APS

Materials performs its plasma spraying in an inert atmosphere chamber. This is done for cooling and to prevent the titanium powder used in many of its coating applications from becoming oxidized thus forming brittle coatings. APS Materials also uses helium, in the spray gun, as well as to adjust the heat level and arc length.

PROCESS DESCRIPTION

Original Process

In the APS Biomedical Parts Division, the company primarily coats cobalt/molybdenum parts and titanium parts with a porous titanium alloy. In order to achieve a strong and adhesive coating, the parts were cleaned with TCA or methanol (TCA for cobalt/molybdenum and methanol for titanium). TCA is more economical than methanol but weakens titanium over time. The cleaning process consists of several steps. The parts undergo visual inspection, tape masking, and grit blasting. After the grit blast has been completed, the masking tape is removed. The part is then immersed in a small pail containing TCA or methanol. The pail is placed in an ultrasonic bath containing warm water for 15 minutes. Solids from grit blasting, oil and grease from the manufacturing and handling of the parts, and any adhesive residuals from the masking tape are removed in this cleaning process. Waste TCA and methanol were being generated at the rate of 1/2 barrel per month each. After the ultrasonic bath, a graphite masking suspension is applied to the part on surfaces where the plasma spray coating is not wanted. The part is then plasma sprayed and cleaned again to remove excess titanium and the graphite mask.

As a check system, APS runs small one-inch diameter disks of the same composition as the part to be coated - called "test buttons" - through the same cleaning and coating process. The test buttons are placed on a tensile strength testing machine which measures the tension required to separate the coating from the substrate as a quality control measure.

Description of Initial Bench Scale Experiments

For this test, DuSQUEEZE (DuBois Chemicals, Inc.) was the product used to determine substitution feasibility. DuSQUEEZE is a blend of surfactants containing 25% limonene. Limonene was selected as a possible substitute for TCA and methanol because of its disposal qualities. Disposal of dilute solutions of DuSQUEEZE could be accomplished by flushing it to a sanitary or industrial sewer, according to local sewer use permit requirements. The feasibility of substituting a dilute, terpene-based cleaner (DuSQUEEZE) for TCA and methanol was determined by assessing the tensile strength of the plasma coating bonds made after cleaning with dilute DuSQUEEZE solutions. Five tests were performed, four on plasma-coated test buttons to assess the tensile strength of bonds made after cleaning with the DuSQUEEZE solutions, as compared to the tensile strength of bonds made after cleaning with methanol and TCA, and one test to determine if any limonene remained on the buttons after being cleaned. In the first test, four titanium test buttons were placed in a stainless steel beaker containing a 20:1 dilute solution of DuSQUEEZE and water. The solution was agitated for 20 seconds. The test buttons were then placed in a stainless steel beaker containing deionized (DI) water which was agitated for 20 seconds. The test buttons were then blow-dried and plasma- sprayed. The tensile strength of the bond between the plasma arc coating and the substrate was measured using a Tinius Olsen tensile tester.

In the second test, four titanium buttons were placed in an ultrasound bath containing a 50:1 dilute solution of DuSQUEEZE for 10 minutes. Next the buttons were placed in a stainless steel beaker containing DI water for 30 seconds. The titanium buttons were blow-dried for 60 seconds and then plasma-sprayed. Then tensile strength of the bonds were tested in the same manner as the first test. The third test followed the same procedure as test two, using a 100:1 dilute solution of DuSQUEEZE. In the fourth test, the buttons were cleaned by the same process as the third test, but the buttons were analyzed for residual limonene and were not plasma-sprayed and tensile-tested. In the fifth test, cobalt/molybdenum buttons were used instead of the titanium buttons with the test protocol identical to the third test.

Modifications to Existing System

APS purchased a heated ultrasonic bath with a timer for the conversion. However, when this ultrasonic bath malfunctioned, a heater was added to the old ultrasonic bath. The TCA/methanol cleaning system did not require a DI water rinse, so a DI water system was purchased along with a stainless steel bath and immersion heater. With the new cleaning system, the parts took longer to dry, so a heat gun was purchased to speed-up the drying process.

SAMPLING AND ANALYSIS

The overall purpose of the sampling and analysis project at APS Materials was to support a purely qualitative judgement of the cleaning capabilities of the substitute cleaning solution (i.e., limonene). The sampling and analysis protocol for this project was set up in three parts; sampling spent solutions of methanol and TCA, sampling the terpene-based cleaning solution after modifications were made to the cleaning system, and developing data for a comparative analysis of plasma-coating bond strengths between the coatings of test buttons that were cleaned with methanol/TCA prior to coating and the coatings of test buttons that were cleaned with the terpene-based solution prior to coating.

Pre-Modification Sampling

The first part of the sampling process was performed prior to any modifications. This sampling was performed in order to determine the type and amounts of contaminants found in the cleaning solvents. Samples of the methanol and TCA cleaning solutions were taken and analyzed for oil and grease, dissolved solids, suspended solids, titanium metal and cobalt metal. This sampling also established the baseline performance for methanol and TCA. The samples were taken by mixing the material in a plastic bucket and then pouring a sample from the bucket through a glass funnel into a glass bottle. The data derived from this sampling served as a bench mark for the ensuing substitution sampling.

Post-Modification Sampling

The second part of the sampling scheme was performed after the modifications were made to the system in order to determine the effectiveness of the terpene-based solvent in cleaning the parts. Sampling of the cleaning solution was performed throughout a typical operating cycle. Samples were recovered at the beginning of a bath cycle (i.e., when the tank contents were completely replaced with fresh cleaning solution) to establish baseline concentrations. A second sample was taken midway through the effective life of the cleaning solution. A final sample was recovered prior to removing the spent solution from the dip tank.

In addition to taking samples of the cleaning solution, wipe samples were taken of the cleaned parts. Wipe samples were taken to evaluate the cleaning efficiency of the solution over time by analyzing for residual contaminants (oil and grease) on the parts. One wipe sample was taken from the cleaned metal parts during each sampling interval to determine if there was a residual of oil and grease. The wipe sample was performed using sterile, uncontaminated cloth. Sterile gloves were worn to prevent contamination of the cloth with oil and grease. The wiping procedure was consistent for each sample. A glass container of sufficient volume was used to hold the cloth after sampling. Three wipe samples were taken over the life of the limonene cleaning solution, to coincide with the three liquid samples described above.

Analysis for metals was performed using inductively-coupled plasma atomic emission spectroscopy (ICP). Oil, grease and dissolved/suspended solids were analyzed using gravimetric analytical techniques. Spikes and replicate analyses were also done to check for accuracy and precision and to identify the presence of any matrix effects associated with sample preparation or measurement. Data were then combined and statistically evaluated.

The analysis of plasma-coating bond strength compared current data collected by APS Materials regarding the strength of coatings applied after parts were cleaned with dilute solutions of DuSQUEEZE and historical data of bond strength resulting from parts cleaning with TCA and methanol. Data generated two months prior and two months following the conversion to the limonene solution were used for this comparison.

RESULTS AND DISCUSSION

Bench Scale Experiments

The before and after tensile strength results were comparable. Overall, the bonding strengths were actually slightly better for the dilute limonene cleaner (see Table 1). No residual limonene was detected (detection limit 1 ppm) for cleaner at 100:1 dilution.

Analyses For In-plant Operations

The initial tests for contaminants in methanol and TCA used for cleaning yielded the results shown in Table 2. The samples for these analyses were taken when the baths were considered spent, just prior to being dumped.

The amounts of oil and grease found in the wipe samples, shown in Table 3, were very low at about 1 mg or less. The increase in oil and grease from the bath dump, as compared to the fresh bath, was very small for one sample and was less than the fresh bath in the second bath dump sample. This latter result could have resulted from the wiping technique. The parts seemed to be cleaned just as well at the time the bath is dumped as when the bath is fresh.

Table 4 shows results from GC/MS method 8270 (SW-846) analyses for residual limonene on the parts. Limonene was not detected in the rinse samples, thus indicating that all of the limonene was removed during dragout and subsequent drying of the parts.

The results in Table 5 indicate that dissolved solids and oil and grease were much higher in the fresh bath and the bath used to clean parts only prior to plasma spraying (Dump#1), than in the bath used also for cleaning after plasma spraying (Dump#2). The reverse was true for the suspended solids. The graphite in the bath may affect the DuSQUEEZE cleaning solution to create these differences.

A comparison of the DuSQUEEZE cleaning solution with the previous methanol and TCA samples, reveals that the oil and grease levels in the DuSQUEEZE are much lower than the other cleaning solvents. Suspended solids for the DuSQUEEZE are lower than the previous solvents, except for the sample containing graphite, which is roughly equivalent. Dissolved solids for DuSQUEEZE are much higher than the other solvents. The higher dissolved solids may reflect the fact that the DuSQUEEZE is an emulsifying agent which converts the oil and grease to dissolved solids. This would explain the lower oil and grease levels for DuSQUEEZE.

Although the data generated by the sampling and analysis program, shown in Table 6, indicates that the terpene-based cleaner adequately cleaned the parts for this process, since wipe samples were not taken for the original process, a statement of comparison between the former and present cleaning techniques is not feasible.

Economic Analysis

Although the old ultrasonic bath was in use at the time of the test, economic analysis is shown for the system that APS is now operating. A summary of the economic analysis is found in Table 7.

CONCLUSIONS

In summary, it has been determined that a terpene-based cleaner can adequately clean metal parts without adversely affecting the performance of the plasma-arc coating application. The use of a terpene-based cleaner in place of methanol and TCA, has proven to be an environmental and economic success. Elimination of the disposal problems associated with methanol and TCA coupled with the maintenance of plasma-arc coating quality, make the use of terpene-based cleaners more attractive than other plasma spray coating processes, as well as other metal cleaning/coating operations. The annual cost savings, as well as the short payback period, also make the cleaner attractive from an economic standpoint.

REFERENCES

1. Herman, Herbert, "Plasma Spray Deposition Processes", MRS Bulletin, p. 60 - 68, December 1988.

2. Test Methods for Evaluating Solid Waste. SW-846, Third Edition, U. S. Environmental Protection Agency, November 1986.

3. Environmental Monitoring and Support Laboratory. Methods for Chemical Analysis of Water and Wastes. EPA-600/4-29-020, U.S. Environmental Protection Agency, Cincinnati, Ohio, March 1983.

TABLE 1. TENSILE STRENGTH TEST RESULTS FOR BENCH SCALE EXPERIMENTS

Test Buttons	Cleaning Agent	Tensile Strength (psi)
titanium	methanol	6300+/-1260
titanium	DuSQUEEZE*	7000+/-570
cobalt/molybdenum	TCA	5150+/-1990
cobalt/molybdenum	DuSQUEEZE*	5400+/-1290

* Tensile strengths measured for test button cleaned with various dilutions of DuSQUEEZE showed no trends or statistical differences, so values shown include all measurements.

TABLE 2. RESULTS OF ANALYSES OF SOLVENT SAMPLES FOR CONTAMINANTS

Test	Methanol (mg/l)	TCA (mg/l)
Dissolved solids	1	29
Suspended solids	33	9
Oil and Grease	911	141
Metals		
Cobalt	-	ND*
Titanium	0.021	-

* Method detection limit is 0.01

TABLE 3. RESULTS OF ANALYSES FOR OIL & GREASE ON PARTS CLEANED WITH 100:1 DILUTE SOLUTION DUSQUEEZE

Test	Oil and Grease Total Mg
Wipe Sample, Fresh Bath	1.0
Wipe Sample, Mid-life Bath	0.4
Wipe Sample, End-life Bath	1.2
BLANK	ND*

* Method detection limit is 0.3

TABLE 4. RESULTS OF ANALYSES FOR RESIDUAL LIMONENE ON PARTS CLEANED WITH 100:1 DILUTE SOLUTION DUSQUEEZE

Test	limonene concentration Total Ug/sample
Rinse Sample, Fresh Bath	ND(<0.3)
Rinse Sample, Mid-life Bath	ND(<0.65)
Rinse Sample, End-life Bath	ND(<0.3)
BLANK	ND(<0.2)

TABLE 5. RESULTS OF ANALYSES OF 100:1 DILUTE DUSQUEEZE SOLUTION FOR CONTAMINANTS

Test	Fresh Bath mg/L	Dump#1 mg/L	Dump#2 mg/L
Dissolved solids	3650	3010	887
Suspended solids	ND*	ND*	19
Oil and Grease	37.0	30.8	15.1
Metals			
Cobalt	0.019	0.18	0.081
Titanium	ND#	ND#	1.65

* Method detection limit is 2
# Method detection limit is 0.047

TABLE 6. TENSILE STRENGTH TEST RESULTS FOR IN-PLANT OPERATIONS

Coating/Substrate	Cleaning Agent	Tensile Strength (psi)
titanium / titanium	methanol	5560+/-600
titanium / titanium	DuSQUEEZE	7180+/-610
titanium / cobalt-moly	TCA	5820+/-370
titanium / cobalt-moly	DuSQUEEZE	5330+/-1560

Table 7. ECONOMIC ANALYSIS

Capital Expenditures

Item	Cost
Ultrasound with heater	$1425
5 gallon stainless steel rinse vessel	38
Immersion heater	105
Heat Gun	75
DI water system installation	150
TOTAL	$1793

Annual Operating Costs

Item	Gal/Yr	Cost
DuSQUEEZE usage	7.8-11.8	$150
DI Water usage	1825-2920	700
TOTAL		$850

Annual Cost Savings

Item	Amount	Cost
Avoided TCA Purchases	330 gal/yr	$1650
Avoided Methanol Purchases	120 gal/yr	1000
Avoided Waste Disposal	6 barrels/yr	3000
TOTAL		$5600

Net Cost Savings $4800

Payback Period: $1793/$4800 = 0.37 year, 4.5 months

ALTERNATIVES TO CFCs IN PRECISION CLEANING: A NEW HCFC BASED SOLVENT BLEND

R.S.Basu and P.B.Logsdon
Engineered Materials Sector
Allied-Signal Inc.
Buffalo Research Laboratory
Buffalo, New York

and

E.M.Kenny-McDermott
Guidance Systems Division
Allied-Signal Aerospace
Teterboro, New Jersey

INTRODUCTION

Recent findings of ozone layer depletion has prompted United Nations Environmental Program (UNEP) to amend the Montreal Protocol, for an accelerated phase-out of chlorofluorocarbons (CFCs), carbon tetrachloride and halons by the year 2000 and phase-out of methyl chloroform by the year 2005. In addition, U.S. Congress has passed the Clean Air Act which puts similar phase-out dates for CFCs and methyl chloroform. Because of these regulations, a search for replacements for all the regulated CFC molecules has been intensified. In addition, global warming is also emerging as another major environmental problem making the search more and more complex.

In this article, we are going to discuss the performance characteristics of a new stratospherically safe alternate to trichlorotrifluoroethane (CFC-113) as a precision cleaning solvent. The alternates are based on hydrochlorofluorocarbons, the selected ones are 1,1-dichloro-1-fluoroethane (HCFC-141b) and 1,1-dichloro- 2,2, 2-trifluoroethane (HCFC-123) with blends based on these compounds. These compounds are formed by the addition of hydrogen to CFCs to make them shorter lived in the troposphere and therefore less harmful to the stratospheric ozone. The Montreal Protocol has also put restrictions on the use of these substances because of their low but non-zero ozone depletion potential. HCFCs are considered interim replacements with phaseout dates starting from 2020 and complete phaseout no later than 2040. The U.S. Clean Air Act assigns earlier phase-out dates for HCFCs, starting with a freeze in production in the year 2015 with complete phase-out by 2030.

SOLVENT SELECTION

In this section we are going to discuss how the solvent for precision cleaning application is selected. Precision cleaning encompasses manufacturing aerospace components, gyroscopes in missile guidance systems, medical devices, computer disks, silicon wafers, etc. These parts to be cleaned contain various metals and plastics and CFC-113 is the universal choice in these applications. CFC-113 is non-flammable, non-toxic, stable, has good solvency characteristics and is compatible with these materials.

Solvent selection depends on a number of environmental factors, along with the ozone depletion and greenhouse warming potential mentioned before. The potential of contributing to smog in the lower atmosphere and pollution to ground water by way of water effluents are also important considerations. The objective of solvent selection is not to trade one environmental problem with another. There are federal and state laws prohibiting the emission of volatile organic compounds (VOCs). These materials break down in the lower atmosphere and contribute to smog pollution. Over and above these, the molecules have to be relatively non-toxic so that the workplace remains relatively safe. Effluents from plants may also pose a problem which is of a lesser concern for solvents than for aqueous systems. The energy requirement is in general lower for solvent cleaning as compared to aqueous cleaning. However, energy consumption also has to be considered in detail because of its contribution to the greenhouse effect.

Among hydrochlorofluoroethanes HCFC-123 (1,1-dichloro-1,2,2-tri-fluoroethane) and HCFC-141b (1,1-dichloro-1-fluoroethane), with boiling points between 80 and 90 F, are being considered as alternates to CFC-113 in solvent applications. The rest of the hydrochlorofluoroethanes are not found to be suitable because of their toxicity, flammability or other undesirable characteristics. These two HCFCs are far less ozone depleting than the current CFCs. Molecular structure of these compounds are shown in Figure 1. The lifetimes, ozone depletion potentials (ODP) of these chemicals and also the greenhouse warming potentials (GWP) are given in Table 1. The ODP and GWP are measured relative to CFC-11 which is assigned a value of 1.0. The ODP and the GWP numbers for the molecules are still being revised by AFEAS and values are expected to be finalized by the end of the year. These two HCFCs are not considered VOCs by US Environmental Protection Agency and therefore, they do not come under that regulation.

Our tests have shown that since HCFC-141b or HCFC-123 alone are not equivalent to CFC-113, a blend of the two is required. HCFC-123 lowers both the ODP and the flammability characteristics of the blend. HCFC-123 is more aggressive towards the plastics and the presence of HCFC-141b makes the blend more compatible to plastics. So blends are preferred to the pure components both from an environmental and performance standpoint.

The preferred blend for precision cleaning is a new azeotrope-like non-segregating blend consisting of HCFC-141b and HCFC-123. The composition of the blend is 80 percent by weight HCFC-141b and 20 percent by weight of HCFC-123. Some of the physical properties of the blend as compared to CFC-113 are shown in Table 2. The blend is commercially available from Allied-Signal under the tradename Genesolv®2020.

As we compare the properties we find that the major difference in physical properties between the CFC-113 and HCFC based blend is their boiling point. This would normally indicate that Genesolv 2020 would evaporate faster than CFC-113 resulting in higher loss rates of the solvent but in reality the evaporation rates of the solvents at room temperature are equivalent. This is due to the fact that Genesolv 2020 has higher heat of vaporization. This property makes it an acceptable substitute for CFC-113 despite its lower boiling point.

Another important use of CFC-113-based blend is in printed circuit board cleaning. The azeotrope- like blend containing HCFC-141b, HCFC-123 and methanol marketed by Allied-Signal under the tradename Genesolv 2010 is the solvent of choice for that application. We are not going to discuss that application in this article. However, we would like to refer the readers to several other articles detailing the use of this blend in defluxing [2,3,4].

SOLUBILITY AND CLEANLINESS STUDIES

In selecting a new solvent, solvency and boiling point are chosen as major physical properties. The solvency of various light oils is determined by measuring their solubility in these solvents. A boiling point range of 25° to 75°C has been chosen as the range for these solvents and various solubility models are used as a tool to select solvents on the basis of solvency. Cleaning tests were performed to finalize the selection.

In the cleaning tests, metal coupons are soiled by various types of oils and heated to 200°F. This is done to partially simulate the temperature attained while machining and grinding in the presence of these oils. The metal coupons thus treated, are degreased in a vapor phase degreaser machine. The coupons are held into the vapor and are vapor rinsed for a period of 15 seconds to 2 minutes depending upon the oils chosen. A short time period is selected so that the solvents can be compared easily. The blend mentioned here is compared to CFC-113 in cleaning performance. Cleanliness of the coupons after cleaning is determined by carbon coulometer which detects amounts of carbon physically adsorbed on the surface of the metals. Since most oils are hydrocarbon-based this method detects the amount of hydrocarbon left on the surfaces.

The overall cleanliness test results with various types of oils are shown in Table 3. The results show that the performance of azeotrope-like blend Genesolv 2020 is equal to or better than CFC-113 in cleaning various oils. The cleaning results shown in Table 3 are in percentages of oils removed from various substrates.

PRECISION CLEANING STUDY OF GYROSCOPE COMPONENTS

High precision gyroscopes contain some mechanical assemblies with tolerances on the order of 1-5 micrometers. As a result, they are extremely sensitive to particulate contamination as well as to very low levels of foreign fluids and polymeric films. Accuracy, precision and reliability of gyroscopes used for military and aviation applications depend on a cleaning process being able to remove these deleterious contaminants without degrading the materials of construction. Gyroscope components are fabricated using a variety of materials, including light metals (i.e. beryllium and aluminum), plastics, adhesives and elastomers. CFC solvents and blends have been eminently suited for this application. Finding a suitable alternative cleaning agent or process will be a difficult task. Extensive material compatibility and cleaning efficacy studies must be conducted on any new solvent system to ensure product integrity. Beryllium metal parts preclude the use of aqueous based cleaning or 1,1,1-trichloroethane which hydrolyzes easily and may cause corrosion. Also, any new solvent system must not attack the polymeric materials in the device.

Present cleaning processes include vapor degreasing, pressure cooking, soxhlet extraction, cold spraying in enclosed booths, cold cleaning in ultrasonics and gyroscope flushing with CFC solvents and blends. Parts and sub-assemblies are typically cleaned several times during assembly in order to remove handling and processing contaminants. Typical soils include particulates, silicones, hydrocarbons, finger oils, bromofluorocarbon balancing fluids, flux and excess adhesive.

This work examines a few of the aspects of testing the new HCFC solvents replacement for CFC solvents for precision cleaning of gyroscope parts and assemblies. It is only the beginning of the extensive research program required to implement these new solvents into the manufacturing mainstream. We are going to talk about the cleaning studies in this section. Cleaning efficacy is determined by comparing the HCFC cleaned surface to one cleaned with CFC using Fourier Transform Infrared Micro and Reflectance Spectroscopy, water break test or weight change of the coupon.

Fourier Transform Infrared reflectance spectra of the surfaces of the coupons cleaned in

CFC-113 and Genesolv 2020 are shown in Figures 2 and 3. The spectra show peaks of various heights indicating materials left on stainless steel coupons after cleaning with CFC-113 and Genesolv 2020. The soils used are phenyl methyl silicone and silicone grease. In case of both of these soils, coupons degreased in Genesolv 2020 show fewer peaks indicating better cleaning compared to CFC-113.

TOXICITY AND MATERIAL COMPATIBILITY

Introduction of new compounds in the marketplace requires extensive toxicity studies. Presently both HCFC-123 and HCFC-141b are undergoing thorough toxicological studies. A Program for Alternative Fluorocarbon Toxicity (PAFT) has been set-up by a consortium of a large number of current CFC manufacturers all over the world. Repeated dose toxicity studies have been completed and no significanct adverse effects have been found. Allied-Signal has been given permission by the US Environmental Protection Agency to manufacture and sell HCFC-141b for solvent use. Of the two HCFCs, HCFC-141b has been found to be less toxic. At this point with available information, an interim PEL of 500 ppm is assigned to HCFC-141b and 50-100 ppm is assigned to HCFC-123. These are interim values, final PELs are expected to be assigned following completion of the chronic studies, which are underway.

Due to lower boiling points, loss of HCFC solvent vapors from older degreasers may be too great for reasons of health and safety and also effect the processing costs. This requires tighter or low emission machines. Existing equipment should, therefore, be evaluated in this respect and be either upgraded to new equipment, or perhaps, be retrofitted to maintain safe and healthy work environment.

Hydrolytic stability of the solvent blend is tested by refluxing the solvent in the presence of various metals and water for a two (2) week period. The solvent even without any stabilizer, showed excellent stability in presence of water. Its stability compares very well with CFC-113 and it seems far superior to 1,1,1-trichloroethane. It appears the presence of a fluorine in the molecule and stronger carbon-fluorine bonding has increased the stability of HCFC-141b over 1,1,1-trichloro- ethane. One important point to note is that, HCFC-141b, HCFC-123 and Genesolv 2020 are all compatible with beryllium metal which makes the solvent blend extremely attractive to clean gyroscopes. As mentioned before, various components in a gyroscopes are fabricated using beryllium and its alloys which makes beryllium compatibility a very important factor in solvent selection.

In commercialization of a new product for precision cleaning another important property to study is its compatibility with various elastomers. These elastomers are present in various components cleaned. The elastomer compatibility is done where the elastomers are refluxed in solvent for a two week period. The solvent blend seemed to have reasonable compatibility with a large number of elastomers. A detailed account of material compatibility appears on references [5,6] and is also available from Allied-Signal upon request.

CONCLUSIONS

In this paper we have shown that a substitute for CFC-113 has been found with a much lower ozone depletion potential for use as a solvent in precision cleaning applications. This is an azeotrope-like blend of HCFC-141b, HCFC-123 (Genesolv 2020) and is presently in the process of commercialization. Solubility measurements with various oils have shown that this blend has a very good potential to be used as a solvent for these applications. This is further confirmed by metal cleaning studies where the blend showed equivalent or better performance compared to

CFC-113. Finally application results were done to demonstrate that Genesolv 2020 performs very well in field applications of cleaning gyroscope components. In this application FTIR reflectance spectroscopy has been used and is found to be an excellent method for cleanliness measurement.

The solvent is currently being produced in pilot quantities in the pilot plant working at Buffalo and is marketed under the tradename Genesolv 2020. The actual commercial production will start by the end of 1991 or by early 1992.

REFERENCES

Montreal Protocol, Final Act 1987, UNEP, 1987; London Amendments 1990.

R.S.Basu, E.L.Swan and M.J.Ruckriegel, "Cleaning Alternative for 1990s and Beyond: HCFC Based Solvent Blend and Low Emission Equipment", Proc. of NEPCON EAST 90, pp. 161-178.

R.S.Basu and J.K.Bonner, "Alternatives to CFCs: New Solvents for the Electronics Industry", Surface Mount Technology, Dec. 1989, pp. 34-37.

J.K.Bonner, "Solvent Alternatives for Electronics for the 1990s", Proc. of NEPCON EAST 90, pp. 189-202.

J.K.Bonner, "Solvent Alternatives for the 1990s", Proc. of NEPCON WEST 90, pp. 1601-1608.

R.S.Basu, K.D.Cook and E.L.Swan, "Stability and Compatibility of HCFC Based Solvent Blends for Defluxing", Proc of NEPCON WEST 91, pp. 926-931.

Table 1: ODP and GWP Values for Selected Compounds

Solvents	Formula	B.P.(°F)	Lifetime	ODP	GWP
CFC-11	CCl_3F	75	60	1.0	1.0
CFC-113	$C_2Cl_3F_3$	118	102	0.8	1.3
HCFC-123	$C_2HCl_2F_3$	82	1.5	0.02	0.02
HCFC-141b	$C_2H_3Cl_2F$	89	10	0.12	0.12

Table 2: Physical Properties Comparison

	Genesolv 2020	CFC-113
Ozone Depletion Potential	0.1	0.8
Greenhouse Warming Potential	0.1	1.3
Flash Point	None	None
Boiling Point (°F)	87.8	119
Liquid Density(g/cc)	1.38	1.56
Kauri-Butanol Value	58	31
Solubility Parameter$(cal/cc)^{1/2}$	7.6	7.3
Evaporation Rate(ether=1)	1.2	1.2
Surface Tension (dynes/cm)	18.4	17.8
Heat of Vaporization (Btu/lb)	91.0	63.1

Table 3: Cleaning Results

OIL	SOLVENT	% Oil Removed		
		Substrates		
		Aluminum	**Stainless Steel**	**Mild Steel**
Petroleum Based	CFC-113	100.0	100.0	100.0
	Genesolv®2020	88.8	97.9	98.3
Semi-synthetic	CFC-113	94.5	94.5	98.3
	Genesolv®2020	99.2	99.2	99.0
Synthetic	CFC-113	19.5	14.6	7.6
	Genesolv®2020	40.1	38.1	54.4

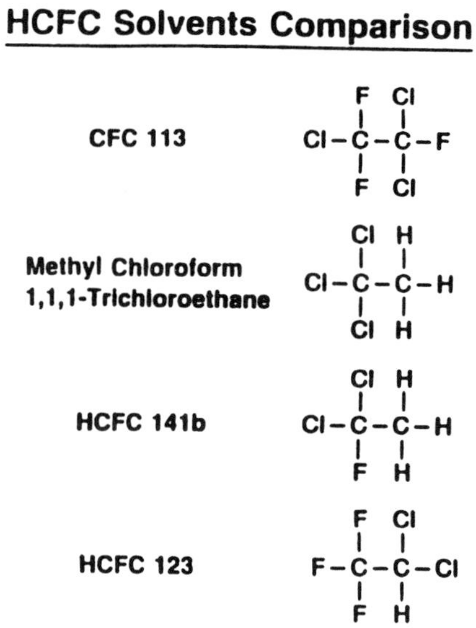
HCFC Solvents Comparison
CFC 113
Methyl Chloroform
1,1,1-Trichloroethane
HCFC 141b
HCFC 123

Fig. 1

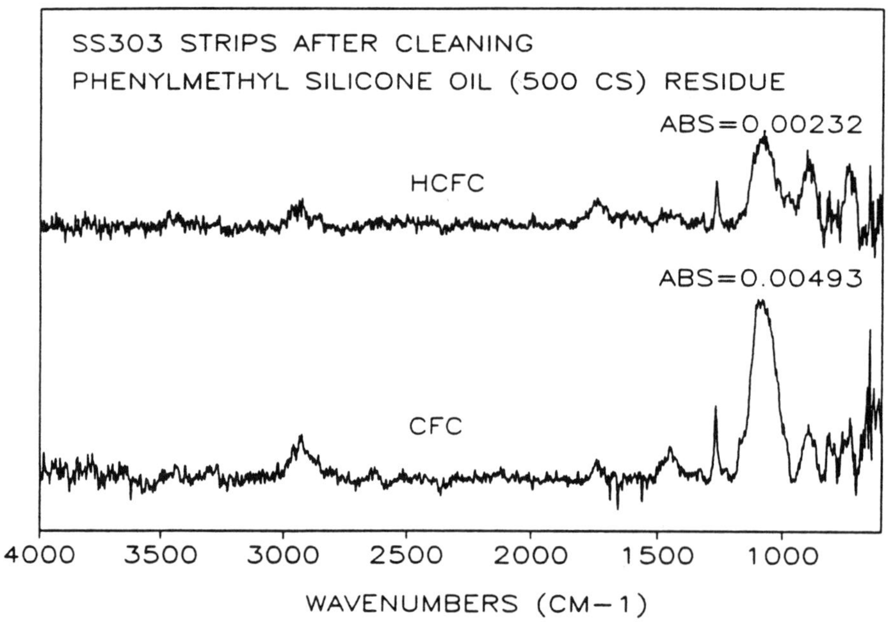
SS303 STRIPS AFTER CLEANING
PHENYLMETHYL SILICONE OIL (500 CS) RESIDUE
ABS=0.00232
HCFC
ABS=0.00493
CFC
4000
3500
3000
2500
2000
1500
1000
WAVENUMBERS (CM-1)

Fig. 2

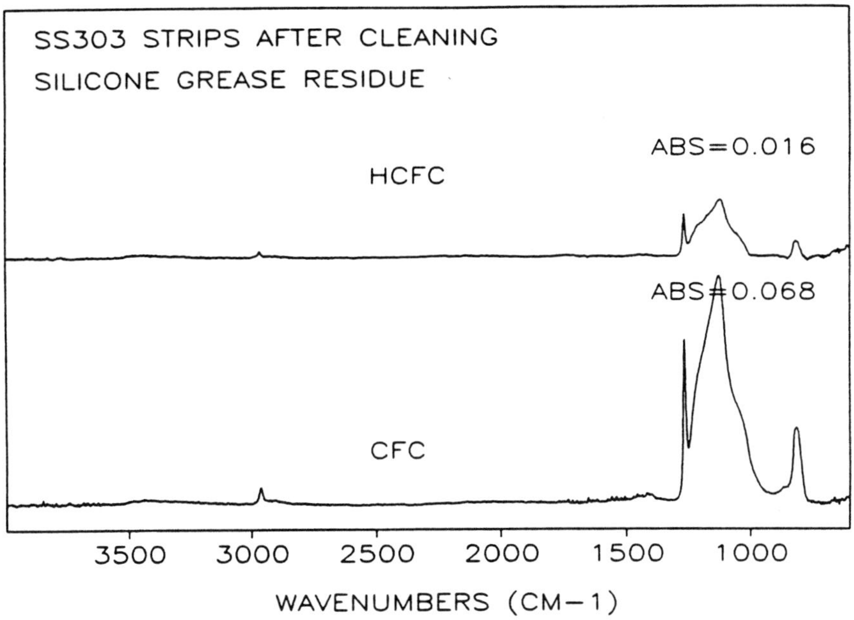
SS303 STRIPS AFTER CLEANING
SILICONE GREASE RESIDUE
ABS=0.016
HCFC
ABS=0.068
CFC
3500
3000
2500
2000
1500
1000
WAVENUMBERS (CM−1)

Fig. 3

Section III

SOLVENT RECOVERY AND RECYCLING

THE SUCCESSFUL IMPLEMENTATION OF A SOLVENT RECOVERY PROGRAM

Marcanne Lynn Burrell
Waste Minimization Company
Bellevue, Washington

ABSTRACT

This paper provides a step-by-step approach for obtaining the technical background and program support necessary for the successful implementation of a solvent recovery program based on an existing program at a Boeing Aerospace and Electronics (BA&E) site in the Puget Sound Area.

INTRODUCTION

It has become apparent that many of the solvents being used in our manufacturing processes today are not only harmful to the environment, but to humans as well. Presently, there are two options to choose from when addressing the problem of environmentally harmful solvents. The first, and more preferred option, is to replace these solvents and paints with substitutes which are as effective, but are not as harmful (eg., replace organic-based materials with aqueous-based materials). Unfortunately, substitutes have not been developed for all processes and may not be for many years. In the interim, harmful chemicals are being released into the atmosphere and disposal costs for these spent solvents are escalating. Until substitutes are available, recycling remains an option that minimizes the generation of hazardous waste and results in considerable cost savings.

PROCEDURE

Step One - Research Background

The first step is to determine the solvent which will best suit your company's needs and evaluate the technical and economic feasibility of recycling this solvent.

Conduct a shop survey. Find out which solvents the shop(s) are currently using. Ask the following questions:

- Would the shop(s) prefer using a substitute that works better, but may not be economically feasible at this time?
- Would it be feasible to change to the more desirable solvent if it was being recycled?
- If the shop(s) are generating different types of solvents, are they willing and able to switch to an universal solvent? (This will avoid the generation of small quantities of different types of spent solvents).

Review historical records (i.e., purchasing and disposal records). This data will help to determine what quantities were consumed and to illustrate trends. Historical records can also be used to verify the information supplied by the shops and other personnel.

After the type of solvent(s) to be recycled has been determined, obtain information on recycling equipment that will be used. Perform a literature search. Interview vendors and survey other organizations and companies that have processes similar to yours to find out if they already have a program in operation. Capitalize upon their past experience. This can save time and money by preventing future problems and duplication of work.

Investigate the regulations and restrictions that may effect this program. This should be done at the federal, state, local and company levels. Agencies at the federal level include the Environmental Protection Agency (EPA), Occupational Safety and Health Administration (OSHA) and National Fire Protection Association (NFPA). The state regulations may be more stringent than regulations on the federal level. Local restrictions would include city and county codes enforced by the fire department, building inspectors, the local air and water regulatory agencies and local transportation criteria. These codes may require permits for the equipment, the location, transportation of the materials and building additions or modifications. Larger companies may have their own fire department, fire protection engineering, safety, industrial hygiene and other governing organizations that will have some prerequisites for this program. Be sure to consult these organizations early and keep them informed.

A pilot scale study is recommended. This can be accomplished by renting, leasing or borrowing equipment from a vendor for several weeks to months depending upon the complexity and magnitude of the program. The pilot study will test the logistics of the entire system, determine the division of labor, and identify and troubleshoot any unexpected problems prior to the purchase of equipment. Samples should be taken during the pilot study to verify the quality of the recycled solvent. The pilot study will indicate the size of equipment necessary for full scale operation. When sizing equipment, it is recommended that the equipment be oversized to allow for additional streams in the future, errors in data and downtime due to maintenance and repairs.

Step Two - Develop Support

The second step is to develop support for the project to ensure its success. First, gain management support by demonstrating the economic and other benefits gained from the program. Show management that the process fits within the company structure, policy and philosophy. Management support is needed to obtain the necessary resources such as capital, labor, and space. Management can also mandate shop participation.

It is also important to continue working with the regulatory agencies. Work with the fire agencies to insure that equipment, procedures, installation and location are all acceptable. Use OSHA regulations or work with OSHA representatives within your company to develop safe operating procedures for the equipment and handling of recycled and spent solvent. Safety, hygiene and toxicology personnel can also be helpful in developing a Material Safety Data Sheet (MSDS).

Develop a good rapport with the shop(s) and its management. Make the project a team effort by keeping the shop(s) informed from the beginning, soliciting their suggestions and making them understand that it is ultimately their program. Shop support is crucial for the program's success.

Step Three - Implementation

Implementation of a program consists of coordinating and troubleshooting. The location of the equipment must now be determined. Ensure that the equipment will be accessible. The location should be acceptable to all parties involved (eg., the Fire Department, Fire Protection Engineering, the site environmental group, plant engineering, transportation, layout, the using organization and the organization(s) generating the spent solvent).

Training of personnel will also have to be coordinated. This includes the operator of the new equipment and the shop personnel generating the solvent. The shop management needs to concur on exactly what is expected of their personnel to support the program once it is implemented.

Construct a manual describing the process procedure and assigning responsibilities to each group involved in the program. Identify key personnel to contact in each organization. This will eliminate grey areas. In the future the manual will be a useful reference for new personnel.

Following the guidelines discussed above will minimize problems. However, complications can arise. The equipment may fail. Contact the manufacturer to discuss troubleshooting and repairs should equipment malfunctions occur. Beware of misinformation from the vendor. As a rule, research all equipment associated with the process (i.e., chillers, pumps, heat exchangers, condensers, etc.) and if possible get guarantees on all purchases. If it is not practical to return malfunctioning or inapplicable equipment to the manufacturer, it may be necessary to make changes to the process, location or installation. For example, upon startup of this program it was discovered the chiller should not be operated outdoors or at ambient temperatures below 50°F. This system was located outside and was operated when the temperature was below 50°F. Unfortunately, this information was not known prior to purchase and installation of this equipment. During winter, the chiller began to malfunction and shut down. To solve this problem, the chiller was moved indoors.

Key personnel may become obstacles. The equipment operator may lack enthusiasm about his/her new duties. It may be necessary to spend additional time with the operator and provide information which will explain the purpose and goals of the program. The shop personnel may also cause problems. They may not care to participate or they may misunderstand the procedures. The shop personnel may also require additional training or monitoring during the initial months of the program until the procedures become a part of their daily routine. Changes in key personnel involved with the program may occur from the initial to the final stages of the project. Forming alliances with these new personnel and educating them on the program will require additional time and effort. Regulations change daily and as a result may require adjustments in the program and its procedures. Keep abreast of new or anticipated changes that may effect this program. Production may also change. Flexibility should be built into the program to cushion these production changes. The characteristics of the waste may change. The product quality can be verified by taking samples throughout the program.

By conducting the pilot study, other problems can be avoided. At the Boeing Aerospace and Electronics sites in the Puget Sound area, spent solvent and unused paint were combined and disposed of offsite. This combined waste stream was initially distilled during the pilot study. It was discovered that the concentration of paint in the solvent was too high to obtain a good separation via distillation. Segregation of spent paint and spent MEK prior to distillation was prescribed to solve this problem.

Different types of stills are available on the market. The conventional still is a cylindrical tank with a manhole at the bottom to shovel out the sludge. Removing this sludge is a chore, since it is a combination of waste paint and solvent that has been heated but has not completely evaporated. There are manufactures that have attempted to remedy this situation by adding polypropolene or Teflon bag liners to contain spent solvent and protect the walls of the still pots. This method is convenient as long as the pot is not too deep (bag liners in deeper stills tend to tear when being removed, therefore a shallow wide bowl is recommended for this technology). The waste should not contain constituents that will wear or dissolve the liner. This can be verified with the vendor.

Some vendors may try to incorporate other manufacturers' technology into their equipment. The utilization of bag liners in stills has been adopted by many manufacturers and vendors because it makes the still bottoms much easier to handle, causes less wear and tear on the still and reduces the chances for spills. Although manufacturers and vendors may be correct when stating that their equipment is capable of utilizing another company's technology (eg., bag liners, condensers, etc.), it is recommended to research if there are patents which would prohibit use of the technology or require royalties to the patenting company. It is less complicated to buy from the originator of the idea or patent (check on patents pending).

During the pilot study at BA&E it was determined that the recycled solvent can only be used for cleanup applications because of the stringent specifications for solvents used in production operations. Use of the recycled solvent for production would require costly and time consuming tests on each batch to verify its quality. Fortunately, the quantity of solvent needed for cleanup purposes exceeded the quantity of recycled solvent produced.

CONCLUSION

In organizations where waste solvent is generated, solvent recovery is critical to proactive waste minimization. Thorough research and an organized step by step approach are key to the success of a solvent recovery program. Implementation of a solvent recovery program requires obtaining technical information while gaining positive program support from management, shops and regulating agencies. Incorporating technical and persuasive skills in an organized manner is vital for the success of this program. Remember, the idea won't reduce hazardous waste or save money unless it is successfully implemented.

Case Study Boeing Aerospace and Electronics' Plant II site in Seattle, Washington was consuming 19,000 gallons of Methyl Ethyl Ketone (MEK) per year (based on a 1989 Sara 313 Report). The disposal cost of spent MEK and associated wastes was approximately $330,000 annually. A pilot study found that it was possible to recover up to 90% of the spent solvent. Based on this data, Boeing purchased a distillation unit which was installed at a central hazardous waste accumulation area at Plant II. The installed cost of this unit was approximately $55,000, including a closed-loop chiller. Boeing anticipates an annual savings of $200,000 with a payback period of less than 4 months. Once the program began, minimal problems were encountered. Some obstacles included minor equipment malfunctions, lack of enthusiasm from shop and other associated personnel, skepticism towards using recycled solvent, health related questions and batches of spent solvent that were not reclaimable. BA&E Plant II site is now recycling spent solvent and using this recycled solvent for cleanup operations.

RECOVERY OF WASTE SOLVENTS BY RECTIFICATION, AZEOTROPIC AND/OR EXTRACTIVE DISTILLATION

Lloyd Berg
Chemical Engineering Department
Montana State University
Bozeman, Montana

Solvents are utilized for cleaning, stripping and various other maintenance operations of aircraft parts and equipment. After use, many of the solvents can be recovered and reused. However, several of the solvents currently in use are chlorinated and emit volatile organic compounds, which are toxic to the environment and to operating personnel. Wastes generated from these solvents are regulated by the U.S. Environmental Protection Agency; and soon use and manufacture of these solvents may be restricted.

The objective of the Solvent Recycle/Recovery Task of the DOE Chlorinated Solvent Substitution Program is to minimize hazardous wastes by identifying recycle/recovery techniques for the proposed substitute solvents.

The following new and unused solvents have been received from the following manufacturers.

Company	Product Designation
Chemical Methods	CM-3707
Chemical Solvents	SP-800
Fine Organics	FO-606
Frederick Gumm	Clepo Envirostrip 222
GAF	M-Pyrol
McGean-Rohco	Cee-Bee A-245
McGean-Rohco	Cee-Bee A-477
Patclin	Hot stripper 126

Company	Product Designation
Rochester Midland	PSS-600
Turco	T-5668
EXXON	Exxate-1000
Bio-Tek	Saf-Solv-140
Orange-Sol	De-Solv-It
3D	3D Supreme
Fremont	Fremont-776

Precision rectifications were made on each of the solvents listed above and the composition of the unused solvents determined. The next step in the program will be to obtain samples of these solvents after being used. They will be screened, filtered and rectified to determine what changes occur during use.

The goal of the program will be to return the used solvents to a purity and composition of a quality which will restore their market value. This will be accomplished by the employment of precision rectification, azeotropic and/or extractive distillation. The principal investigator has been awarded 95 patents in the separation of acids, alkalies, alcohols, esters, ketones, amines, glycols, sulfur compounds, nitrogenous compounds, heterocyclics and hydrocarbons.

The principal investigator has been awarded U.S. Patents in azeotropic and extractive distillation to separate the following mixtures:

Ethylbenzene from Xylenes
Ethyl acetate from Ethanol (2)
Isopropyl ether from Acetone (2)
m-Xylene from o-Xylene (7)
n-Butyl acetate from n-Butanol (2)
n-Propanol from Allyl alcohol (3)
Benzene from Non-aromatic hydrocarbons
n-Propyl acetate from n-Propanol (2)
Formic acid from Water (4)
Isopropyl acetate from Isopropanol (7)
n-Amyl acetate from n-Amyl alcohol
t-Amyl alcohol from Isobutanol (2)
Ethanol from Isopropanol (2)
n-Propanol from 2-Butanol
2-Butyl acetate from 2-Butanol
Ethanol from t-Butanol
2-Pentanone from Formic acid
Formic acid from Dioxane (3)

Formic acid from 3-Methyl-2-butanone (2)
Acetic acid from 4-Methyl-2-pentanone
Benzene from Acetone
n-Propanol from t-Amyl alcohol
Ethyl benzene from Styrene (2)
Ethylene glycol from Butanediols
Glycerine from Polyols

Toluene from Non-aromatic hydrocarbons
Isopropyl ether from Methyl ethyl ketone (2)
Acetone from Methyl ethyl ketone
Methanol from Acetone (3)
Methyl acetate from Methanol (5)
Isopropyl ether from Isopropanol (2)
Isopropyl ether from Isopropanol
Ethanol from Water (2)
Isobutyl acetate from Isobutanol (2)
Methyl t-butyl ether from Hydrocarbons
Formic acid from Acetic acid (2)
n-Hexyl acetate from n-Hexanol (2)
Isopropanol from t-Butanol
Propionic acid from Water
Acetic acid from Water
2-Butanol from t-Amyl alcohol
Acetic acid from Dioxane (2)
m-Diisopropylbenzene from p-Diisopropylbenzene (7)
Formic acid from 2-Pentanone
Vinyl acetate from Ethyl acetate (2)
2,3-Butanediol from Propylene glycol
Formic acid from 4-Methyl-2-pentanone
Acetic acid from 4-Methyl-2-pentanone
2-Methyl-1-butanol from Pentanol-1
Methyl, Ethyl, Propyl & Butyl Lactates

RECYCLING ALTERNATIVES

James L. Schreiner
Exxon Chemical Company
Bayton, Texas

With increased awareness of the impact of handling and disposing waste streams, many companies are looking into recycling and recovering materials used in their operations. In areas related to cleaning applications, the recovery and recycle of modern hydrocarbon and oxygenated solvents is important. These fluids are proposed to replace chloroflourocarbons (CFCs) and other chlorinated solvents for a variety of applications.

RECYCLE

To facilitate recycle and for purposes of waste minimization, it is important to have knowledge up-front of the product's composition. That is, the product must be of a quality which will not degrade in use or during standard recycle operations.

For our purposes we will focus on two qualities: 1) a defined and narrow boiling range and 2) a chemistry with limited unsaturation and/or reactivity. The first criterion simplifies as well as enables effective recycling. The latter ensures product quality after recycle, or in other words, eliminate degredation of the product as a result of use or recycle.

It is with these in mind that various hydrocarbon and oxygenated fluids are finding niches in cleaning applications. That is, these products have compositions such that they may be recycled with the reused material having the same properties as the original product.

Product Contamination and Clean-up

Cleaning operations impart many contaminants to the cleaning fluid. Contaminants entering a cleaning fluid were previously characterized in a paper entitled "Reclamation and Reprocessing of Spent Solvents" by Arthur Tarrer, et. al. These contaminants include hydrocarbon oils, such as lubricating oils, greases, transmission oil, fuel oil; asphalts; tars; waxes; paint and varnishes; or soily materials such as clay and silt, cement, soot or lampblack. In addition, some contamination from metallic fines should also be expected in metal cleaning operations. Reclamation of metals will not be addressed in this paper. Information on this subject is available through publications such as Metal Finishing or seminars such as SUR/FIN.

Relating to oil and soil contaminations of hydrocarbon and oxygenated fluids, Exxon Chemical has, in other industries or applications, had similar challenges of product contamination. In these applications, a product has been used in a process becoming sufficiently loaded (typically) with a heavy oil, such as a hydraulic fluid, where the product is no longer effective in the application. At this point, the user has several alternatives: 1) Disposal of the product, incurring significant charges for handling the spent material which will be classified as a hazardous waste under RCRA. 2) Sale of the used product to an independent reclaimer, recovering some portion of the original solvent investment. 3) Establishment of an outlet for reclaiming these products for recycle in its own operations.

As the third alternative has been found to be the more economic and environmentally safe alternative, we frequently consult with companies working to recycle our products. Typically, the reclamation procedure involves distillation, but in some cases carbon adsorption has been used. Most of the conversion companies we have dealt with are members of the Association of Petroleum Refiners or the National Association of Solvent Recyclers.

One technique an analytical chemist may use when viewing a contaminated or spent product is gas chromatography. This technique can be used to identify materials based on the distribution of the components by boiling point. This criteria is needed to determine the effectiveness of distillation as a means of recycle and recovery.

An uncontaminated hydrocarbon or oxygenated fluid will have a discrete narrow boiling range with no indication of components trailing past the final boiling point (FBP). This is indicated by a flat baseline on a gas chromatogram.

One of the fluids that has been contaminated will show peaks or aberations in the gas chromatographic baseline after the final boiling point of the fresh fluid. These baseline movements are indicative of the type of contamination in the fluid. One example is that of a hydraulic fluid contaminating a hydrocarbon fluid. This hydraulic fluid has a boiling range from 600-1040°F and a viscosity of 32 cSt at 100°F. The contamination ratio of cleaning fluid to hydraulic fluid is 96:4.

The use of a controlled batch distillation procedure removed all of the contamination from the cleaning fluid. The quality of this recycled material, as determined by gas chromatography, was identical to the unused fluid. Additional analytical testing techniques also confirmed that the recycled product quality was identical to that of the unused fluid.

Processing conditions to effect this recycle were stated by the recycler in Indiana as being quite complicated. A simplistic and generalized interpretation of the process, however, is one that would usually require only atmospheric capabilities if using a hot oil system or a fired reboiler to heat the material. However, for more difficult distillations or if only high pressure steam is available, mild vacuum and a few theoretical stages will effect an optimal separation. The use of thin film evaporators may also be effective.

Another example of the recycle capabilities involves the removal of contaminants from an oxygenated fluid system. The contamination consisted of carbon, high melt point wax and a 46 cSt oil. The loading level was close to 15 wt%. The carbon was effectively removed using a simple filtration. This filtered material was then distilled to return the solvent to its former quality. It is important to remember that this may only be accomplished if the product's chemistry does not degrade in use or in recycle. The use of vacuum distillation and/or other temperature controls during distillation will help reduce the risk of degredation during reclamation.

Distillation results in a recycled product within the proper boiling range of the original product. Other product quality inspections, such as flashpoint, oxidation level, color, particulates, etc., are necessary to approve a recycled product for reuse. These examples are only a demonstration of a single screening analysis for reviewing the effectiveness of a recycle operation.

CAPTURE

Other considerations for reclaiming product will now be discussed. These are capture techniques which facilitate controlling emissions of products. While some 10-20 technologies are under review by the Department of Energy (DOE) group SITE (Superfund Innovative Technology Evaluation), the following will address only the more common and established methods of capture, especially as they may be directly applied to cleaning equipment modifications. Many of these technologies were recently discussed at the DOE-sponsored Conference on Industrial Solvent Recycling held in October in Charlotte.

The first technology is direct condensation to control loss of vapor laden air. The air is routed over refrigerated coils condensing the liquid for capture and recycle.

Another common capture technology is carbon adsorption. Usually, for efficiency of operation, two adsorbers are installed where one is active in adsorbing vapors from the operation and the other adsorber is going through a regeneration procedure. This regeneration can be accomplished using either steam or inert gas. Using this technology, recovered material is desorbed from the carbon adsorption bed by the steam or inert gas and collected in a receiving vessel.

If steam is being used, the immiscible products can be recovered in a separator while the miscible products are removed by subsequent distillation. The preferred method is desorption using an inert gas if that is available. Inert gas avoids the issues of water quality in the desorption and regeneration phases of the recovery and does not require additional distillative processing. Skid mounted units and special container carbon modules are available from multiple suppliers to assist regeneration of carbon for smaller users or for facilities not having steam or inert gas available.

An emerging technology for large industrial applications is called reverse Brayton cycle. The vapor enters a dehumidifier to remove moisture and then goes to a chiller to reduce the temperature. From here, some fluid may be recovered with the remaining vapor sent to a turbo compressor where pressure is increased dropping the temperature still further. From here, the product is cooled through an interchanger and pumped through the expander of a free-spindle turbo bringing the temperatures to as low as -80°F. The vapor in the gas stream is then recovered as a condensed liquid with the inert gas passed through a pump to the exhaust. The energy resulting from the pressure change through the expander helps drive the compressor for energy savings.

"From and to adsorber" shown on the diagram illustrates the use of the reverse Brayton cycle to desorb a carbon adsorption bed that has been previously saturated with vapors. The hot gas exiting the recovery operation is used for desorbing the volatile organic compound (VOC) and regenerating one adsorber, therefore providing energy savings and operation efficiencies in the process.

Recovery of solvents by membranes is a technology which continues to evolve and warrants close attention over the next few years. All of the other technologies discussed, however, are immediately available and are applicable to recovering and recycling hydrocarbon and oxygenated fluids. A screening model is being prepared by Science Applications International Corporation for the DOE Office of Industrial Technologies. The model will review these technologies to address the most favorable economics for recovering VOC types - considering concentrations and gas flow rates of the process stream.

Many challenges face us with the latest clean air legislation. There are many options available to meet these challenges. All technology and chemical replacements should be reviewed in the pursuit of safe alternatives. Safety should be coupled with energy conservation to minimize waste. As recovery and recycle are steps toward minimization, proper product selection is important to ensure that the product chosen for use in a specific cleaning operation is capable of being recovered and recycled. Previous industrial experience indicates that some hydrocarbon and oxygenated fluids are available which may be recycled effectively, returning product to former quality for reuse by the customer. This recycling may be conducted through outside tolling facilities or in purchased on-site facilities.

Although recovery technologies such as membranes, solidification, and in-situ vitrofication are developing quickly, various recycle and recovery technologies exist today to assist waste minimization efforts, at the same time controlling costs.

Exxon Chemical would like to acknowledge the assistance of Dr. Victor S. Engleman, of Science Applications International Corporation, in the preparation of information and graphics for this presentation.

This talk was prepared by Janet S. Catanach and Kishore K. Chokshi of Exxon Chemical Company.

THIN FILM EVAPORATION FOR REUSE/RECYCLE OF WASTE ORGANIC SOLUTIONS

W. N. Whinnery
Paducah Gaseous Diffusion Plant*
Paducah, Kentucky

ABSTRACT

A thin film evaporator TFE has been used at PGDP to evaluate the feasibility of recovering waste organic solutions as reusable products or to reduce the waste disposal volume. The setup of the TFE and the hot oil heating system has allowed a wide range of solutions to be tested. These solutions include 1,1,1 trichloroethane, trichloroethylene, waste deplating solution, lacquer thinner (paint waste), and a polychlorinated biphenyls (PCB) and uranium contaminated waste oil solution. The recovery rate or waste volume reduction are presented for each solution tested. Cost savings for specific compounds and requirements for the use of recovered trichloroethylene and 1,1,1 trichloroethane are explained. Actual plant reuse of the recovered chlorinated solvents and the lacquer thinner has occurred.

EXECUTIVE SUMMARY

The thin film evaporator (TFE) is being used to evaluate the feasibility of recovering waste organic solutions as reusable products or to reduce their waste disposal volume. The TFE has successfully recovered trichloroethylene, 1,1,1 trichloroethane and lacquer thinner. These chemicals have been returned to a large vapor degreaser, small parts cleaning bath in the pump shop, and the paint shop for field use testing. No problems have been found with the recovered products compared to the virgin solvents.

Testing of PCB (polychlorinated biphenyls) and uranium co-contaminated waste yielded a 28% decrease in disposal volume. The condensate recovered from the evaporation was below 50 ppm PCB, the federally regulated guideline. These test results proved volume reduction of PCB laden waste can be achieved prior to final disposal (incineration).

The technology is applicable to other sites for reduction of hazardous and contaminated waste feeds prior to incineration.

Material Tested To-Date

Materials tested to-date include trichloroethylene, 1,1,1 trichloroethane, nickel stripper solution, co-contaminated oil, and lacquer thinner. Percent solvent recovery is defined as weight of solvent recovered divided by the weight of solvent processed. Maximum recoveries achieved were 96% for trichloroethylene and 98% for 1,1,1 trichloroethane. Recovery rates were found to be a function of:

1. Solids loading
2. Number of passes through the TFE system
3. Percent oil in the waste

Uranium removal has averaged 99% with typical 25 ppm uranium starting levels being concentrated to 350 ppm in the bottoms and the product reduced to 0.02 ppm.

The readdition of stabilizers to the recovered product suppresses the generation and accumulation of HCl during reuse of the

*Operated by Martin Marietta Energy Systems, Inc. for the U.S. Department of Energy under Contract No. DE-AC05-840R21400.

solvent. This prevents corrosive reactions in the steel storage drums which would lead to pinhole leaks. To prevent corrosion, each specific solvent is inhibited with an acid accepting stabilizer. The level of inhibition of a solvent is measured by the non-amine acid acceptance (NAA). The average operating range of the NAA in the degreaser is 0.10 to 0.13% as NaOH. The maximum amount of inhibitor is needed in waste solvent after processing due to the interaction of the used solvent and the oil from dirty equipment. 1,1,1 trichloroethane is stabilized by 1,2 butylene oxide and secondary butyl alcohol. Trichloroethylene is stabilized by 1,2 butylene oxide, secondary butyl alcohol, cyclohexane oxide and diisopropyl amine.

The nickel-stripping solution is composed of two parts. The "A" component is a dry powder composed of sodium meta nitro benzyl sulfonic acid. The "B" component contains 2% ammonia and 20% aliphatic amine. TFE treatment of the nickel-stripper solution has yielded two distinct fractions, an ammonia-water mixture in the distillate and the Metex solid with high boiling liquids in the bottoms. The volume reduction achieved by separation of the ammonia-water mixture from the feed material was 55%.

The co-contaminated waste oil solution was high in suspended solids (75 g/l) and was filtered to 100-micron particle size prior to TFE processing. The PCB level of the initial sample prior to filtration was 6700 ppm. The TFE concentrated the waste in the bottoms solution to 55 g/l solids content. After processing, the bottoms concentration of 5200 ppm PCB's was lower than expected due to the solids removed in the filtration. The solids filtered out had PCB's sorbed to the solid material and produced the lower than expected bottoms concentration. The condensate contained less than 50 ppm PCB, which is below the federally regulated guideline for PCB's. The volume of the riginal solution contaminated with PCB's was reduced by 28% on a single pass through the TFE. Increased volume reduction would have been obtained if a lower initial suspended solids concentration had been present. A 50% recovery of the available volatile and water fraction of the solution was obtained. This proves volume reduction of mixed volatile wastes containing PCB can be achieved prior to final disposal (incineration).

Test Materials Chemical Parameters

The specifications for chlorinated solvents were taken from the manufacturers data sheets and our material specification. Additional parameters of interest were added, such as uranium and total suspended solids. The list of test parameters for the chlorinated waste is shown in Table 1.

Testing of the waste nickel stripper solution was monitored for percent volume decrease. Earlier testing in the laboratory showed the liquid fraction of the two-part, nickel-stripping compound was the active ingredient. The "B" liquid fraction is partially fractionated in the evaporation process. The high boiling liquid in the bottoms fraction could serve as an activator for the stripping process; however, the bottoms does have high suspended solids and separating the liquid was difficult. Recovery of the bottoms liquid for reuse was not feasible. The dried bottoms residue was ineffective as an additive to rejuvenate the stripping solution. The deplating operation would be affected by increased nickel concentrations from additions of the dried bottoms and would reduce the life of the bath rather than extend it. The condensate (product) yields an ammonia-water mixture that was tested in the laboratory and would not deplate nickel plated steel samples. The use of the ammonia-water mixture as a fertilizer is a reuse application of the condensate solution.

The waste oil solution contaminated with PCB's and U was analyzed for the characteristics listed in Table 2.

The lacquer thinner material was tested for the same general properties as the chlorinated solvents. The lacquer thinner waste solution produced a very clear condensate solution. The waste was identified by GC/MS comparison with stores current product of lacquer thinner. The test run recovered 57% of the total waste solution on the first pass. The solution was prefiltered and removed the paint fines that would have coated the TFE jacket walls and reduced heat transfer.

Flow Diagram of the TFE

The flow diagram of the thin film evaporator system is shown in Figure 1. This system includes an oil heater unit to provide heating of the jacketed evaporator section. The upper temperature limit of the hot oil heating system is 550°F. The evaporator is composed of a rotating center section, producing a thin film on the inside wall of the evaporator jacket, a condenser, feed pump, condensate pump, and bottoms pump.

Optimum Machine Conditions

The major parameters of operation that were monitored on all testing runs are listed in Table 3. Amperage load on the rotating center shaft ranged 62 - 67% of full load during all the testing. Increases in the vacuum on the TFE produced quicker product rates. The vacuum setting in processing the chlorinated solvents was five psia. At vacuum conditions above 13.5 psia, the rotating center section will vibrate. The vacuum can be used to draw feed material into the TFE to aid in priming the feed pumps provided the back pressure valve is set low enough to allow this to occur. Optimum conditions for processing the waste tested are listed on Table 4.

The feed rate will vary depending on the solids loading, percentage of oil in the solvent, and type of pump used. All trichloroethylene waste was processed in a single pass. The temperature difference between 1,1,1 trichloroethane and trichloroethylene represents the added heat required to process the higher boiling trichloroethylene. The flow rate will dictate how many passes the solvent should make before all available material is recovered. The time required to process a 55-gallon drum of waste chlorinated solvent was 4 - 6 hours depending on the percent solvent in the feed.

The co-contaminated solution was filtered prior to treatment. The TFE removed excess water and volatile organics from the waste oil mixture. High solids loading resulted in pumping out the bottoms area frequently.

The lacquer thinner solution processed through the TFE was filtered to 30 microns prior to treatment to remove paint particles from the solution. Solution fed very smoothly after the pump was primed and the correct pump feed pressure was established. The temperature in the vapor phase was still too low for single pass complete recovery.

The product cycle time has been plotted in control chart form for each of the materials tested and out of control points can be traced to changes in the feed pump settings.

DISCLAIMER

expressed herein do not necessarily state or reflect those of the United States Government or any agency thereof.

TABLE 1
TEST PARAMETERS FOR CHLORINATED WASTE SOLUTIONS UNDER SPECIFICATIONS FOR MATERIALS

Trichloroethylene	1.1.1 Trichloroethane
Acidity	Acidity
Acidity after oxidation test	Flash point
Alkalinity	Fire point
Residue on evaporation	Non-volatile materials
Moisture	Water
Free halogen	Odor
Color	Color
Copper corrosion	Boiling range
Boiling range	Specific gravity
Specific gravity	

Additional Experimental Test Parameters

Components
Uranium
Total suspended solids
Spectrochemical

TABLE 2

CONTAMINATED OIL SOLUTION ANALYSIS PARAMETERS

PCB
Uranium
Spectrochemical
Specific gravity
Water
Color
Residue after evaporation
Total suspended solids
Acidity
Boiling range

TABLE 3

PARAMETERS MONITORED FOR TESTING OF THE THIN FILM EVAPORATOR

Hot oil heater use time

Heater temperature control (maximum set point)

TFE temperature control (maximum set point)

Process temperatures

Point 1 overhead vapor temperature
2 fed in temperature
3 hot oil inlet temperature
4 hot oil outlet temperature
5 condenser outlet temperature
6 bottoms pump outlet temperature

Systems' Vacuum

Amperage load on the rotor

Product (condensate) cycle time

Feed pump setting

TABLE 4

OPTIMUM MACHINE CONDITIONS FOR TFE TREATED WASTE

	Jacket Temperature °F	Vapor Temperature °F	Vacuum psia	Product Cycle min:sec
Trichloroethylene	315	183	5	1:48
Trichloroethane	250	163	5	1:26
Cocontaminated Oil	350	181	10	3:00
Nickel Stripper	317	212	0.5-21.5	3:15
Lacquer Thinner	355	210	10	1:47

FLOW DIAGRAM OF THE THIN FILM EVAPORATOR SYSTEM

FIGURE 1

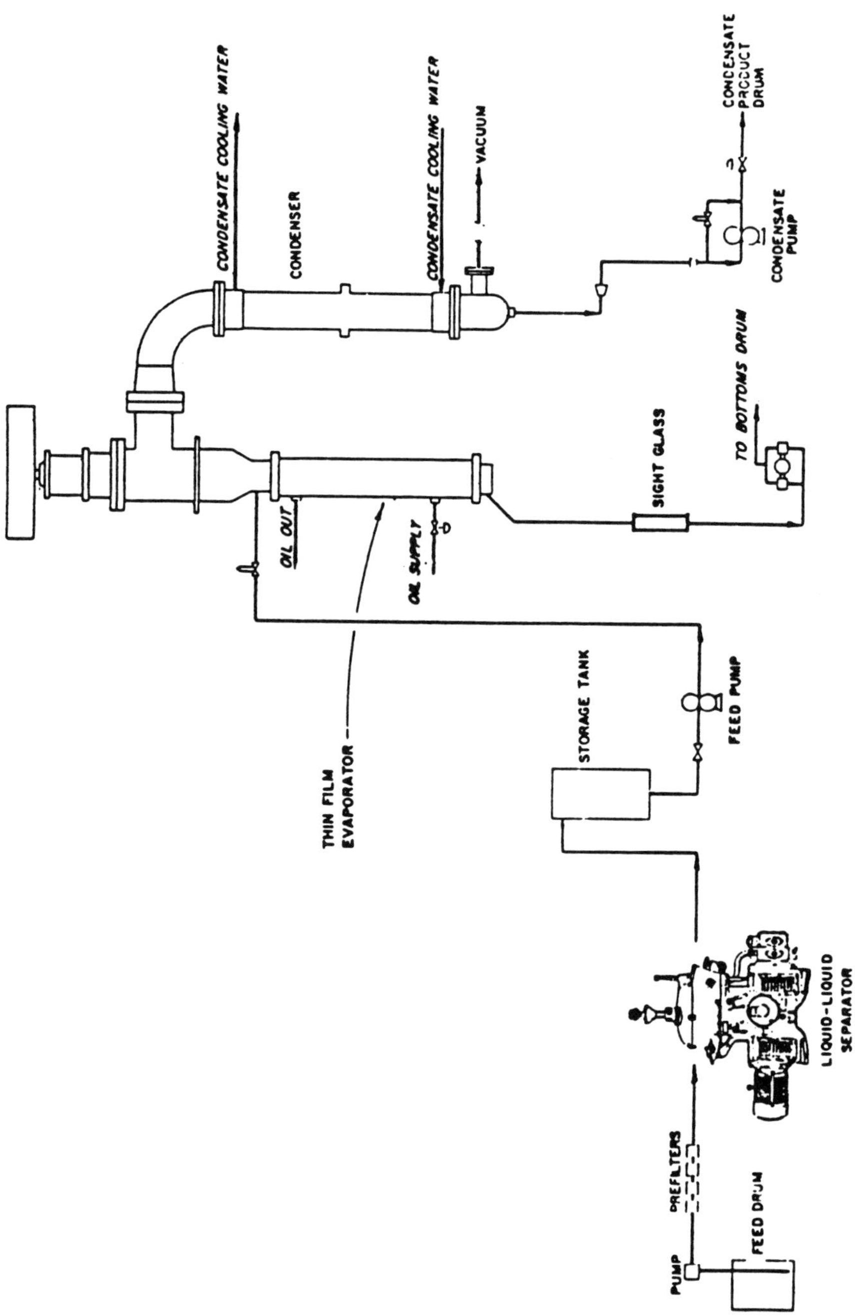

Section IV

DEALING WITH LOW VOCs

ON-LINE MONITORING OF VOLATILE ORGANIC SPECIES

Gregory C. Frye and Stephen J. Martin
Sandia National Laboratories
Albuquerque, New Mexico

ABSTRACT

On-line chemical monitoring systems can help ensure safe, environmentally sound operation of industrial processes using hazardous chemicals. Using polymer-coated surface acoustic wave (SAW) sensors, we have demonstrated monitors that are capable of detecting dilute concentrations of volatile organic species. Using changes in both wave velocity and wave attenuation, the identity and concentration of an isolated chemical species can be determined. A polysiloxane coating has been found to provide unique properties for monitoring chlorinated hydrocarbons (CHCs) such as trichloroethylene: good discrimination of CHCs from most other organic species, rapid and reversible sensor response, and low detection limits. Using this technology, a portable acoustic wave sensor (PAWS) system has been constructed.

INTRODUCTION

A wide variety of volatile organic species are commonly used in industry, especially in cleaning operations such as vapor degreasing. Proper use of these chemicals requires addressing several environmental, safety and health related concerns. For example, chlorinated hydrocarbons such as trichloroethylene (TCE) are often used as cleaning solvents. Their ozone depleting potential makes their emission to the environment a grave concern. This has resulted in severe restrictions in their use, as dictated by the Montreal Protocol and the Clean Air Act. In addition, many CHCs are known or suspected human carcinogens and their concentration in the workplace is regulated by agencies such as OSHA. In some DOE and industrial applications, acceptable alternatives to CHC-based cleaning are not currently available. In these situations, continued operation will require documenting that CHC usage is in compliance with all relevant regulations.

Real-time, on-line monitoring of exhaust stacks and workplace environments can play a critical role in enabling safe, environmentally sound operation of processes using volatile organic species. These monitors can be used to document emissions and verify that concentrations in the workplace do not exceed safety standards. They can be an effective tool in waste minimization and pollution prevention efforts, too. This latter application can be as simple as using the direct chemical information to optimize worker protocols or to evaluate changes in processes or equipment. Alternatively, these monitors can be integrated with on-line process control systems to optimize process operation for improved product quality and yield as well as to minimize waste and prevent pollution.

There are several requirements for useful industrial monitors. First, they need to have sufficient sensitivity to detect the chemical species of interest in dilute concentrations. For many applications, this requires detection limits on the order of one ppm or less. Second, they need to be insensitive to temperature fluctuations and potential chemical interferants, especially the omnipresent water vapor. Sensitivity to these effects can be circumvented by constructing a sensor system which enables the periodic determination of the baseline sensor response by excluding the chemical species of interest from the sensor. Third, for many applications, the monitors must be able to identify the chemical species providing the sensor response to prevent unwarranted reactions, e.g., the emergency evacuation of a workplace due to a misinterpretation of a sensor response. In addition to these requirements, there are many other critical issues which must be considered such as cost, size, reliability, durability, speed of response, ease of integration into the industrial process, and simplicity of operation.

In this paper, we will describe monitoring systems that meet these basic requirements. These systems utilize surface acoustic wave (SAW) devices and can be applied to monitoring a wide variety of volatile organic species, including CHCs. The current system is unable to determine the concentrations of multiple species in mixtures; however, it is able to discriminate between responses due to different isolated chemical species, allowing species identification and quantification using a single SAW sensor.

SURFACE ACOUSTIC WAVE SENSORS

Surface acoustic wave devices consist of input and output interdigitated transducers patterned on a piezoelectric substrate such as quartz (see Figure 1). When an alternating voltage is applied to the input transducer, an alternating mechanical strain is generated in the underlying substrate which launches the wave. This wave travels along the surface and interacts with a thin film formed on the device surface before being converted back into an electrical signal by the output transducer [1]. This SAW/thin film interaction can result in perturbations in both wave velocity and wave attenuation (i.e., loss of acoustic power) in response to perturbations in film properties [2,3].

As shown in Figure 1, an effective method for monitoring changes in wave velocity is to operate the SAW device as the feedback element of an oscillator circuit. In this configuration, relative changes in oscillation frequency (f) can be directly related to relative changes in wave velocity (v): $\Delta f/f_o = \Delta v/v_o$ where f_o and v_o are the unperturbed frequency and wave velocity, respectively. Wave attenuation can be measured with a SAW oscillator system by incorporating a vector voltmeter to monitor changes in the input and output signal levels to the SAW device [2,3]. Alternatively, if the oscillator circuit is operated with constant input power to the device (i.e., amplifier saturation), then wave attenuation can be obtained by monitoring the RF signal level at a single point in the circuit after the device, as shown in Figure 1.

The SAW device used in this study was patterned on an ST-cut quartz substrate and operated at 97 MHz. To demonstrate the extreme sensitivity of this device, a 1 Hz frequency change corresponds to a change in surface mass of only 80 pg/cm^2. The film used in this study was a polysiloxane polymer film formed by plasma-assisted chemical vapor deposition of hexamethyldisiloxane. The film thickness, based on profilometry, was approximately 1 μm. Since no oxygen was added to the chamber during deposition, the polymer contains a large number of methyl groups, making the film hydrophobic in nature. The resulting affinity of this polymer for many volatile organic species, with comparatively low affinity for water, makes this film ideal for monitoring organics such as chlorinated hydrocarbons.

With this film, there are two film properties which are altered by absorption of species and which result in detectable sensor responses. First, the increase in the mass of the film results in a velocity decrease with no change in the attenuation. This "mass loading" response is the most common interaction used in SAW sensor applications [4]. Second, the viscoelastic properties of the film are altered by the plasticizing action that occurs upon dispersal of species between polymer chains. In contrast to the mass loading interaction, this viscoelastic interaction results in changes in both the velocity and the attenuation.

DUAL OUTPUT SENSORS FOR MOLECULAR IDENTIFICATION

Figure 2 shows the response of the polysiloxane-coated SAW device to various concentrations of TCE in a dry nitrogen stream. The relative velocity changes were obtained from frequency changes, while the attenuation changes were determined using an HP 8505A vector voltmeter. Vapor concentrations were varied using computer-operated mass flow controllers to vary the relative flow rates of a "carrier" stream (saturated with TCE by passage through a bubbler) and a dry nitrogen "mix-down" stream. Using this system, partial pressures (P) from 3 to 97% of the saturation vapor pressure (P_{sat}) can be obtained (see references 3 or 5 for details). Due to the large responses observed with TCE, the concentration scan had to be cut short at 16% of saturation; excessive attenuation makes device operation impractical.

The advantage of monitoring both velocity and attenuation responses can be demonstrated by plotting the attenuation response vs. the frequency response as shown in Figure 3 (the gas phase concentration determines the position along the curve for each species). It is clear that a unique curve is traced out in this plot for each chemical species; thus, these two sensor responses are independent. This demonstrates the capability for distinguishing species on the basis of these two sensor responses; In fact, the relative magnitudes of these two responses indicate the chemical species providing the sensor response. With the chemical species identified, the concentration of that species can be determined from calibration curves for that species (see Fig. 2). Thus, a single SAW device can be used to both identify and quantify an isolated chemical species [5,6].

The results of studies with this polysiloxane coating have demonstrated several other advantages of this coating: rapid response due to fast diffusion, large response to volatile organic species (1000 ppm frequency response to TCE at 15% of saturation), reversible response, and relatively small sensitivity to water vapor (50 ppm frequency response with 90% relative humidity). Based on the tests shown in Figure 3, along with tests with a variety of other species (acetone, 3-methylpentane, dodecane, ethanol, isopropanol and water), a trend was observed in the data for all species except those with significant hydrogen bonding capability (e.g, water and the alcohols). The velocity shift at a given value of attenuation was found to be proportional to the liquid density of the absorbing species (density values are listed in Fig. 3). Since chlorinated hydrocarbons have significantly higher densities (1.4 to 1.6 g/cm^3) than typical organic solvents (0.6 to 1.0 g/cm^3), this trend provides a distinct set of responses for CHCs. Thus, this polysiloxane film is ideally suited for discriminating between chlorinated hydrocarbon species as compared with most other volatile organic solvents. Studies with polybutadiene and polybutadiene/polystyrene films showed this same dependence on liquid density [3]. This trend appears to be due to the manner in which these solvents plasticize these films, specifically that the amount of plasticizing action correlates with the volume of species absorbed into the film [3]. In contrast, a polyimide film has also been tested and, in this case, the responses appear to correlate with the molecular weight of the absorbing species [5]. This suggests a "mass spectrometer on a chip"; however, the slow diffusional properties of this film make it impractical for sensing all but a few small species.

Differences in response trends between polymers can be used to choose a polymer which will optimize the sensitivity and the discriminating capability for a given sensing application (i.e, a given set of chemicals to be sensed along with potentially interfering species). In addition, for analyzing mixtures, arrays of SAW devices with different coatings could be used along with pattern recognition schemes [7]. Since each sensor provides two independent responses, the number of sensors required to quantify a mixture can be reduced by a factor of two as compared with single output sensors. This simplifies the often difficult task of finding a sufficient number of oatings which provide independent responses to the chemical species of interest [7].

PORTABLE ACOUSTIC WAVE SENSOR SYSTEMS

In order to demonstrate the practical application of these dual output SAW sensors, development has begun on a prototype portable acoustic wave sensor (PAWS) system using this technology. A schematic of the PAWS system is shown in Figure 1. The sensor module contains a coated SAW sensor in a gas test fixture, RF electronics to operate the device, and gas handling equipment. The RF electronics consist of: (1) two modular amplifiers (Pasternack) having a combined maximum gain of 55 dB, (2) a coupler after the output transducer to split a portion of the signal to send to an external frequency counter (HP 5384A), and (3) a second coupler between the amplifiers which splits power off to an RF detector (HP 8471A: converting RF power level to a DC voltage) which is connected to an external DC voltmeter (HP 3457A). The frequency counter and voltmeter are monitored using a computer (HP 9816).

The PAWS gas handling equipment consists of a pump (Romega) to draw a gas sample across the SAW sensor from the environment to be tested, an activated carbon scrubber and a three-way teflon solenoid valve (Valcor Scientific) to direct the gas flow through the scrubber upon request. Since the activated carbon effectively removes volatile organic species without removing water vapor, periodically passing the stream through the scrubber reestablishes sensor baseline. The prototype PAWS module was assembled in a metal box with dimensions of 30 x 30 x 10 cm.

Using the polysiloxane-coated SAW device, preliminary testing of the PAWS module has been initiated. Baseline response with no vapor present has been characterized in order to evaluate noise levels and sensor drift. Significant drift in the signals is observed during experiments performed over many hours. By monitoring the temperature in the box using a thermocouple, it was determined that this drift is due almost solely to changes in the device temperature. Since one reason for choosing the ST-cut of quartz for SAW devices is due to the small temperature coefficient around room temperatures, this temperature-induced SAW sensor drift is probably dominated by changes in the viscoelastic properties of the polysiloxane film. Increasing temperature acts to soften the film much like what happens with solvent plasticization [3]. Since these viscoelastic properties form the basis for the independence of the two sensor responses, this sensitivity to temperature cannot easily be removed. However, it can be circumvented by the periodic use of the activated carbon scrubber to reestablish sensor baseline or by temperature control of the SAW test case.

The sensor drift makes it difficult to establish an estimate of the noise level. Since this drift was approximately linear over short times, the rms error between a linear least squares fit and the data was calculated to estimate the short-term noise. For frequency (used to indicate velocity), the noise level for one and five minute intervals was only 0.8 and 1.8 Hz, respectively. For attenuation (calculated by calibration of the DC voltage from the RF detector), the noise level is 0.00037 and 0.00043 dB at one and five minute intervals, respectively. This value is smaller than that obtained using the more expensive and bulky vector voltmeter.

Estimating the lower limit of detection with this film requires comparing device sensitivity (i.e., slope of the response vs. concentration curve) to these noise levels. Since the saturation vapor pressure of TCE at 20 °C is 57 Torr (a concentration of 75,000 ppm with an ambient pressure of 760 Torr), a 5.4 Hz frequency shift (three times the five minute noise level) represents a gas phase detection limit for TCE of 0.5 ppm. For attenuation, a 0.0013 dB shift corresponds to 1.6 ppm, indicating that the frequency response is slightly more sensitive for detecting TCE. These low detection limits will only be valid for situations where the species concentration changes over a short period, e.g., in the application of these devices as gas chromatographic detectors. For on-line monitoring applications, the ability to account for sensor drift and to accurately establish sensor baseline determines the detection limits.

Studies to date on the response of the prototype module to vapors have focused on the response of the polysiloxane-coated SAW device to trichloroethylene vapors. To simulate "real-world" conditions, these challenges were performed by probing the head space in small vials containing the solvent. As shown in Figure 4, rapid, reversible responses in both wave velocity and wave attenuation are observed upon challenging with TCE. The response is large (e.g., responses of over 70 kHz and 10 dB), as expected based on the previous work with this polysiloxane coating. Due to the dynamic nature of this experiment, TCE concentration is continuously changing with time. The sensor follows these concentration changes, as indicated by the varying signal levels. The frequency response is delayed with respect to attenuation due to the one second gate time used for the frequency counter.

The ability of the activated carbon scrubber to provide sensor baseline has also been evaluated. With no vapor challenge, a small (approximately 100 Hz frequency shift) sensor response is observed when the scrubber is activated. This response is reversible and the signal level corresponds to that expected for a TCE challenge at a concentration of about ten ppm, indicating the practical detection limit when the scrubber is being used to provide sensor baseline. As shown in Figure 4, when a vapor is being passed over the sensor, the activation of the scrubber results in return of the sensor responses to near their "unchallenged" values in about ten seconds. The ability of the scrubber to remove about 99% of the TCE from the gas phase (based on the level of return of the signals in Fig. 4) should be sufficient to reestablish sensor baseline. In conclusion, these preliminary studies demonstrate the basic capabilities of the PAWS module: real-time detection of volatile organic species such as TCE, two sensor responses for species identification, and insensitivity to sensor drift due to changes in temperature and relative humidity based on the use of the activated carbon scrubber to periodically reestablish sensor baseline.

ACKNOWLEDGEMENTS

We gratefully acknowledge helpful discussions with L. Gilliom and A. J. Ricco and the technical assistance of B. L. Wampler, A. K. Hays, G. C. Cordes, and T. V. Bohuszewicz, all of Sandia National Laboratories (SNL). A special thanks to L. Casaus of SNL for his work in assembling the prototype PAWS module and to H. Wohltjen of Microsensor Systems, Inc. for useful suggestion on the configuration of the PAWS module. This work was performed at SNL, supported by the U.S. Department of Energy under contract no. DE-AC04-76DP00789.

REFERENCES

1. D. P. Morgan, Surface-Wave Devices for Signal Processing, Elsevier, New York, 1985.

2. S. J. Martin and A. J. Ricco, "Effective Utilization of Acoustic Wave Sensor Responses: Simultaneous Measurement of Velocity and Attenuation," in Proc. 1989 IEEE Ultrasonics Symp., IEEE, New York, 1989, p. 621-625.

3. S. J. Martin and G. C. Frye, "Surface Acoustic Wave Response to Changes in Viscoelastic Film Properties," Appl.

Phys. Lett., 57 (1990) 1867.

4. M. S. Nieuwenhuizen and A. Venema, "Surface Acoustic Wave Chemical Sensors," Sensors and Materials, 1 (1989) 261.

5. G. C. Frye and S. J. Martin, "Dual Output Surface Acoustic Wave Sensors for Molecular Identification," Sensors and Materials, 2 (1990) 187.

6. G. C. Frye and S. J. Martin, "Dual Output Acoustic Wave Sensor for Molecular Identification," U. S. Patent Application, Serial No. 07/592,383, Filed Oct. 3, 1990.

7. S. L. Rose-Pehrsson, J. W. Grate, D. S. Ballantine, Jr. and P. C. Jurs, "Detection of Hazardous Vapors Including Mixtures Using Pattern Recognition Analysis of Responses from Surface Acoustic Wave Devices," Anal. Chem., 60 (1988) 2801.

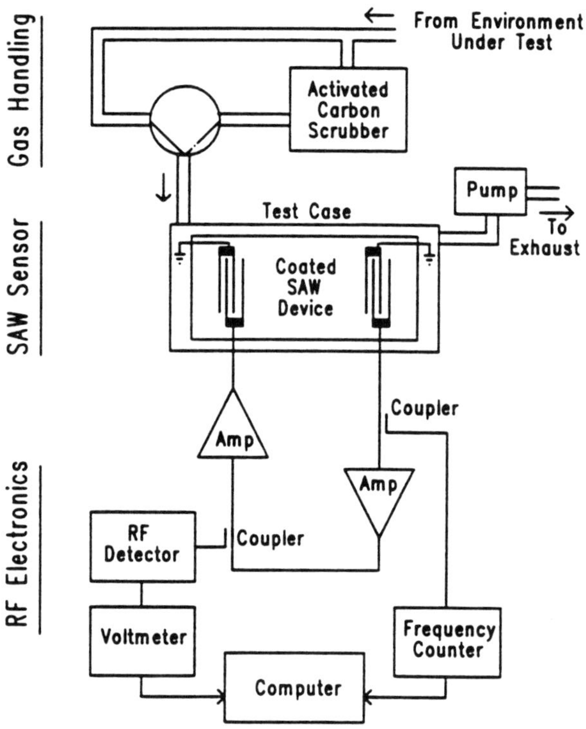

Fig. 1: Schematic of chemical monitoring system consisting of a coated surface acoustic wave sensor, a gas handling system to draw in a gas sample, and electronics to operate the device and monitor changes in wave velocity and attenuation.

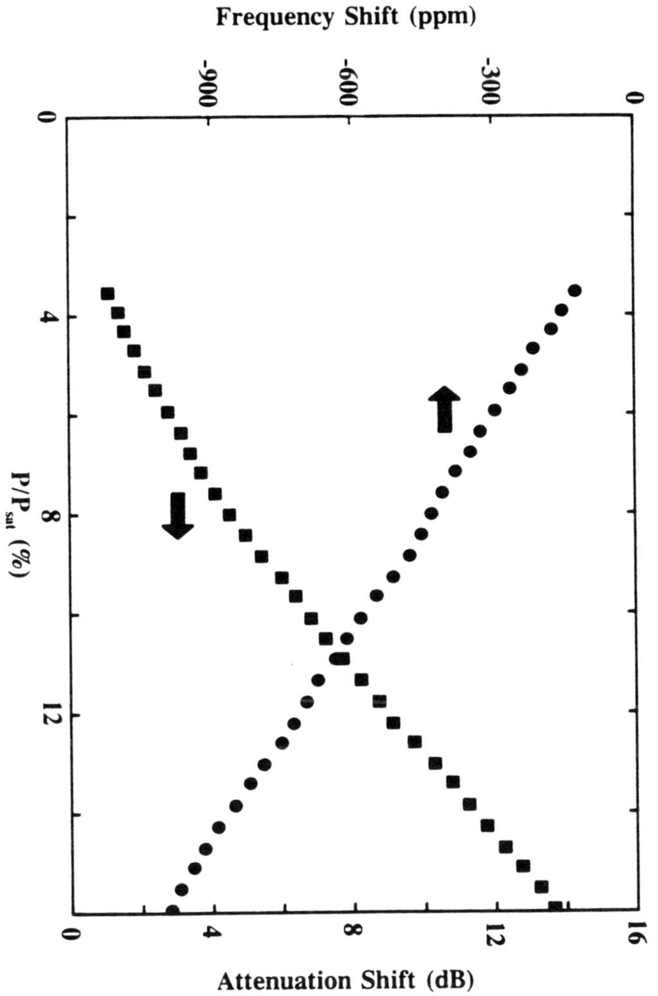

Fig. 2: Changes in frequency and attenuation for a polysiloxane-coated SAW device vs. relative partial pressure of trichloroethylene vapors. Either of these curves can be used as a calibration curve to determine vapor concentration if it is known that the chemical providing the sensor response is TCE.

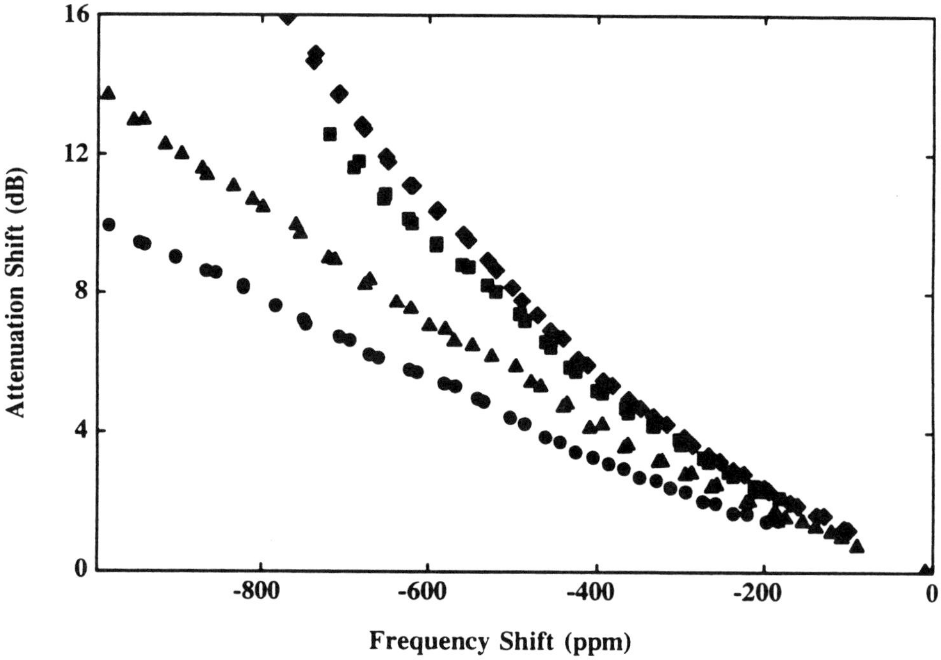

Fig. 3: Plot of attenuation response vs. frequency response for a polysiloxane coated SAW device showing discrimination of chemical species based on differences in the relative magnitudes of these two sensor responses: (◆) hexane (liquid density ρ = 0.660 g/cm^3), (■) toluene (ρ = 0.867), (▲) trichloroethylene (ρ = 1.464), and (•) dibromomethane (ρ = 2.497).

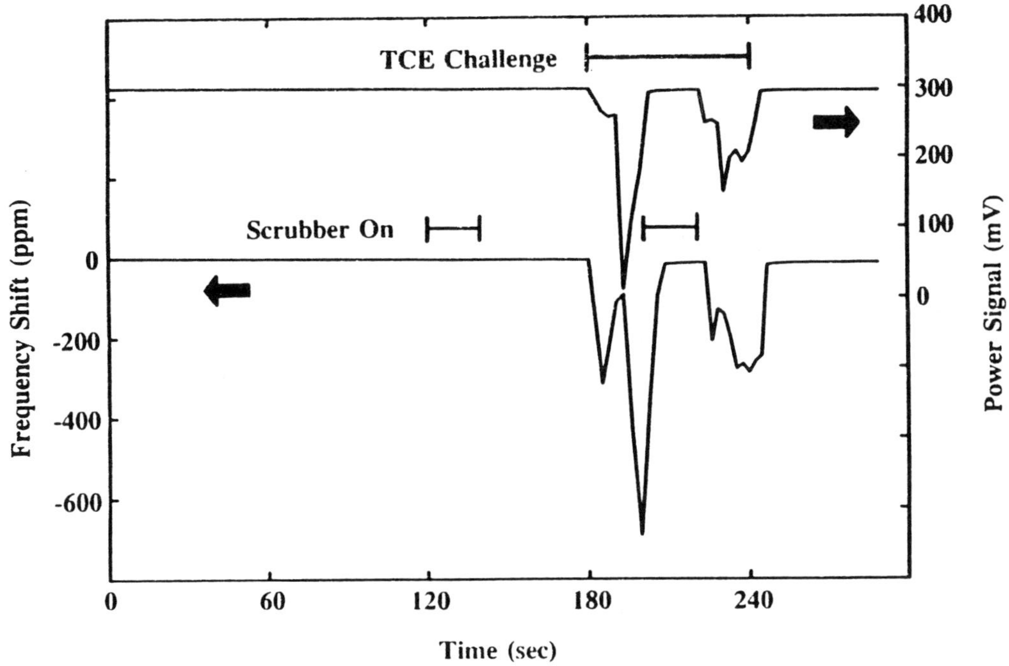

Fig. 4: Response of PAWS module to a trichloroethylene challenge (180 to 240 sec) obtained by probing the head space in a vial containing TCE. Scrubber activation with a vapor challenge (200 to 220 sec) shows the ability to reestablish sensor baseline. Scrubber activation with no vapor challenge (120 to 140 sec) shows a small response (undetectable on this scale).

EVALUATION OF LOW VOC MATERIALS AT THE BOEING COMPANY

Linda H. Hsu and Judith A. Werner
Metals and Finishes
Boeing Commercial Airplane Group
Seattle, Washington

INTRODUCTION

Over the past several years, the regulatory emphasis has been directed toward control and reduction of smog-forming solvents, sometimes referred to as volatile organic compounds or VOCs. The Boeing Company has devoted considerable effort and resources to the development of environmentally acceptable materials and processes.

The projects described below are generally those in which significant research or production scale up efforts have been made. Many of the Boeing-developed processes are currently of a proprietary nature. Information given in these areas is general and has been provided to indicate where environmentally acceptable alternates are probably going to be available in the near future in the area of solvent substitution.

LOW VOC CHEMICAL RESISTANT PRIMERS AND TOPCOATS

The development efforts of low VOC chemical resistant primers and topcoats for metals and composites have been focused on materials subject to regulation in the South Coast Air Quality Management District (SCAQMD). The regulatory maximum levels for the VOC contents of chemical resistant primers and topcoats are 350 and 420 grams/liter, respectively. VOC levels of conventional chemical resistant primers and topcoats are approximately 650 and 600 grams/liter, respectively. Four methods have been utilized by coating manufacturers to achieve low VOC contents:

1. Use of chlorinated solvents (Exempt, non-smog forming solvents)
2. Use of water as carrier (Water Reducible)
3. High solids
4. Electrodeposition

Overall Boeing has evaluated more than 200 coating alternative candidates. Extensive testing requirements are imposed on all the candidates. The main tests these primers and topcoats must pass include salt spray corrosion resistance, chemical resistance (including Skydrol, fuel, lube oil, solvents), water, sealant adhesion, condensing humidity, and primer/topcoat compatibility. In addition, the urethane compatible and corrosion resistant primers must pass filiform corrosion and rain erosion tests.

Interior Structural Primers

The interior structural primers comprise approximately 35 to 40% of the total paint used on Boeing commercial aircraft. They are the primary corrosion preventive coating used on airplanes today. Of the numerous candidates evaluated, there are only three primers which have met Boeing specification requirements. Two acceptable primers are chlorinated solvents containing chemical resistant primers. These two coatings were implemented in 1988 and 1989 in the Los Angelos basin areas. However, these recently qualified chlorinated solvent primers are not used for any kind of wet installations, to avoid the possibility of corrosion from trapped residual chlorine. The third qualified primer is the water reducible version of the chemical resistant primer. Currently a version of the water reducible primer is being used for military applications by the Boeing Aerospace and Electronics Company. Boeing Commercial Airplane Group is currently conducting a manufacturing scale-up study for commercial aircraft applications.

Work is continuing on water reducible and high solids formulations to replace the conventional solvent based primers. Several of the otherwise promising candidates are faced with the problem of topcoat incompatibility. The alternative primers are sensitive to the combination of substrate surface film conditions and enamel overcoat. A major effort has been devoted to development and manufacturing evaluations of these primers.

Composite Finishing Primer

Organic finishes for composite surfaces comprise approximately 5 to 15 % of paint used on Boeing commercial aircraft. Similar to all other primers, the regulatory target is 350 grams/liter for the SCAQMD. The advantage of the composite finishes over metal finishes is that the formulation does not contain chromates for corrosion protection. Currently a water reducible, non-chromated primer is under manufacturing suitability study for commercial aircraft applications on composite parts.

Fuel Tank Primer

The development of low VOC fuel tank primer alternates represents a significant engineering challenge. The fuel tank primer contains stringent engineering requirements because of its dual function of corrosion protection and sealant adhesion necessary for fuel containment. The combination of required properties has resulted in a time-consuming material development program. The fuel tank primer comprises about 10% of the total paint volume on a commercial aircraft. The regulatory target to be met in 1993 is at 420 grams/liter. Currently Boeing is conducting preliminary screening tests on water reducible and high solids alternatives submitted by suppliers.

Interior Structural Topcoat

The interior structural topcoat usage is about 15 to 20% on a commercial aircraft. The regulatory VOC content is targeted at 420 grams/liter or less. Of the numerous candidates that Boeing has evaluated, two high solids materials have passed the qualification tests. They will be implemented by this summer at subcontractors in the South Coast Air Quality Management District jurisdiction.

Abrasion Resistant Topcoat

The abrasion resistant topcoat is another coating containing difficult engineering requirements for material qualifications. The abrasion resistant topcoat is used on trailing edge flaps and other rub areas. The usage is about one percent of the total paint volume on an aircraft. Currently the Boeing Commercial Airplane Group is evaluating primer/topcoat compatibility with two high solids, low VOC candidates. Material qualification is in progress for the two alternatives.

Interior Cabin Finish System

Testing is in progress on a one-coat candidate to replace the existing primer/topcoat interior decorative enamel system. The major criteria for interior decorative coatings are stain resistance and flammability. A high solids, low VOC candidate is being evaluated to meet the 420 grams/liter requirement.

Concurrently to research and development of alternative materials, the Boeing Company is also implementing environmentally compliant application equipment to reduce the VOC emissions. High Transfer Efficiency (HTE) spray equipment has been evaluated and is in use with a variety of coatings and applications within the Boeing facilities. Electrostatic equipment is currently being used at the Boeing Fabrication Division Auburn facility. In addition, the electrostatic equipment is also used widely in many paint shops and paint hangars. High Volume Low Pressure (HVLP) spray guns have also been implemented for military applications in the Boeing Seattle sites and at the Boeing Helicopter division in Philadelphia. Several commercial facilities in the Puget Sound area are currently evaluating the HVLP spray guns for a variety of applications.

ENVIRONMENTALLY ACCEPTABLE SOLVENT CLEANERS

Regulations on the VOC content of cleaning solvents generally control the vapor pressure of the VOCs. The usual requirement is that VOC vapor pressure not exceed 45 mm Hg at 20C. Engineering and manufacturing evaluation of cleaning solvents with low VOC vapor pressures is in progress.

Low VOC solvent research and development activities primarily include the replacement of Methyl Ethyl Ketone (MEK), naphtha, toluene, and other high vapor pressure volatile commercial cleaners. Several low vapor pressure cleaning solvents and mixtures have been qualified for use on commercial and military hardware. These materials are VOCs, but, emissions are reduced because the materials do not have vapor pressures as MEK or toluene. Since these alternatives do not evaporate as quickly as the traditional cleaning solvents, the residual cleaning materials are wiped off from the hardware and captured.

The material qualification for general-use solvents is an on going effort for the Boeing Company. In addition to general cleaning applications, several low VOC materials are being qualified to meet the stringent engineering requirements of cleaning before painting, sealing, and adhesive bonding. Because of the non-conventional nature of some of these candidates, each solvent candidate is tested for compatibility with aircraft metals, plastics, coatings, and sealant. This requirement includes corrosion resistance, degradation testing of plastics, elastomers and paint, and hydrogen embrittlement testing.

CORROSION INHIBITING COMPOUNDS

As a part of the corrosion control program for the airplane fleet, the application of corrosion inhibiting compounds on most in-service airplanes has expanded in the international aviation industry in the recent year. The areas of application required by design on Boeing airplanes includes most of the interior metal aircraft structure. As a result, the increased volume of the corrosion inhibiting compounds used in the factory has presented environmental, personal health, and production-rate concerns. The corrosion inhibiting compounds are not regulated at this time. However, as a part of the Boeing effort to reduce the total VOC emissions, low VOC candidates are being evaluated.

The current corrosion inhibiting scheme uses a duplex system (one thin coat plus one thick coat) for high corrosion-susceptible areas. Candidates with reduced VOC contents are being evaluated to replace the duplex system. In addition, Boeing is currently developing a single-coat, heavy-duty system, which will be equivalent in the performance requirements of the duplex system.

DRY FILM LUBRICANTS

Dry film lubricants are used to facilitate the installation of fasteners. Currently the development activities on dry film lubricants have been aimed at qualifying water reducible low VOC alternatives to solvent-based cetyl alcohol fastener coatings. A manufacturing feasibility study is scheduled and the alternative coatings will be implemented prior to the regulatory effectivity date. The regulatory target for the SCAQMD is at 250 grams/liter.

ADHESIVE BONDING PRIMER

The structural adhesive bond primer is a critical component of bonded aircraft structures. The material qualification contains stringent engineering requirements and therefore exhaustive testing programs are set up for new candidates. Two batches of materials have passed the qualification testing; the manufacturing evaluation is being conducted for the commercial airplane applications. The regulatory target for SCAQMD is at 250 grams/liter, which is a dramatic reduction from current 850 grams/liter.

CONCLUSION

The regulatory emphasis has been on reduction of VOCs in coatings, i.e. primers and topcoats. A more recent shift in the regulatory focus has been towards lowering vapor pressures of the cleaning solvents. Local and U.S. government have set aggresive goals in reducing environmentally hazardous chemicals. Material research and development is a long-term involved process. Boeing will continue to devote major efforts to develop environmentally acceptable materials and processes.

WATER -REDUCIBLE POLYURETHANE ENAMELS: CANDIDATE LOW VOC AEROSPACE TOPCOAT FORMULATIONS

David J. Swanberg
Mechanical Systems Technology
Boeing Defense and Space Group
Seattle, Washington

ABSTRACT

Recently, air quality regulations have prompted significant efforts by coatings resin producers and coatings manufacturers to develop new low Volatile Organic Compounds (VOC) coatings. One approach has been to develop water-borne polyurethanes having performance approaching that of solvent-borne coatings.

Several water-borne resins were tested in white topcoat formulations versus solvent-borne controls (MIL-C-83286 and Deft 1-COAT). Selected chemical resistance and flexibility tests were performed to determine whether solvent-borne performance could be achieved in a water-borne formulation. Coatings were formulated with and without crosslinkers to determine whether chemical resistance could be improved without sacrificing flexibility.

In general, chemical resistance of the water-bornes was not as good as that of the solvent-borne controls; all formulations failed 7 days' immersion in Skydrol 500B. Crosslinking tended to improve chemical resistance but in some cases decreased flexibility of the coatings. An aziridine-crosslinked formulation performed best but is undesirable because of potential health risks associated with aziridines. With additional development, water-borne coatings should be able to satisfy performance requirements of the military specification for high-solids polyurethane topcoats, MIL-C-85285.

INTRODUCTION

In recent years, regulations that restrict allowable emissions of Volatile Organic Compounds (VOCs) from coating processes have affected virtually every area of the coatings industry. Regulatory agencies in some states are requiring significant reductions in solvent content of organic coating materials, whereas others are limiting VOC emissions from spray booth stacks. In either case, regulations have prompted significant development efforts by coatings resin manufacturers and innovative new products from coatings formulators. All efforts are intended to help reduce VOC emissions below regulatory levels while still providing equivalent coating performance.

Reducing VOC while maintaining performance presents a significant challenge to the aerospace industry. Aerospace coatings often provide specialized functions in addition to protection of the substrate and are routinely subjected to extreme environments. While performance is a primary consideration, other factors will also influence the success or failure of low VOC replacement coatings. These include application characteristics, appearance, cost, and scale of coating operations (i.e., painting large assemblies).

The state of the art in performance aerospace topcoats for many years has been the solvent-borne, two-component polyurethane meeting Military Specification MIL-C-83286. In terms of emissions, this material contains typically 600 grams/liter VOC. High-solids polyurethane technology (MIL-C-85285) is available at about 420 grams/liter VOC with similar performance. The goal of our work is to provide equivalent performance in a liquid coating with VOC emissions approaching zero. Water-soluble or 100% solids coatings can achieve this. Our work to date has concentrated on water-borne polyurethanes and polyesters (1,2) with potential for use in two-component formulations that can be externally crosslinked in the applied film.

The long-term goal of a solvent-free aerospace topcoat is consistent with the concept of minimizing hazardous waste by eliminating hazardous components from manufacturing processes. Aside from the obvious benefits to the health of workers and the environment, major incentives for eliminating organic solvents from coatings include the following: 1) avoided cost of purchase, operation, and maintenance of emissions control equipment, 2) reduced hazardous material inventory, and 3) reduced cost and liability of hazardous waste disposal.

The practical objective of the water-reducible polyurethane enamel project is to develop an ultra-low VOC performance topcoat that can be spray-applied with high transfer efficiency equipment and cured at ambient conditions. Formulations must be applicable for original finish processes as well as for re-finish operations and touch-up in the field. The coating should be available in a range of colors and reflectance ranging from high gloss to very low gloss (camouflage applications). The resin backbone should be aliphatic for exterior durability and should contain functionality that can be crosslinked for enhanced chemical resistance. Finally, once applied, the coating should be strippable by available means with minimal hazard to workers and the environment.

EXPERIMENTAL APPROACH

The current phase of this work involved testing fully pigmented water-borne coatings formulations for selected chemical resistance and flexibility requirements of MIL-C-83286 and MIL-C-85285. In coatings development, flexibility and chemical resistance are often considered mutually exclusive since the higher crosslink density required to improve chemical resistance tends to decrease flexibility of the coating. Details of performance requirements and test methods used in this work appear in Table I.

Materials

The water-borne resins used in this study were supplied as single-component materials suitable for formulation of lacquer-type coatings. The water-borne resins tested are described in Table II. All contained carboxylate functionality to impart water solubility and allow dispersion of the polymers in an aqueous phase. Water-compatible crosslinkers that were tested are shown in Table III. All of the experimental coatings were formulated both with and without crosslinkers to determine whether increased crosslinking could significantly improve chemical resistance without sacrificing flexibility.

Two solvent-borne polyurethane coatings were also used for comparison with water-borne coating performance. They were: 1) MIL-C-83286 two-component polyurethane and Deft 1-Coat (see Table IV). The 1-Coat material was included since it represents a recent development in aerospace coatings technology whereby corrosion resistance is incorporated into the topcoat eliminating the need for the traditional epoxy primer (3). In addition, this material is based on high-solids polyurethane technology with a VOC of less than 420 grams/liter.

Formulation of Water-Borne Coatings

The water-borne coatings used in this study were formulated by first dispersing titanium dioxide pigment in a portion of the resin. This dispersion was then let down with additional resin to the desired ratio of pigment to binder (usually 0.75 by weight). Additives such as dispersing aids, leveling agents, and defoamers were rarely used since earlier work showed that they tend to cause severe cratering, crawling, and fisheye even at very low addition levels.

After the letdown, formulations were thinned with de-ionized water to a viscosity of about 30 seconds on a #2 Zahn cup and the pH was adjusted to between 8 an 9 with ammonium hydroxide. Formulations were usually set aside overnight to de-aerate prior to application. When used, crosslinkers were added just prior to application and pH and viscosity were adjusted as reqired.

Sample Preparation

Test panels were typically 3 x 5 x 0.020 inch 2024 aluminum alloy pre-treated with a chromate conversion coating (Alodine 1200). Low temperature flexibility specimens were bare, deoxidized 2024 substrates only.

The test panels were primed (with the exception of the 1-Coat) with either MIL-P-23377 or MIL-P-85582 epoxy primers to dry film thicknesses of 0.008 to 0.0012 inch. These were then topcoated with the various test formulations to dry film thicknesses of 0.0015 to 0.0027 inch. After topcoating, the test panels were allowed to cure for 7 days at ambient conditions prior to testing.

RESULTS

General coating properties, flexibility, and chemical resistance results for all formulations tested appear in Table V. Water-borne formulas are listed by resin type and resin/crosslinker combination. Where crosslinkers were used, the addition ratio is listed in percent based on resin solids of the base resin. VOCs for water-bornes were calculated values, disregarding the volume of water in the coating (4). VOC values followed by an asterisk (*) indicate that organic coalescing solvent was added to the formulation, contributing additional VOC over that present in the base resin. For comparison purposes, results of tests on solvent-borne polyurethanes appear at the bottom of Table V.

The results of these evaluations indicate that in general the water-bornes did not develop gloss, hardness, and chemical resistance of the solvent-borne coatings they hope to replace. Skydrol resistance was particularly low, even with crosslinkers added. Although Skydrol resistance is not a requirement for some aerospace applications, it is an excellent indicator for comparing resin system performance. The best Skydrol result for a water-borne in this study was the XW110 with CX100 (6B after 7 days).

Crosslinking water-bornes with XL25 carbodiimide showed interesting results when compared to the uncrosslinked coatings made from the same base resins. Both XW110 and XW121 urethanes showed lower flexibility and greater sensitivity to deionized water immersion with XL25 added. Skydrol resistance was very slightly improved but still below specification requirements. R-9637 and R-9000 urethanes both showed slightly increased Skydrol resistance with XL25 added. R-9637 flexibility increased slightly while the result for the R-9000 was mixed. Please note that the XL25 addition levels of 17% were well above the manufacturer's recommendation of 5-10% on resin solids. The higher addition level was chosen based on equivalent weights of diimide functionality in the XL25, to carboxyl functionality in the base resins.

Uncrosslinked 72-7230 polyester was not listed in Table V since the material did not fully cure at ambient conditions. Both crosslinked versions, however, had excellent gloss and good flexibility. Chemical resistance was lower than the water-borne urethanes. High crosslinker additions may have been partially responsible for the water sensitivity. This material has excellent poential for good performance, high gloss appearance, and very low VOC with optimized crosslinking.

Three fairly low molecular weight diamines

were also tested as crosslinkers for the R-9000 polyurethane. All showed slightly improved Skydrol resistance without significant reduction in flexibility. An attractive advantage of the amine crosslinkers is that they do not contribute VOC to the formulation. The R-9000/BAPP formula showed overall performance as good as any of the water-borne materials tested, with a VOC below 100 grams/liter.

The results for the solvent-borne coatings were excellent in all categories with the exception of Skydrol resistance. The MIL-C-83286 material showed Skydrol resistance somewhat lower than expected while Skydrol resistance of 1-Coat was slightly better than that of water-borne polyurethanes with crosslinkers added.

Other results, related this work but not listed in Table V, indicate that baked cure conditions for water-bornes can enhance performance over ambient cure. A white topcoat formulation using XW121 polyurethane with 10% XL25 added was baked for one hour at 300°F after solvent flash-off. This formed a hard coating (3H) that showed good Skydrol resistance (2B after 7 days) and also passed the low temperature flexibility test. In another example, two of the amine-crosslinked R-9000 formulations showed improved low temperature flexibility after baking for one hour at 150°F. These were very low VOC formulations with little organic coalescing solvent present. It is possible that improved low temperature flexibility was related to improved polymer coalescence and film formation at the elevated temperature.

CONCLUSIONS

In general, low VOC water-borne coatings with performance appraoching aerospace requirements can be successfully formulated using available materials. Although Skydrol resistance was low for all water-borne formulas, overall performance of crosslinked polyurethanes was nearly as good as the high-solids, solvent-borne polyurethane (1-Coat). All water-borne formulations appeared to satisfy the MIL-C-85285 hydraulic fluid resistance requirement with no significant degredation of the films. Full evaluation of the best water-borne formulations per MIL-C-85285 requirements should be pursued.

An aziridine crosslinked polyurethane showed the best Skydrol resistance of all water-bornes tested relative to MIL-C-83286 requirements. This formulation however, will not be recommended unless health risks associated with aziridines are demonstrated to be minor. One approach would be to find a relatively innocuous analog to the ethylene-imine ring.

A water-borne polyester was tested that, when crosslinked, produced coatings with performance similar to the water-borne polyurethanes. The main advantage of this material is the ability to formulate at very low VOC and produce glossy, decorative coatings with substantial chemical resistance. Such coatings may have aerospace applications for non-flight hardware.

A very promising water-borne system evaluated in this study was R-9000 polyurethane crosslinked with amines. The distinct advantage of this system is the ability to provide overall performance as good as any of the water-bornes tested, but in very low VOC formulations (white topcoat < 100 grams/liter). It is likely that the chemical resistance of this polyurethane could be improved even further by increased crosslinking.

In general, crosslinked water-borne polyurethanes and polyesters show potential to provide durable, chemically resistant coatings for aerospace applications. In addition, they have potential to provide this performance at very low VOC and should continue to be devloped per aerospace requirements.

REFERENCES

1. R.E. Tirpak and P.H. Markush, "Aqueous Dispersions of Crosslinked Polyurethanes," Proceedings of the 12th Water-Borne Higher-Solids Coatings Symposium, University of Southern Mississippi, 1985.

2. M. Glavas, M. Hage, and A. Heitkamp, "Waterborne Dispersion Resins for Very Low VOC Coatings," Proceedings of the 17th Water-Borne Higher-Solids Coatings Symposium, University of Southern Mississippi, 1990.

3. C.R. Hegedus, "Development of a Primer/Topcoat and Flexible Primer for Aluminum," Report No. NADC-87016-60, Naval Air Development Center, Warminster, PA, March 20, 1987.

4. ASTM D3960-89, "Practice for Determining Volatile Organic Compound Content of Paints and Related Coatings," 1990 Annual Book of ASTM Standards, Vol. 6.01, American Society for Testing and Materials, Philadelphia, PA, 1990.

Table I. Selected Coating Performance Requirements

Test	Test Method	Criterion/Goal Water Borne	MIL-C-83286 paragraph number	MIL-C-85285 paragraph number
60° Specular Gloss	ASTM D523	Gloss: 90+ Camouflage: ~5	3.7.1.2	3.7.5
Film Hardness (Pencil Scale)	ASTM D3363	H - 2H	not specified	not specified
Wet Tape Adhesion	ASTM D3359 FTMS 141C-6301.2	< 5% loss	3.7.2.1	3.7.7
Impact Flexibility	ASTM D2794	96 in-lb direct & reverse	3.7.2.2	3.7.8
Low Temperature Flexibility	ASTM D522	no crack, 1" mandrel @ -65 F	3.7.3.4	3.7.8
Immersion Tests				
De-ionized Water	ASTM D870 ASTM D714	7 days, room temp. no blistering	3.7.3.5	not specified
Skydrol 500B	ASTM D1308	7 days, room temp. film intact, hardness change < 2 units	3.7.3.5	not specified
MIL-H-83232	ASTM D1308	24 hrs, 150 F. film intact, min. stain, hardness-no change	not specified	3.8.1

Table II. Water-Borne Resins

Product	Type	Manufacturer	VOC (gm/L)
XW110	Polyurethane	Mobay	360
XW121	Polyurethane	Mobay	345
R-9637	Polyurethane	ICI	245
R-9000	Polyurethane	ICI	35
72-7230	Polyester	Cargill	85

Table III. Water-Compatible Crosslinkers

Product	Type	Manufacturer	VOC (gm/L)
XL25 SE	Carbodiimide (self-emulsified)	Union Carbide	500
CX-100	polyaziridine (100% active)	ICI	0
EDR-148	Di-functional amine	Texaco	0
Bis-Aminopropyl Piperazine (BAPP)	Di-functional amine	Texaco	0
Dimethylamino Propylamine (DMAPA)	Tertiary amino alkylylamine	Texaco	0

Table IV. Solvent-Borne Coatings

Coating	Type	Color	VOC (gm/L)
MIL-C-83286	Conventional 2-comp. Polyurethane	Gloss Grey	590
1-Coat	High-Solids 2-comp. Polyurethane	Gloss White	418

Table V. Coating Evaluation Test Results

Formula (crosslinker wt%)	VOC (g/L)	60° Spec Gloss	Pencil Hardness	Wet Tape Adhesion	Low Temperature Flexibility	ImpactFlex in-lb direct/rev	De-ionized Water 7 days, room temp.	Skydrol 500B 7 days, room temp.	MIL-H-83282 24 hours, 150° F
XW110	315	70	H	5B 0% removed	no cracking	160 160	no blistering hardness: n.c.	severe softening <6B mars easily	light stain hardness: +1 unit
XW110/ XL25 (17%)	345	63	H	5B 0% removed	hairline cracks	96 64	blisters, #8 Med hardness: n.c.	severe softening <6B	light stain hardness: -1 unit
XW110/ CX100 (10%)	290	70	2H	5B 0% removed	no cracking	160 120	no blistering hardness: n.c.	softened 6B	no stain hardness: n.c.
XW121	295	66	H	5B 0% removed	hairline cracks	80 80	no blistering hardness: n.c.	severe softening <6B mars easily	very light stain hardness: n.c.
XW121/ XL25 (17%)	325	50	H-2H	5B 0% removed	slight crazing	24 32	lt blisters, #8 Few hardness: n.c.	severe softening <6B loss of adhesion	very light stain hardness: n.c.
72-7230/ XL25 (25%)	175	91	B	5B 0% removed	no cracking	60 24	no blistering hardness: n.c.	severe softening <6B mars easily	lt. stain, gloss: -20% hardness: -1 unit
72-7230/ CX100 (15%)	80	87	B	5B 0% removed	slight crazing	160 160	blisters, #6 Dense, film wrinkled	coating removed	very light stain hardness: n.c.
R-9637	210	62	H	5B 0% removed	peeled	40 40	blisters, #8 Dense hardness: n.c.	severe softening <6B loss of adhesion	light stain hardness: +1 unit
R-9637/ XL25 (17%)	265	62	2H	4B <5% removed	hairline cracks	64 44	blisters, # 8 Dense hardness: n.c.	severe softening <6B	very light stain hardness: n.c.
R-9000	225*	64	B	4B <5% removed	hairline cracks	96 80	blisters, #6 Few hardness: -1 unit	severe softening <6B loss of adhesion	very light stain hardness: n.c.
R-9000/ XL25 (17%)	335*	66	H	5B 0% removed	cracks	160 120	no blistering hardness: n.c.	severe softening <6B	very light stain hardness: -1 unit
R-9000/ EDR148 (3%)	260*	54	HB	4B <5% removed	no cracking	120 120	no blistering hardness: n.c.	severe softening <6B	very light stain hardness: n.c.
R-9000/ BAPP (10%)	65*	66	HB	5B 0% removed	hairline cracks	160 160	no blistering hardness: +1 unit	severe softening<6B	very light stain hardness: n.c.
R-9000/ DMAPA(2.3%)	195*	65	HB	5B 0% removed	hairline cracks	160 160	no blistering hardness: n.c.	severe softening<6B	very light stain hardness: n.c.
MIL-C-83286	590	92	3H	5B 0% removed	no cracking	92 160	no blistering hardness: H-2H	softening 4B gloss: - 12%	very light stain hardness: no change
1-Coat	418	93	3H	5B 0% removed	no cracking	160 160	no blistering hardness: F-H	severe softening <6B gloss: -85%	very light stain hardness: no change

LOW VOC COATING ALTERNATIVES

Mark D. Smith
Materials Engineering
Allied-Signal Inc.
Kansas City, Missouri

INTRODUCTION

Allied-Signal Inc., Kansas City Division, (KCD) is a prime contractor for the Department of Energy. KCD is one of the manufacturing sites for the Department of Energy, (DOE) integrated weapons complex. KCD manufactures various components, many of which require surface coating in one way or another. Recent changes in DOE directives and application of state and local air quality regulations caused coating operations to become a concern.

AREAS OF SURFACE COATING CONCERN

The two main areas of surface coating concern at KCD are the painting operations, (both product related and painting done for maintenance or non-production applications) and the application of dry film lubricants. As is detailed later, both operations were, prior to this work, using materials that were high in volatile organic compounds, (VOCs). This will cover the two operations separately, mainly because the regulations treated them differently.

PAINTING

Status Prior to July, 1989

Prior to July of 1989, we were vaguely aware of state and local air quality regulations. Some effort was being made to develop a long range "site plan" to address the situation as it was understood at KCD. We were totally unaware of the existence of low VOC paints. A cursory investigation into this area had been done a few months earlier regarding the currently used Military Specification paints. At this time KCD used approximately 25 paints on a routine basis with internal Material Specifications existing for another 250-300 that could be used at any time that they might be needed. The VOC content of none of these paints was known. Our paints in use were a mixture; some were Mil. Spec. paints and some were not, depending on what our customer wanted. In addition, at this time all of the painting VOCs were being released to the atmosphere through the paint booth stacks.

REGULATION IMPOSED

The air pollution regulation imposed in July 1989 was "Missouri Air Pollution Rule 10 CSR 10-2.230, Control of Emissions from Industrial Surface Coating Operations." This regulation applies to locations emitting more than 6.8 kilograms per day or 2.7 tons per year of VOCs. KCD was regulated under the provisions for painting "Miscellaneous Metal Parts" since our products were not considered to be covered by any of the other classifications in the regulation. This classification of products had an emission limit of 3.5 pounds of VOCs per gallon of coating as applied.

Compliance Actions, July 1989

On July 7, 1989, KCD voluntarily halted all spray painting operations. The process of developing a "site plan" for obtaining state and local EPA approval was accelerated drastically. A survey was started of the currently used paints to determine their VOC contents. Figure 1 shows the results of that survey with the individual paints averaged into groups by chemical type. The results clearly show that the paints being used at that time were all significantly higher in VOC content than the 3.5 pounds per gallon allowed by the regulations.

A comprehensive search for paints "compliant" to the regulations was started. To our pleasant surprise, a large percentage of the currently used paints had low VOC compliant versions available or in development. Some paints were being replaced by the paint industry with other types of paints which were equivalent or better in performance. Samples or small batches of these compliant and replacement paints were obtained and VOC-testing begun on them. Figure 2 shows the VOC content of selected replacement paints as claimed by the vendors. Actual test data confirmed these values within experimental error. Along a parallel path, while the new paints were being tested for VOC content, samples were being test sprayed and their physical properties checked to determine suitability for KCD production applications.

On July 25, 1989, a meeting was held with other DOE integrated weapons contractors to appraise them of the situation and determine possible alternate sources for production painting within the weapons complex. The prime consideration in this meeting was the ability of the other contractors to meet their own state and local air quality standards. It was not considered feasible to send DOE contracted work to a location that was also not in compliance with air quality regulations. Production schedules and other restrictions eliminated the possibility of sending parts to other DOE contractors for painting.

An exhaustive survey was also conducted of painting industry contacts to determine possible alternate sources of production painting outside the weapons complex. Over 500 companies were contacted. The prime consideration in these contacts was also their ability to meet their state and local air quality standards. As with the DOE contractors, it was not considered feasible to send DOE contracted work outside the complex to a location that was also not in compliance with air quality regulations. This survey found less than five contacts able to meet local regulations, fewer that would take on KCD work and fewer still that could meet the paint quality standards required for the product.

A small activated carbon filtering system for the production painting area was obtained but local EPA approval was needed before it could be installed and tested. It was small enough that it would not handle all production painting but could be used on a limited basis through a partitioned section of an existing paint hood. Start-up of this unit was included in the proposed site plan which awaited EPA approval.

CONCERNS IN SWITCHING TO REPLACEMENT PAINTS

The mixture of Mil. Spec. and non-Mil. Spec. paints used previously was not an optimum situation for the type of work done at KCD considering the customer for our products. Since changes had to be made in the types of paints being used anyway, it was considered preferable to strive for a paint inventory of all Mil. Spec. paints.

The VOC regulations were written, in general, to reduce the amount of chemicals expelled into the air that cause the formation of ozone and smog at low altitudes. For this reason, several chlorinated and fluorinated solvents are considered by the regulations to be "exempt" or compliant for use in paint formulations. Many of the new low VOC compliant paints were formulated using 1,1,1-trichloroethane as the principal thinning solvent, (shown on Figure 2 with an asterisk beneath the paint type), in place of the high VOC hydrocarbon thinners such as methyl ethyl ketone or xylene. The solvents considered exempt for the VOC regulations are in fact considered ozone depleters in the upper atmosphere by other regulations. Significant effort is being expended at KCD, as at other DOE contractor sites, to minimize or eliminate the use of chlorinated and fluorinated solvents for that reason. In this case, two major efforts to reduce air pollution were essentially incompatible. Therefore it was also considered preferable to strive for a paint inventory as free from chlorinated and fluorinated solvents as possible.

There were a number of paints for which there were no direct low VOC replacements, including lacquers, wash primers and Teflon®-based paints. It was decided that in most cases this impediment to changing paint systems could be worked around since the number of products requiring these paints was small. In addition some concern existed over the need for new painting techniques and equipment to properly apply the new paint systems. This was also considered a small obstacle to overcome since without the new low VOC paints, painting operations could not resume.

PRESENT KCD PAINTING STATUS

Most spray painting applications are now using low VOC paints. In rare instances where substitute paints are not yet available, limited use of high VOC materials is made within the special spray booth, the stack emissions being routed through activated carbon filters as approved in the new EPA site plan. This filtering system has proved to be greater than 95% efficient in removing VOCs from that air stream and has not required changing over the past year.

The number of paint systems now in routine use has been significantly reduced from 25 high VOC versions down to approximately 5 low VOC versions. A larger percentage of the paints now being used are also based on Mil. Spec.'s and Federal Color Standard numbers. Both of these factors eases our purchasing and qualification testing of the paints. In addition, the inventory is easier to maintain, both in size and complexity and from the standpoint of shelf-life requalification of fewer paints. So, along with reducing our VOC emissions, the amount of waste from unused, out of date paints will, in the future, be reduced.

ANTICIPATED FUTURE PAINT WORK

KCD will continue to emphasize the use of high-solids, low VOC polyurethane paints. Those meeting Mil-C-85285 have worked well in replacing high VOC urethanes as well as other types of paints such as epoxies and acrylics. The reduction or elimination of chlorinated solvent based coatings will be continue to be pursued. In one case, a chlorinated solvent based, zinc chromate primer was replaced with a Mil. Spec. waterborne epoxy primer with satisfactory results. Other chlorinated solvent based paints have been replaced with the Mil-C-85285 urethanes.

One major program remains to be changed over from a high VOC system of primer, aluminum pigmented epoxy and a clear epoxy top-coat to low VOC waterborne primer and low VOC urethane in place of the epoxy. Problems with the clear urethane continue to be worked through.

In anticipation of tighter EPA regulations and Allied-Signal corporate directives, consideration is being given to the activated carbon filtration of all the paint booths, including those using low VOC paints, instead of just the special booth now using high VOC paints. This would allow more flexibility in painting and reduce the emissions closer to the "as low as reasonably attainable" level desired by the corporate policies.

Alternate coating methods are being evaluated including powder painting and coating by wet chemical electrophoresis. A joint study of powder coatings applied by an outside contractor is underway with Sandia National Laboratories. A laboratory-scale, batch-type powder facility is being set-up at KCD to serve the DOE weapons complex.

DRY FILM LUBRICANTS

Present Dry Film Lubricant Status

Most of the dry film materials presently used at KCD are high VOC, ranging from 6.42 to 8.12 pounds per gallon as applied, (Figure 3). The current Missouri regulation regarding emissions from surface coating operations does not regulate the application of dry film lubricants. However, considering the dynamic nature of air quality regulations and the corporate policies to reduce emissions, considerable work is underway to evaluate low VOC dry film lubricants or alternative lubrication methods that would be suitable for our applications.

Concerns in Switching Dry Film

The application techniques and equipment required to apply some of the dry films being considered are significantly different than what is now is use and will require substantial changes in operations. Some of the technologies being investigated require licensing and some are performed only by the vendors. In certain cases, such as classified or sensitive parts, this would be prohibitive.

The performance requirements of the present materials are not well defined, therefore it will be hard to define what properties a suitable replacement should have. Many of the dry films were put into use to solve a particular problem on a particular part years ago and their use has been continued without the problem and solution being fully understood.

Some of the new low VOC versions of existing dry films are relatively unproven in actual use, similar to some of the low VOC paints, which causes some trepidation in adopting their use in an ongoing production environment. Substantial testing will be required in various modes of lubrication for several different types of assemblies.

KCD DRY FILM ACTIONS

A joint Sandia National Labs./KCD group has begun studying dry film lubricants. The dry film market is being surveyed for possible low or zero VOC materials and technologies. Possible alternatives identified at this point include: Dicronite®, sputtering of MoS_2, sputtering MoS_2 followed by ion treatment, electrophoretic application of MoS_2 rich coatings, increased hardness coatings, Microseal® and low VOC versions of the presently used E/M materials. A database of the dry films being considered is being developed for future reference including vendor data and manufacturer's Material Safety Data Sheets.

The study is also attempting to define dry film performance requirements for existing and planned assemblies. This covers hundreds of individual parts and assemblies and various different modes of lubrication needed, some within the same assembly. In addition, the study will need to determine objective inspection techniques for qualifying and comparing the dry films that will be tested. Several short term tests are being made on individual parts or assemblies and are generating useful data. However, the large number of parts that use dry films suggests that a broad study that will yield consistency in the future use of dry films is in order.

CONCLUSIONS

Although a great deal of progress has been made at KCD in the effort to reduce VOC emissions from surface coating operations and the state and local regulations are being complied with, work has not and will not stop in trying to further this effort. Future EPA regulations concerning air quality and Allied-Signal corporate policies of trying to reach emissions levels of "as low as reasonably attainable" assure that this work is not completed.

Figure 1. VOC's of Selected High VOC Paint Systems, (as applied)

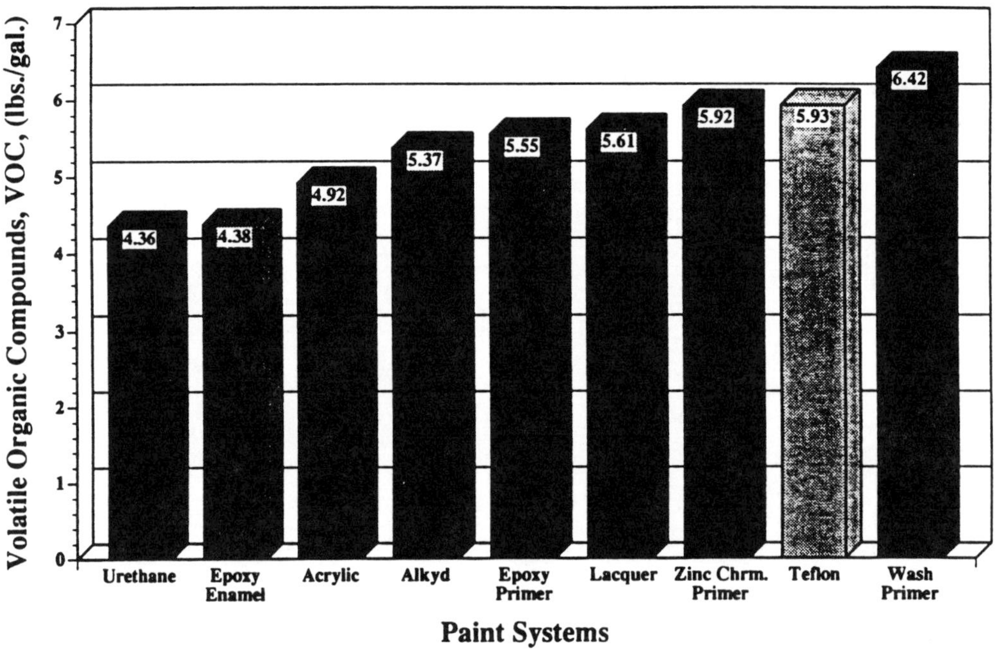

Figure 2. VOC's of Selected Replacement Paints, (vendor data)

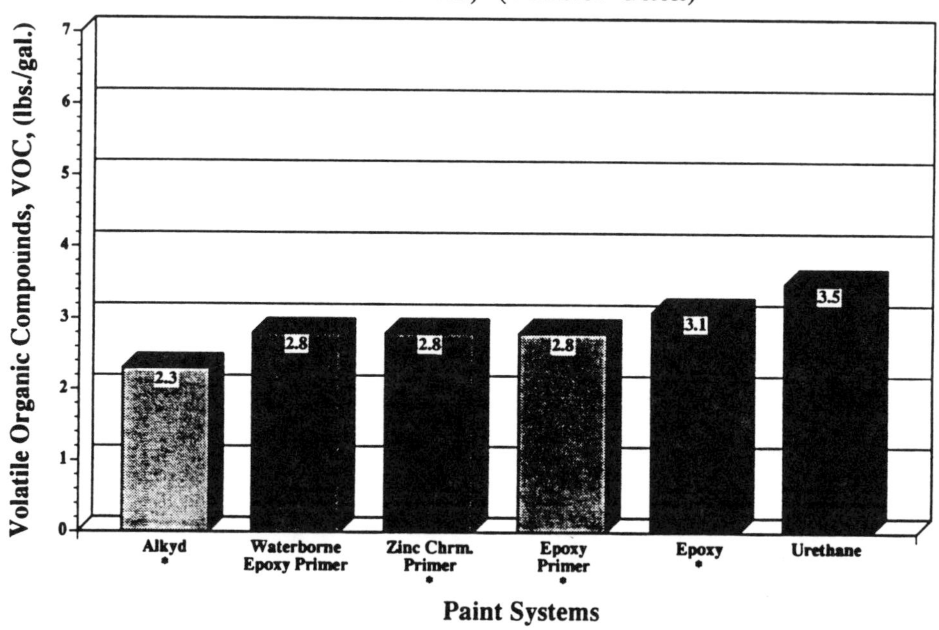

Figure 3. VOC's of Selected E/M Corp. Dry Films, (as applied)

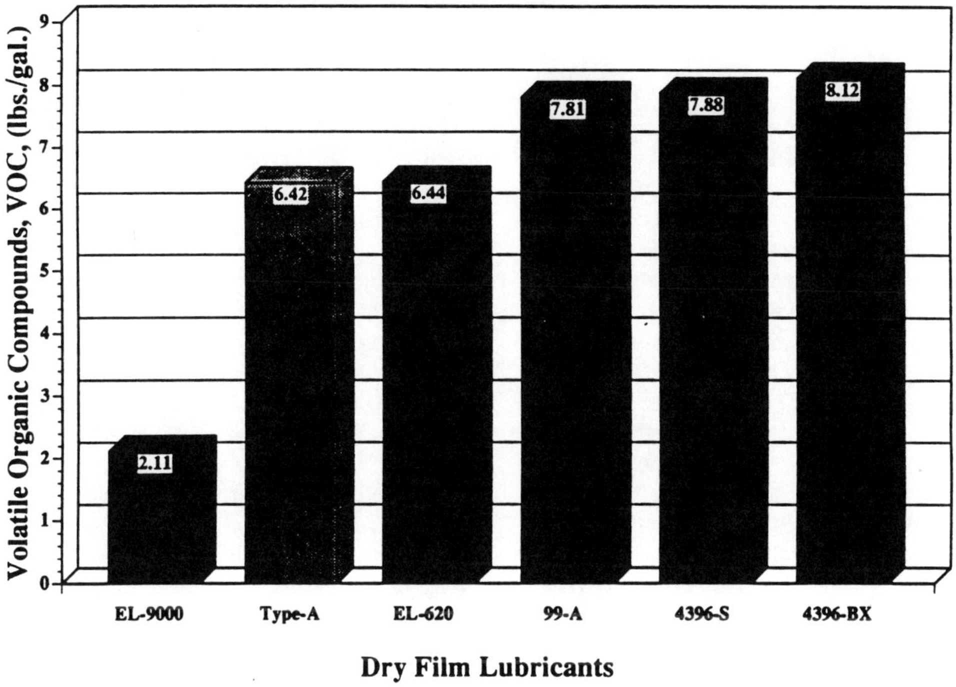

DUAL CURE PHOTOCATALYST SYSTEMS

Steven J. Keipert, Ph.D.
Corporate Research Laboratories
3M Company
St. Paul, Minnesota

The 3M Dual Cure process is a method for producing radiation cured polymer compositions via a solventless coating process. The process involves the simultaneous cure of two different monomer types. The first type may be either an epoxy monomer, or the polyol and polyisocyanate precursors to a urethane polymer. These are polymerized by use of an iron based organometallic photocatalyst which gives a Lewis acid upon irradiation. The second monomer type consists of acrylate monomers, which are polymerized via a free radical photoinitiator to give acrylate polymers. The two polymerization types procede separately to give substantially independent, interpenetrating polymer networks. The properties of the resulting Dual Cure compositions are often found to be superior to either component alone.

The work described in this paper was developed under a 3M/U.S. Department of Energy (DOE) cost share contract titled "Industrial Gaseous Waste Reduction" funded by the DoE Office of Industrial Technologies. The research was performed in the 3M Corporate Research Science Research Laboratory, in cooperation with other 3M laboratories. The contract consists of three phases. Phase I, which is complete, involved demonstration of technical feasibility on a laboratory scale. Several catalyst combinations were evaluated in a model curable composition, and the physical properties of the resulting cured polymers were analyzed and compared. Phase II, which is nearing completion, consisted of a pilot scale demonstration, primarily involving protective metal coating applications. Phase III will involve the demonstration of the technology in full scale commercial applications in cooperation with internal and external industrial partners.

The objective of the Dual Cure program is to eliminate solvents from coating compositions, while simultaneously maintaining or improving the physical properties of the coating and the processing conditions used to prepare them. This will lead to reductions in solvent use, with a resulting reduction in energy consumption and solvent emissions.

Projections for the use of various coating technologies in the future show a reduced market share for solvent coating techniques. Most of this will be replaced by high solids, waterborne, and powder coating methods. Although decreasing as a percentage of the total, market growth means that, in absolute terms a substantial amount of solvent coating will remain in absolute terms. Radiation coating techniques offer an alternative replacement for solvent based systems. Projections show this sector growing, but never gaining a substantial market share. We feel that a major reason for this trend is the inability of currently existing U.V. curing processes to deliver the materials performance required by many coatings applications. We believe that the 3M Dual Cure process, which allows an expansion of the number of radiation curable monomer types, will enhance the production of high performance coatings. This should facilitate the increased use of radiation curing, and a further reduction in solvent use. Preliminary projections of energy savings resulting from the adoption of this technology show savings of 1.5×10^{13}

btu/year by the year 2010.

We are primarily interested in two types of Dual Cure compositions. The first results from the combination of epoxies with acrylates to give epoxy/acrylate compositions. The second results from the combination of a urethane and an acrylate to give a urethane / acrylate. These are not to be confused with the acrylated urethanes, which are photocurable, acrylate functionalized urethane oligomers which polymerize by a free radical acrylate type polymerization. In the Dual Cure process, both polymers are formed directly from their respective monomers, with limited interconnection of the two polymer networks formed.

Five different catalyst combinations were investigated in phase I of the contract. Systems 1 through 3 use a cationic iron catalyst. These compounds are relatively inert in the dark, and are air-and-moisture stable. Upon irradiation, they produce a coordinatively unsaturated, Lewis acidic iron species through the initial loss of the arene ligand. This is an active thermal catalyst for the polymerization of epoxies and urethanes. They are also able to initiate free radical polymerizations, although the mechanism for this is not clear at present. These iron catalysts were used alone as in catalyst system 2, or in combination with other radical photoinitiators such as benzil dimethylketal in catalyst system 1, and oxidants such as diphenyliodonium salts in catalyst system 3. A neutral binuclear iron species was also investigated in catalyst system 4. Although it was found to be an active photocatalyst, its reactivity in the dark limited its usefulness and this system was abandoned early in phase I. Catalyst system 5 consisted of the controls. For epoxy/acrylates this consisted of a combination of a triarylsulfonium salt, which gives a protic acid upon photolysis, in combination with a radical photoinitiator. For urethane/acrylates, the control consisted of dibutyltin dilaurate, a commonly used thermal catalyst for urethane polymerization, in combination with a radical photoinitiator.

The urethane portion of the urethane/acrylate Dual Cure formulations are formed from a mixture of a polyisocyanate, preferably an aliphatic polyisocyanate, and a polyol. Prior to the discovery of the Dual Cure catalysts, these materials could only be polymerized by thermal processes. The two components will polymerize alone, but the rate of polymerization is enhanced greatly by incorporation of a catalyst such a dibutyltin dilaurate. Thermal polymerizations of this type have several disadvantages which will be discussed later. The acrylate portion of the urethane/acrylate Dual Cure formulations is generally a mixture of mono and polyfunctional acrylates. The optimum ratio of urethane to acrylate for protective coating applications has been found to be approximately 70:30. The high viscosity of urethane formulations has often limited their processability without the addition of thinning solvents. Incorporation of the acrylate results in reduced viscosity, and improved processability.

Because of the reactivity of the urethane precursors even in the absence of catalyst, these materials require storage as two components, which are mixed prior to use. The Dual Cure compositions do offer processing advantages in comparison with thermally catalyzed urethane polymerizations. Figure 2 shows the effect of the different catalyst compositions on the cure speed of a urethane/acrylate formulation. These are the cure times to give a tack-free film (set to cotton) at 100 degrees Celcius after U.V. irradiation. Catalyst concentrations were adjusted to give comparable cure speeds for all systems. The iron catalyst containing compositions (1-3) contained 300 ppm catalyst, and the control (5) contained 40 ppm of dibutyltin dilaurate. Free radical initiators and oxidants were present at 0.1% in the appropriate systems. Figure 3 shows the potlife of these same formulations at room temperature in the dark. The useful potlife is defined as the time required to double the coating viscosity of the formulation. These data show that although all compositions gave comparable cure speeds, the compositions based on the Dual Cure catalysts (1-3) had potlives in excess of 12 hours, comparable to uncatalyzed compositions. In comparison, the control composition containing dibutyltin dilaurate had a potlife of less than 1/2 hour. This demonstrates an advantage in using the photoactivated Dual Cure catalyst in urethane systems. The necessary tradeoff between cure speed and potlife found when using thermal catalysts is eliminated. With Dual Cure, cure speed may be adjusted as desired, with no effect on the potlife.

Synergistic effects are often observed with respect to the physical properties of Dual Cure compositions. Figure 4 shows a tensile strength comparison between a urethane/acrylate composition and the identically cured urethane and acrylate portions alone. The acrylate has a higher tensile strength than the urethane. The Dual Cure composition is seen to retain most of the tensile strength of the stronger acrylate portion, even though it consists of only 30% acrylate. Figure 5 shows a comparison between the elongation to break behavior of the same three compositions. The urethane has a much higher elongation to break than the strong but brittle acrylate. The Dual Cure composition retains most of the elongation properties of the urethane portion. Figure 6 shows a comparison of the toughness (energy to break) of the three materials as measured by the area under the stress/strain curve. Since toughness is a function of both high tensile strength and high elongation to break, it is not unexpected that the Dual Cure composition is found to be substantially tougher than either of its components separately.

The second class of Dual Cure materials which were investigated are the epoxy/ acrylates. The preferred epoxies are the difunctional cycloaliphatic epoxies such as ERL-4211. These are combined with a mixture of mono and polyfunctional acrylates. The most useful monomer ratio for protective metal coating applications was found to be 60:40. Since neither of these monomers will polymerize without activated catalyst or initiator, they are stable one-component materials as long as they are stored in the dark. These compositions are generally of low viscosity and easily coated.

The synergistic effects observed in the urethane/acrylate system are also seen with the epoxy/acrylates. Figure 7 shows a comparison of the tensile strength of the Dual Cure composition with the separate epoxy and acrylate portions. The epoxy/acrylate is seen to maintain a significant portion of the tensile strength of the stronger epoxy component. Figure 8 shows a similar comparison of the elongation to break data. In this case the Dual Cure composition outperforms both of the separate components. The toughness data is shown in figure 9. Once again, the Dual Cure material is found to be significantly tougher than either component separately.

In summary, it has been shown that the Dual Cure process offers several advantages with respect to materials, properties and processing. The urethane/acrylate system, when compared to conventional urethanes, gives a lower viscosity, more easily coatable formulation, and avoids the cure time/potlife tradeoff. In addition, it has been demonstrated that both the urethane/acrylate and epoxy/acrylate Dual Cure systems exhibit beneficial synergistic effects, retaining the best physical properties of each component. Phase III of the contract will extend this technology into full scale commercial demonstrations.

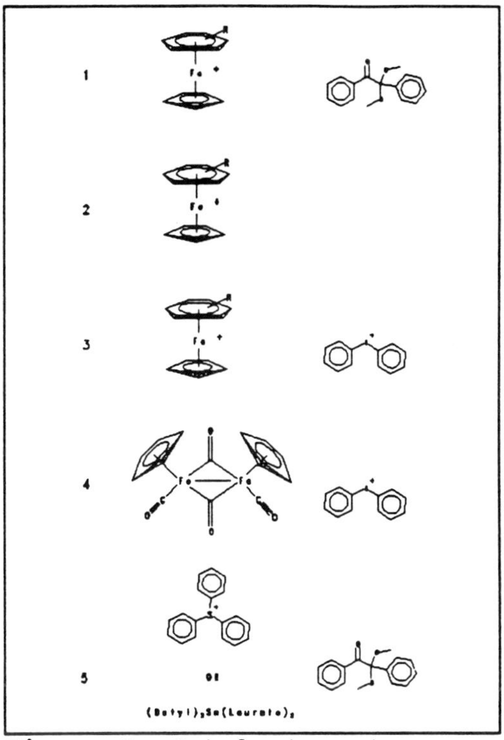

Figure 1: Catalyst systems.

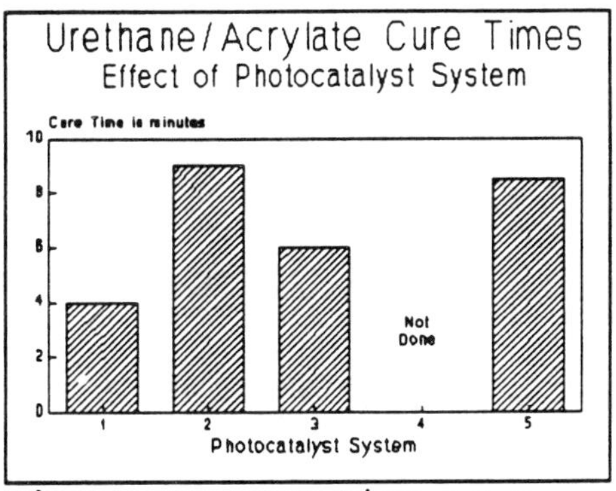

Figure 2: Cure time as a function of catalyst system.

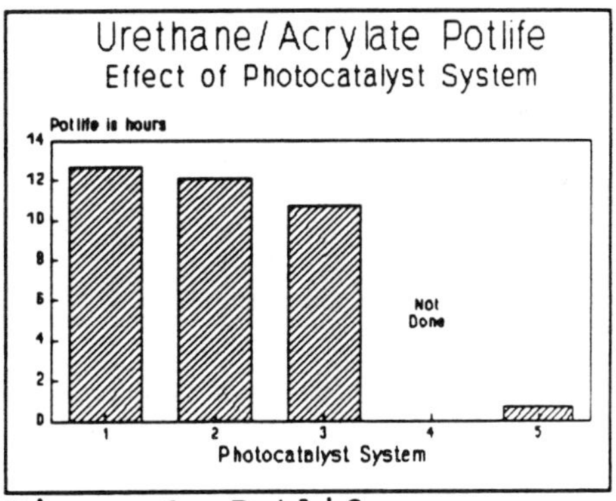

Figure 3: Potlife as a function of catalyst system.

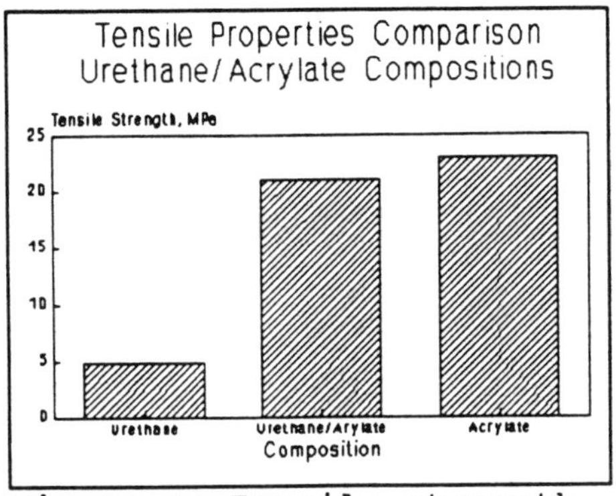

Figure 4: Tensile strength comparison with pure components.

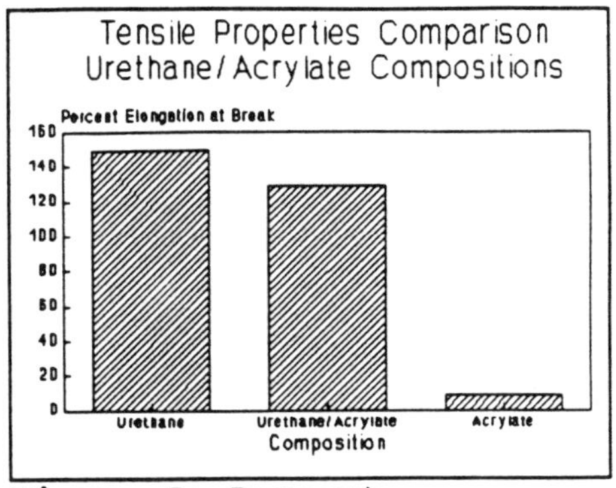

Figure 5: Percent elongation to break comparison with pure components.

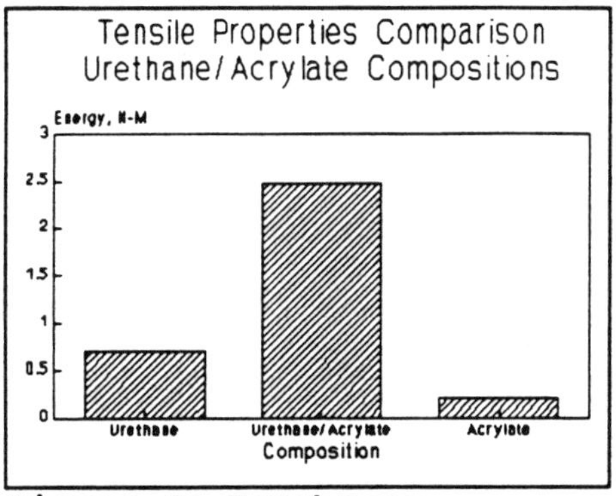

Figure 6: Toughness comparison with pure components.

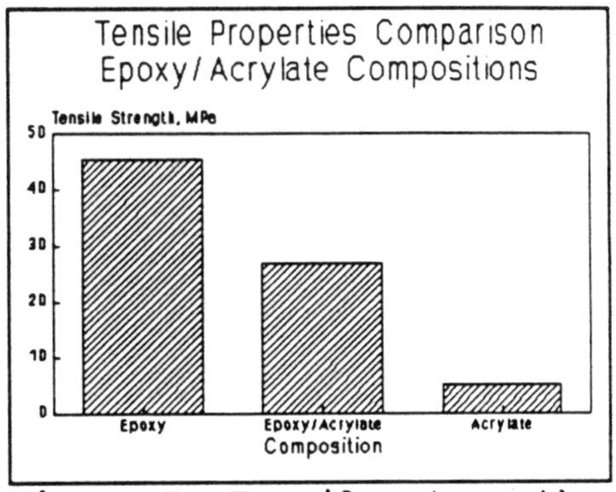

Figure 7: Tensile strength comparison with pure components.

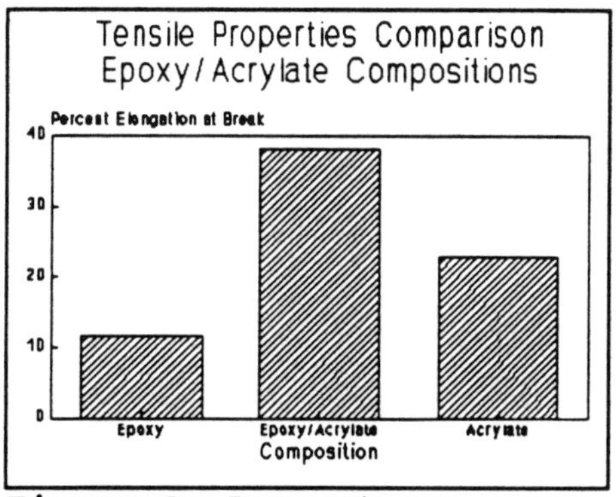

Figure 8: Percent elongation to break comparison with pure components.

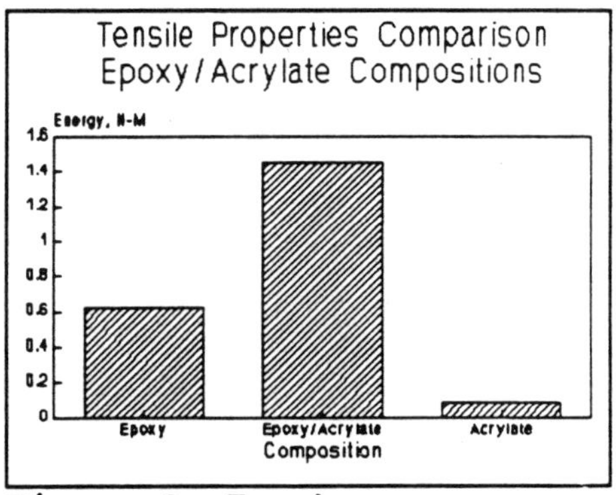

Figure 9: Toughness comparison with pure components.

SCREENING OF VOC CONTROL TECHNOLOGIES: TECHNOLOGY OPTIONS AND COMPARATIVE COSTS

Dr. Victor S. Engleman
Science Applications International Corporation
San Diego, California

Emissions of volatile organic compounds (VOCs) can be controlled by recovery methods and destruction methods. Recovery methods include adsorption, absorption, condensation and separation options. Destruction methods include incineration and biodegradation. A screening model has been developed to perform preliminary evaluations on feasibility and relative costs of VOC control technologies. The combinations evaluated by the model are as follows:

- carbon adsorption with off-site regeneration/destruction
- carbon adsorption with on-site steam regeneration
- carbon adsorption with on-site inert gas regeneration and reverse Rankine solvent recovery
- carbon adsorption with on-site inert gas regeneration and reverse Brayton solvent recovery
- carbon adsorption with decoupled inert gas regeneration and reverse Brayton solvent recovery
- closed cycle Rankine condensation
- closed cycle Brayton condensation
- open cycle Brayton condensation
- cryogenic liquid condensation (partially implemented)
- absorption by heavy organic liquids (partially implemented)
- membrane separation (partially implemented)
- thermal incineration without heat recovery
- thermal incineration with heat recovery
- catalytic incineration
- regenerative incineration
- biodegradation/biofiltration (partially implemented)

Some of the parameters that are considered by the model that influence the cost of these control technologies are:

- chemical composition of the waste stream
- concentrations of the compounds
- recovery value of the solvents
- flow rate of the stream
- temperature of the stream
- pressure
- operating time (hours/day, days/week, weeks/year)
- percent recovery required
- utility and chemical unit costs

Volatile organic compounds presently included in the model are the following:

- acetic acid
- acetone
- benzene
- carbon tetrachloride
- chlorobenzene
- p-dichlorobenzene
- ethylene
- ethylene dichloride
- formaldehyde
- methyl ethyl ketone
- methyl isobutyl ketone
- methylene chloride
- perchloroethylene
- toluene
- trichloroethylene
- trichlorofluoromethane
- vinyl chloride

- heptane
- methanol
- methyl chloride
- methyl chloroform
- trichlorotrifluoromethane
- m-xylene
- o-xylene
- p-xylene

Recovery technologies represent an opportunity to save both energy and money, since solvent recovered may be reused in the same or a similar application. Each specific case requires separate evaluation because of the specific requirements of the individual application. In some cases, it may not be possible to use recovery technologies because of technical impediments.

The screening model has been used for the initial evaluation of control technologies for specific cases. The results of calculations from this screening model are presented below, in graphic form, to illustrate the regions where control technologies are cost-competitive for the specific case of toluene emissions. Figures 1-3 represent the regions in which each of the control technologies represent the least-cost control option, given the input used for this particular test (the most important of which is the value for the recovered solvent of $0.20 per pound). Annualized costs were used as the basis for these figures.

Figure 1 indicates the regions in which adsorption-based technologies are the least-cost control method. Adsorption with off-site regeneration is most favorable at low concentrations and low flow rates. Adsorption with on-site steam regeneration is most favorable at high flow rates and low to medium concentrations. Adsorption with mobile inert-gas regeneration fills the gap between the first two methods. Adsorption with on-site Brayton regeneration overlaps the region for steam regeneration, but extends to higher concentrations. The region of higher concentrations is also covered by adsorption with on-site Rankine regeneration. Since the Brayton system can achieve lower temperatures more readily than the Rankine systems, it tends to be more favorable for streams requiring very low condensation temperatures.

Figure 2 indicates the regions for incineration technologies. Because of the good recovery value used for toluene and the fact that toluene is readily recoverable, even from steam regeneration, incineration technologies were among the least-cost control option only in a limited range. In the vicinity of 10,000 cubic feet per minute and 100 ppm, regenerative incineration was among the least-cost control options. If technical factors in the system had made adsorption technically unfeasible, incineration would have covered with a wider range. Incineration is technically feasible across a broad range.

Figure 3 indicates the regions for condensation technologies. Both direct Brayton (also called open-cycle Brayton) and indirect Rankine (closed-cycle Rankine) condensation technologies were the least-cost control options at high concentrations across the entire range of flow rates. The direct Brayton does become limited in applicability above 10 or 12% because it becomes difficult to condense the VOCs by direct expansion.

The graphs shown do not apply universally. Higher or lower boiling points, difference in reactivity, and differences in recovery costs, among other factors, influence the regions in which control technologies are most favorable.

The model outputs include capital costs (equipment and installation), operating costs (labor, energy, maintenance, supplies, taxes/insurance, overhead and solvent recovery credit) and total annualized costs. Since this is a screening model, results are expressed as ranges rather than as absolute numbers. The ranges represent the uncertainty of the calculations, based on site-specific factors not considered.

The model is still under development and interested parties are encouraged to contact the author at the above address or by phoning (619) 587-9071, ext. 169. Input for cost information, suggestions for inclusion of additional features, case studies for calibration and general interest are solicited.

EXAMPLES OF RANGES FOR CONTROL TECHNOLOGIES TOLUENE (1 OF 3)

OTHER FACTORS AFFECTING VOC CONTROL SELECTION

- o VALUE
- o # COMPONENTS
- o CONDENSATION TEMP
- o REACTIVITY
- o WATER CONTENT
- o MISCIBILITY

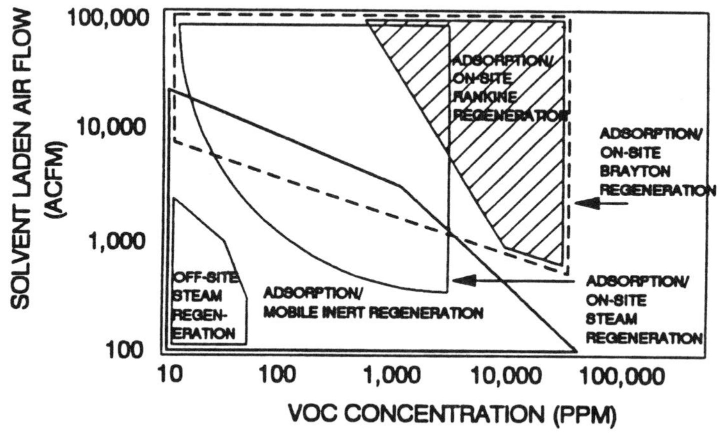

FIGURE 1

EXAMPLES OF RANGES FOR CONTROL TECHNOLOGIES TOLUENE (2 OF 3)

OTHER FACTORS AFFECTING VOC CONTROL SELECTION

- o VALUE
- o # COMPONENTS
- o CONDENSATION TEMP
- o REACTIVITY
- o WATER CONTENT
- o MISCIBILITY

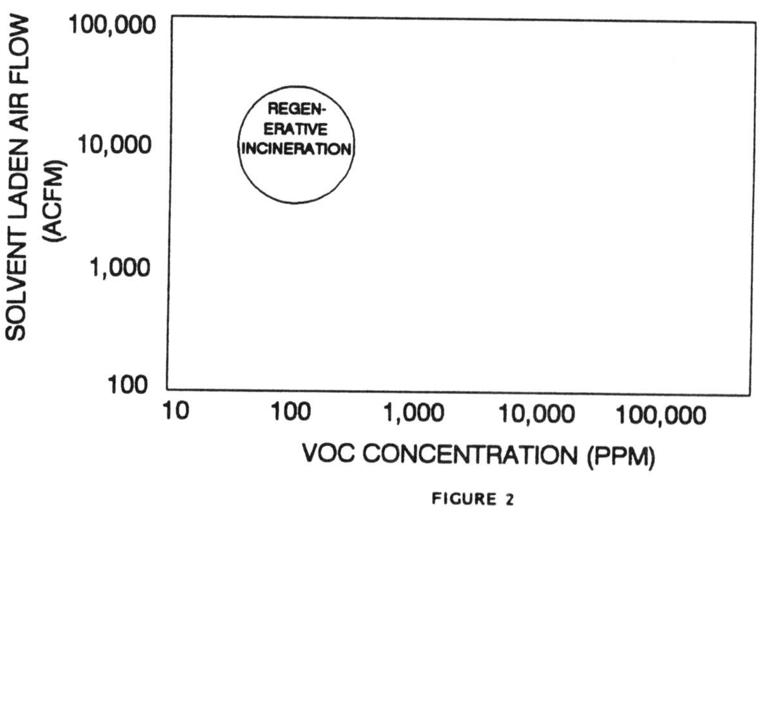

FIGURE 2

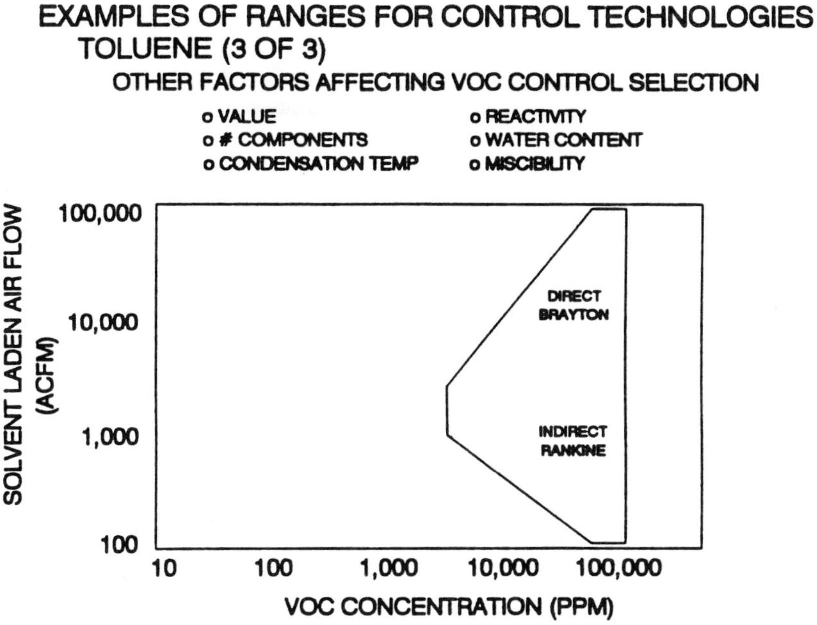
EXAMPLES OF RANGES FOR CONTROL TECHNOLOGIES
TOLUENE (3 OF 3)
OTHER FACTORS AFFECTING VOC CONTROL SELECTION
o VALUE
o # COMPONENTS
o CONDENSATION TEMP
o REACTIVITY
o WATER CONTENT
o MISCIBILITY
SOLVENT LADEN AIR FLOW (ACFM)
100,000
10,000
1,000
100
DIRECT BRAYTON
INDIRECT RANKINE
10
100
1,000
10,000
100,000
VOC CONCENTRATION (PPM)

FIGURE 3

Section V

TREATMENT FOR ENVIRONMENTALLY SAFE DISPOSAL OF TOXIC SOLVENTS

GENERAL OVERVIEW OF HAZARDOUS WASTE INCINERATION

Philip C. Lin, Ph.D.
Risk Reduction Engineering Laboratory
Office of Research and Development
U.S. Environmental Protection Agency
Cincinnati, Ohio

ABSTRACT

Improper disposal of hazardous materials generated by industries in the U.S. in the past several decades has prompted the Congress to enact a series of environmental laws to govern and cleanup hazardous waste. Conventional methods of waste disposal, such as landfilling and deep-well injection, arebeing restricted. Waste minimization, recycling, and treatment are being vigorously pursued. Although there are many different treatment methods, such as physical/chemical treatment and stabilization/solidification methods, incineration is the most universally applicable. Incineration, which is capable of the highest degree of waste destruction, is considered to be able to destroy the broadest range of hazardous waste. However, incineration may produce residuals in the ash and emit undesired trace amounts of unburned hazardous waste, incomplete combustion by-products, metals, and particulates. This paper is to review the current technology available and practice of incineration of halogenated hydrocarbons; regulations and standards for incinerators, boilers, and furnaces; and results from field evaluations of the incineration of various hazardous wastes in the early 1980's. Recent incineration research by the USEPA is also discussed.

INTRODUCTION

Improper disposal of hazardous materials generated by industries in the U.S. in the past several decades has prompted the Congress to enact a series of laws to control the hazardous waste. The first federal standards for control of incineration emissions were enacted under provisions of the Clean Air Act (CAA) in 1970. Only particulate emissions from incineration sources were regulated. In 1976, the Resource Conservation and Recovery Act (RCRA), which defined and identified hazardous waste and provided provisions for controlling the storage, transport, treatment, and disposal of hazardous waste, was enacted. Incineration of polychlorinated byphenyls (PCBs) was controlled in May 1979 under rules established under the Toxic Substances Control Act (TSCA) of 1976. The rules prohibited further manufacture of PCBs after July 2, 1979, set limits on PCB use in commerce, and established regulations for proper disposal. In 1980, a national fund to assist the clean-up of uncontrolled waste sites created by poor disposal practices was established under the Comprehensive Environmental Response, Compensation, and Liability Act (CERCLA). On June 24, 1982, the final incinerator standards of performance were published in the Code of Federal Regulations (CFR) under 40 CFR 264.343. The Hazardous and Solid Waste Act (HSWA) of 1984 amended and reauthorized RCRA to establish a strict timetable for restricting untreated hazardous waste from land disposal. By 1990, most wastes were restricted and required to be pretreated by Best Demonstrated Available Technology (BDAT) before disposal. The 1986 Superfund Amendments and Reauthorization Act (SARA) reauthorized the Superfund programs and greatly expanded the provisions and funding of the initial Act. SARA also emphasized the need to select clean-up technologies that would result in a permanent decrease in the toxicity, mobility, or volume of hazardous materials. The impact of these various statutes will be a significant modification of waste management practices.[1]

REGULATIONS AND STANDARDS

Environmental regulations and standards promulgated under various federal statutes are chronologically listed as follows:

Regulations and Standards for Incinerators

1. Clean Air Act (CAA, 1970)

   Particulate concentration emitted from all incinerators constructed after August 1971 must not exceed 0.08 grains/dscf corrected to 12% CO_2 in the stack gas.

2. Resource Conservation and Recovery Act (RCRA, 1976)

   The final incinerator performance standards were published on June 24, 1982, listed in the Code of Federal Regulations (CFR) under 40 CFR 264.343. In order to receive a RCRA permit to run the facility, it must attain the following performance levels:

   a. Principal organic hazardous constituents (POHCs) designated in each waste feed must be destroyed and/or removed to a destruction and removal efficiency (DRE) of 99.99% or better.

   b. DRE is defined by the following formula:

      DRE (%) = 100 x (Win- Wout)/Win

      where Win = POHC feed rate to the incinerator
      Wout = POHC emission rate from the incinerator

   c. Gaseous hydrogen chloride (HCL) emissions must either be controlled to 4 lbs/hr or less, or be removed at 99% efficiency.

   d. Particulate emissions must not exceed 180 mg/dscm corrected to 7% O_2 in the stack gas. The measured particulate concentration is multiplied by the following correction factor (CF):

      CF = (21 - desired O_2) /(21 - measured O_2)

   The POHCs are listed in RCRA Appendix VIII which consists of 390 organic and inorganic compounds first published in the May 19, 1980 Federal Register and updated semiannually in 40 CFR 261. The Appendix VIII constituents, which are in the highest concentration in the waste feed and are most difficult to incinerate, are to be selected as POHCs in the trial burn. The heat of combustion of the POHCs was initially suggested as a measure of compound incinerability, but this has been replaced by an EPA-generated Incinerability Ranking Index.

   To obtain a RCRA permit, a trial burn to meet these requirements must be conducted. The standards also specify the requirements for waste analysis, operation, monitoring, inspection, and procedures by which permits will be granted.

3. Toxic Substances Control Act (TSCA, 1976)

   Whenever disposal of PCBs is undertaken, they must be incinerated, unless the PCB concentration is less than 50 ppm. If the PCB concentration is between 50 and 500 ppm, the waste can be used as a fuel in a high-efficiency boiler. For PCBs exceeding 500 ppm, they must be incinerated with a combustion efficiency no less than 99.9% and a

DRE of 99.9999% and meet a number of specific incinerator operating conditions, such as combustion temperature, residence time, stack oxygen concentration.[1] The combustion efficiency (CE) is the ratio of the following:

$$CE = CO/(CO+CO_2)$$

Further manufacture of PCBs was prohibited after July 2, 1979.

4. Dioxin Rule[2] (promulgated on January 14, 1985 under RCRA)
   The incinerator must be capable of achieving 99.9999% DRE for chlorinated dioxins or similar compounds.

5. Superfund Amendments and Reauthorization Act (SARA, 1986)

   a. Reauthorized the Superfund program and expanded the provisions and funding.

   b. Emphasized the need to select clean-up technologies that will result a significant decrease in the toxicity, mobility, and/or volume of hazardous materials.

Proposed Regulations on Boilers and Furnaces

On May 6, 1987, EPA proposed rules to control the burning of hazardous waste in boilers and furnaces. They were promulgated in December, 1990. The following performance standards must be met:

1. 99.99% DRE for each selected POHC in the waste feed.

2. 99.9999% DRE for dioxin and other extremely toxic substances.

3. Risk-based emission limits for carcinogenic metals (arsenic, beryllium, cadmium and hexavalent chromium) must meet the ratio of 1 out of 100,000.

4. Risk-based emission limits for noncarcinogenic metals (antimony, barium, lead, mercury, silver and thallium). must meet minimal levels.

5. Risk-based emissions limit for HCL is set to 20 ppm or less in the volume of flue gas.

6. A 100 ppm limit is required for CO emissions, corrected to 7% O_2 and based on a 60-minute average, in order to minimize the amounts of PICs.

The proposed regulations for boilers and furnaces cover 12 categories including:

1. Aggregate kilns, lime kilns, cement kilns, phosphate kilns.

2. Blast furnaces, halogen acid furnaces.

3. Smelting, melting and refining furnaces.

4. Coke ovens.

CURRENT PRACTICE AND TECHNOLOGIES

Incineration

Incineration is a process that employs decomposition via thermal oxidation at high temperature to destroy the organic waste. In 1981, EPA estimated that the annual hazardous waste generation was about 250 million metric tons (MMT) which was confirmed in separate studies by the Congressional Office of Technology Assessment (OTA) in 1983[3] and the Congressional Budget Office (CBO) in 1985.[4] Of this, approximately 47 MMT per year could have been incinerated. However, the CBO projected that only 2.7 MMT were actually incinerated in 1983. A 1981 EPA study also estimated that approximately 3.8 MMT of hazardous waste was disposed in over 1300 industrial boilers and furnaces.[5] These numbers are illustrated in Figures 1 and 2. This estimate does not include wastes generated from uncontrolled hazardous waste sites. The implementation of the HSWA (1984) land disposal restriction regulations and generator concern for long-term liability will result in increased utilization of incineration for ultimate disposal.

Thermal Destruction Systems

Three types of thermal destruction systems (incinerators, boilers, and furnaces) are discussed below.

1. Incinerators

Various types of incinerators are available for handling different physical forms of hazardous waste. An incinerator typically includes a primary and a secondary combustion chamber. Pollution control devices for reducing particulate, hydrogen chloride, sulfur oxides, metals and other emissions may be added. The most common types of incinerator designs are as follows:

a. Liquid injection

Liquid wastes are blended and then pumped into the combustion chamber through atomizing devices.

b. Rotary kiln

Rotary kilns generally consist of a rotating kiln and an afterburner. Afterburners are used to ensure complete combustion of flue gases before their treatment for air pollutants. The rotary kiln incinerator can generally be used for the destruction of solid organic hazardous waste.

c. Fixed hearth

Fixed hearth is also called a controlled air, a starved air, or a pyrolytic incinerator. The starved air condition causes most of the volatile fraction to be destroyed in the primary chamber with 50-80% of stoichiometric air and chamber temperature at 1200-1800F. The resultant smoke and pyrolytic products pass to the secondary chamber where excess air is injected to complete the combustion reactions. This type of incinerator generally has a smaller capacity than liquid injection or rotary kiln incinerators.

d. Fluidized bed

Fluidized beds are either circulating or bubbling bed designs. Operating temperatures are maintained in the 1400-1600F range and excess air requirements range from 100 to 150 percent. It is generally used for sludges or shredded solid materials. Fluidized bed incinerators offer high gas-to-solid ratios, high heat transfer, and uniform temperatures through out the bed.

e. Fume incineration

Fume incinerators are used to destroy gases or fume wastes. Wastes are injected by pressure or atomization through the burner nozzles.

2. Boilers and Furnaces

Boilers are constructed to produce steam. The concept of disposing of hazardous wastes in boilers has centered around industrial boilers. Wastes are cofired with conventional fuels in these boilers. Furnaces used include cement kilns, blast furnaces and smelters.

Air Pollution Control Devices (APCD)

Combustion gases generally need to be further treated in an air pollution control system. The presence of chlorine and other halogens in the waste requires a scrubbing or absorption step to remove HCl and other halo-acids. Particulate emissions require collection devices of moderate efficiency to meet the RCRA emission standard (0.08 grains/dscf). The most common system used for the air pollution control is a quench, followed by a venturi scrubber (particulate removal), a packed tower adsorber (acid gas removal) and a demister.

1. Wet Scrubber Systems

These may include spray towers, centrifugal scrubbers, and venturi scrubbers, etc.

a. Venturi scrubbers

Venturi scrubbers involve the injection of a liquid, usually water or a water/caustic solution, into the exhaust gas stream as it passes through the throat. The fine-atomized liquid entrains fine particles and a portion of the absorbable gases from the gas stream. They are reliable and simple to operate but they often require significant pressure drop across the throat (60-120 in. of water)[1] which represents a significant percentage of total cost of operation.

b. Packed bed scrubbers

These are vessels filled with packing material such as polyethylene saddles. The liquid is fed to the top of the vessel, with the gas flowing countercurrent to it. The liquid wets the packed material to remove the acid gas from the stack gases.

c. Plate tower scrubbers

They are similar to packed bed scrubbers, relying on absorption for the removal of contaminants. They are mostly used with liquid injection incinerators for absorption of soluble gaseous pollutants such as HCl and SOx. For rotary kiln or fixed hearth facilities with high ash feeds, venturi scrubbers are also used in series with packed bed scrubbers.

2. Dry Sorbent Injection (DSI)

Dry alkali sorbents are injected into the flue gas downstream of the combustor outlet and upstream of the particulate matter control device to form salts. The removal efficiency depends on flue gas temperature, sorbent type and feed rate, and the extent of sorbent mixing with the flue gas.

3. Wet Electrostatic Precipitation (ESP)

4. Ionizing Wet Scrubbers (IWS)

5. Fabric Filters

ESP, IWS, and fabric filters are used

for small particle removal.

6. Spray Dryer (SD)

Lime or limestone slurry is injected into the SD, the water in the slurry evaporates to cool the flue gas and the lime/$CaCO_3$ reacts with acid gases to form salts that can be removed by a particulate matter control device. The key design and operating parameters that affect SD performance are SD outlet temperature and lime-to-acid gas ratio.

Residuals and Ash Handling

Ash is frequently accumulated on-site prior to hazardous waste land disposal. It may be dewatered or may be chemically (fixation/stabilization) prior to disposal. Residuals generated by the APCD contain particulates, absorbed acid gases which have reacted to become salts and small amounts of organic contaminants. The suspended contaminants can be concentrated before disposal. The waters may either be returned to the process or treated and discharged to sewers as necessary.

PERFORMANCE TESTING OF THERMAL DESTRUCTION FACILITIES IN THE EARLY 1980's

EPA conducted performance testing at several thermal destruction facilities in the early 1980's. Complete test reports have been published for 8 incinerators, 11 industrial boilers, and 8 furnaces. These data as well as trial burn results from 14 RCRA applicants for incinerators have been summarized in an EPA report, "Permit Writer's Guide to Test Burn Data - Hazardous Waste Incineration."[7] Results of these studies (on the DREs, etc) have been summarized, and compared with the hazardous waste incinerator standards and proposed boiler and furnace regulations, by Dempsey and Oberacker of the USEPA in 1988.[5] Their comparisons are reproduced and shown in Figure 3 through Figure 16. It was concluded that properly designed thermal destruction systems equipped with suitable air pollution control devices can meet or perform better than the requirements set by the regulations.

RECENT STUDIES

Recent studies (1988) on hazardous waste cofired in industrial boilers under nonsteady or offset conditions by EPA have demonstrated that industrial boilers can provide adequate thermal environments for hazardous waste destruction, achieving an average DRE of 99.998% for RCRA toxic organics.[8,9] Results from a pilot-scale boiler cofiring test (1988) to investigate nonsteady effects on DRE have also revealed that unburned POHCs and PICs could be adsorbed on soot deposited on boiler surfaces during cofiring and desorbed back into the combustion gases after waste cofiring ceases, an effect which has been termed hysteresis.[10] The impact of this hysteresis on the DRE for POHCs was further tested in a full-scale hysteresis study on a watertube package boiler in 1990 by EPA to determine if the hysteresis effect actually exists, and if so, to evaluate its effect on DRE measurements. Results from POHC cofiring tests - Trichloroethylene (TCE) and Monochlorobenzene (MCB)- under sooting and nonsooting conditions indicated that DREs were generally low during the sooting operations, only three to four "nines". The hysteresis effect does exist: however, it does not significantly affect the DRE performance.[11] During 1988 and early 1989, other halogenated hydrocarbons wastes were incinerated.[12] Over 300,000 gallons of EDB (ethylene dibromide) pesticide were incinerated at commercial hazardous waste incineration facilities. Sulfuric acid was fired into the kiln to prevent the release of bromine to the atmosphere. EDB ($C_2H_4Br_2$) is a liquid halogenated hydrocarbon which was registrated as a pesticide in 1948 and is carcinogenic, mutagenic, and has adverse reproductive effects. The level of bromine in the stack gas was non-detectable and particulate emissions were 0.0081 to 0.0123 grains/dscf corrected to 7% CO_2. Figure 17 shows that the DREs of the EDB were at least 6 nines. It is obvious that properly operated incinerators are able to destroy a wide range of hazardous wastes.

CONCLUSIONS

Incineration is capable of the highest degree of waste destruction and is able to destroy the broadest range of hazardous waste. It can greatly reduce the solid waste volume, recover heat energy from the combustion process, and permanently destroy the waste. If the incinerators are properly designed and operated, incineration remains the most efficient and available technique for disposal of most organic wastes. More research is needed to study on how to reduce the trace amounts of unburned hazardous wastes, products of incomplete combustion, haloacids, halogen gases, metals and particulates in stack gases and residuals in the ash.

REFERENCES

1. E. T. Oppelt, "Incineration of Hazardous Waste: A Critical Review," the Journal of the Air Pollution Control Association (JAPCA), Vol., No. 5, 1987.

2. U.S. EPA, "Dioxin Rule," Federal Register, January 14, 1985.

3. U.S. EPA, "Standards Applicable to Owners and Operators of Hazardous Waste Treatment Facilities; Interim Final Rule and Proposed Rule," Federal Register 47(122, Part V):27516-27535, June 24, 1982.

4. U.S. Congress, Congressional Budget Office, "Hazardous Waste Management: Recent Change and Policy Alternatives,"U.S. Governmental Printing Office, 1985.

5. C. Dempsey and D. Oberacker, "Overview of Incineration Performance," Presented at the Engineering Foundation's Conference on "Hazardous Waste Management Technologies," Mercersberg, Pennsylvania, August 7-12, 1988.

6. U.S. EPA, "Polychlorinated Biphenyls (PCBs) Manufacturing, Processing, Destruction in Commerce, and Use Prohibition," Federal Register, Vol. 52, No. 87, 1987.

7. U.S. EPA, "Permit Writers Guide to Test Burn Data - Hazardous Waste Incineration," EPA-625/6-86/012, 1986.

8. M. Wool, C. Castaldini, and H. Lips, "Engineering Assessment Report: Hazardous Waste Cofiring in Industrial Boilers Under Nonsteady Operating Conditions," Acurex Summary Report TR-86-103/ESD, U.S. EPA/RREL, Cincinnati, Ohio, July 1989.

9. H. B. Mason, et al. "Pilot-Scale Boiler Cofiring Tests to Investigate Nonsteady Effects," Proceedings of the 14th Annual EPA Research Symposium on Land Disposal, Remedial Action, Incineration and Treatment of Hazardous Waste, Cincinnati, Ohio, May 9-11, 1988, EPA/600/9-88/021, 7/88, pp.332-345.

10. H. B. Mason, J. A. Nicholson, M. Chan, R. J. Derosier, and R. Gale, "Pilot-Scale Testing of Boiler Waste Cofiring-Hysteresis Effects," Midwest Research Institute report, U.S. EPA, ORD, EPA Contract No. 68-03-3241, August 1988.

11. G. D. Hinshaw, S. W. Klamm, G. L. Huffman and P. C. L. Lin, "Sorption and Desorption of POHCs and PICs in a Full-Scale Boiler Under Sooting Conditions", Presented at the 16th Annual research Symposium on Hazardous Waste, Cincinnati, Ohio April 3-5, 1990.

12. D. Oberacker and C. Stangel, "Incinerating Ethylene Dibromide and Dinoseb Stocks," Presented at the U.S. EPA's 15th Annual Research Symposium on Remedial Action, Treatment and Disposal of Hazardous Waste, Cincinnati, Ohio, April 10, 1989.

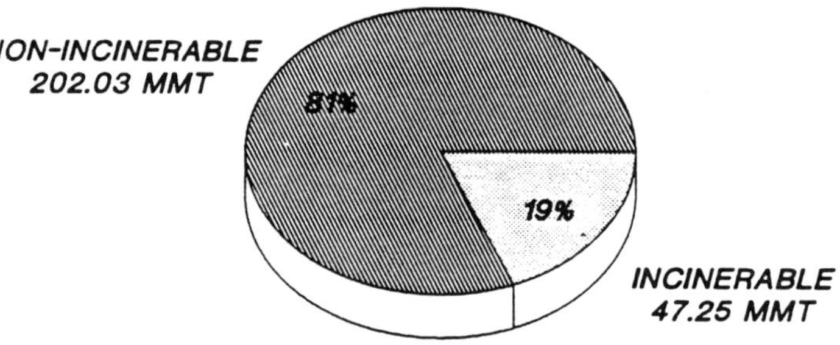

Fig. 1. Annual Hazardous Waste

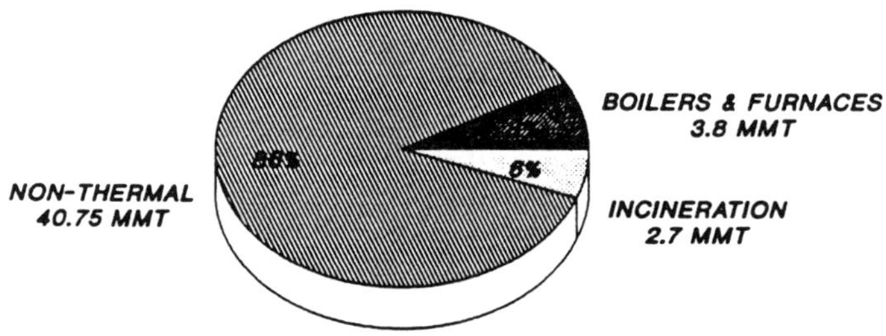

Fig. 2. Incinerable Hazardous Waste

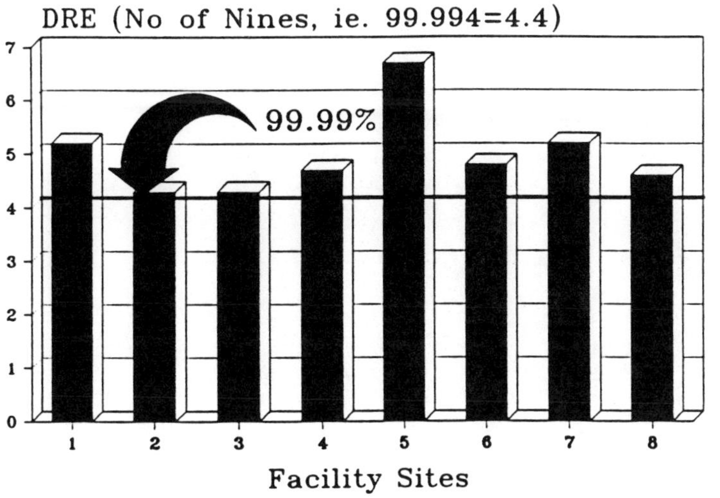

Fig. 3. Avg. DREs from 8 EPA Tested Incinerators

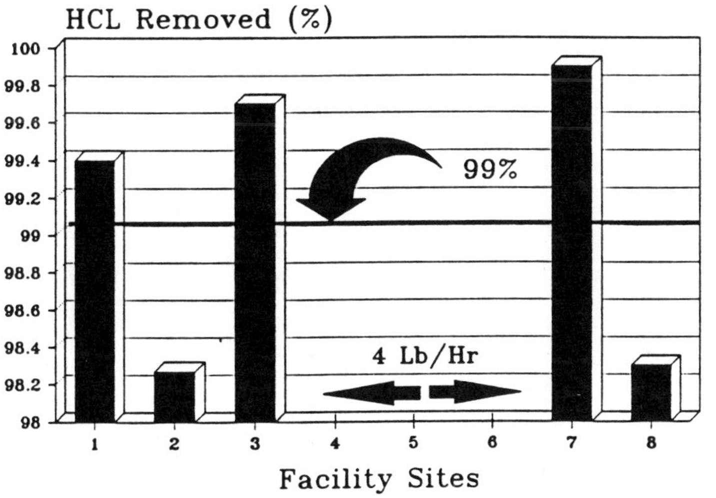

Fig. 4. Avg. HCL Removal from 8 EPA Tested Incinerators

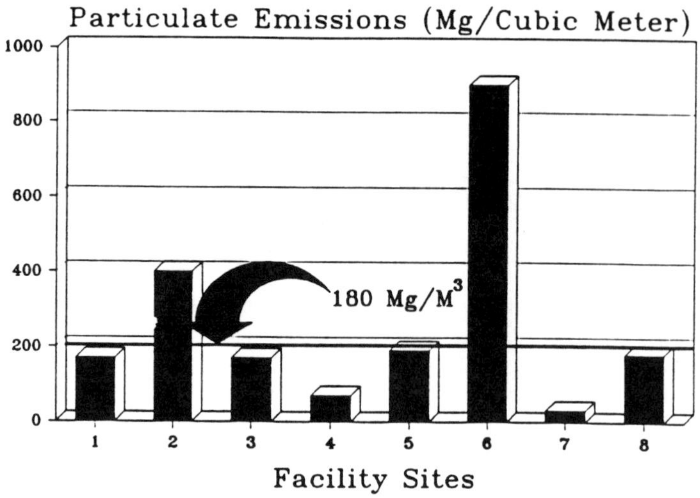

Fig. 5. Avg. Particulates from 8 EPA Tested Incinerators

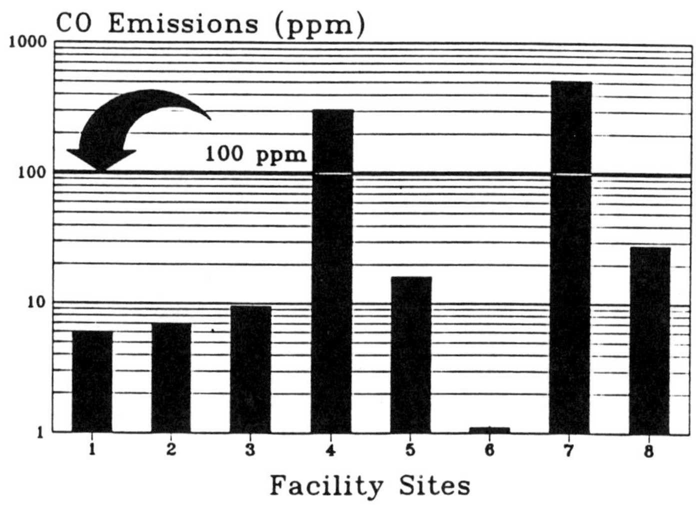

Fig. 6. Avg. CO Emissions from 8 EPA Tested Incinerators

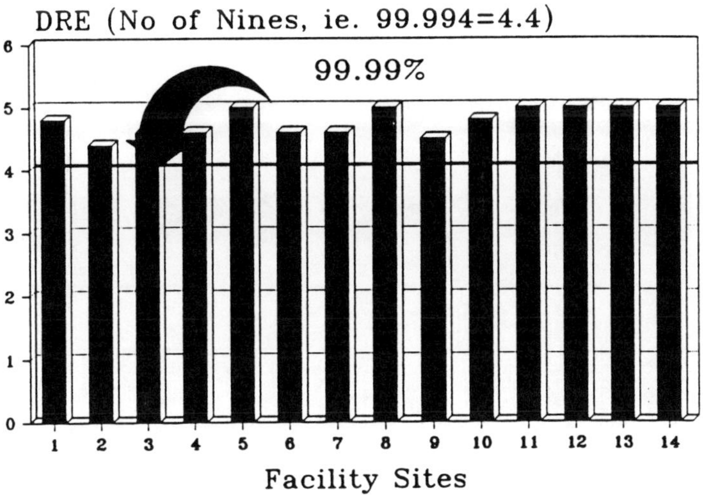

Fig. 7. Avg. DREs from 14 RCRA Applicants

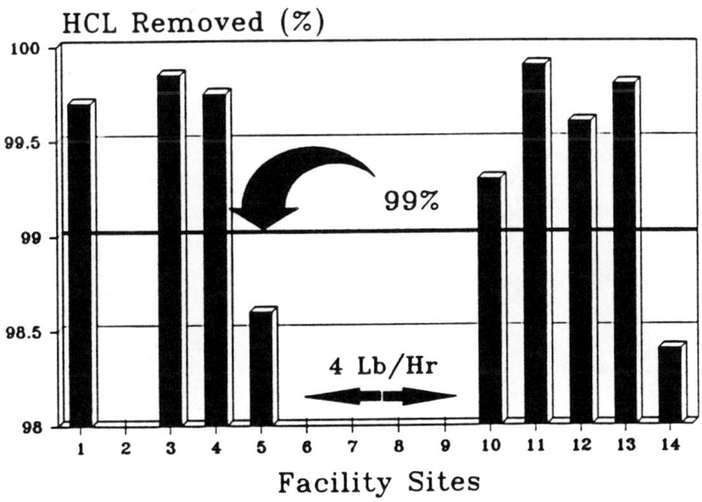

Fig. 8. Avg. HCL Removal from 14 RCRA Applicants

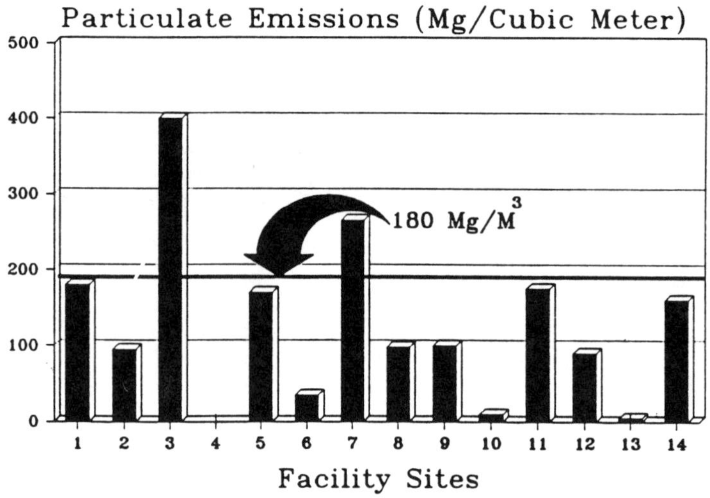

Fig. 9. Avg. Particulates from 14 RCRA Applicants

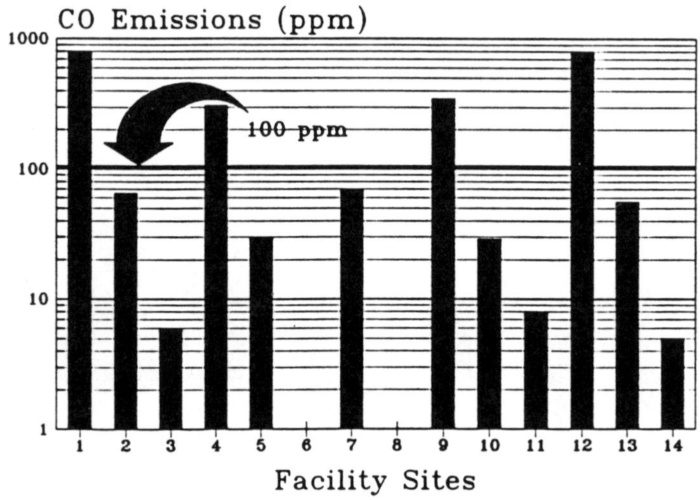

Fig. 10. Avg. CO Emissions from 14 RCRA Applicants

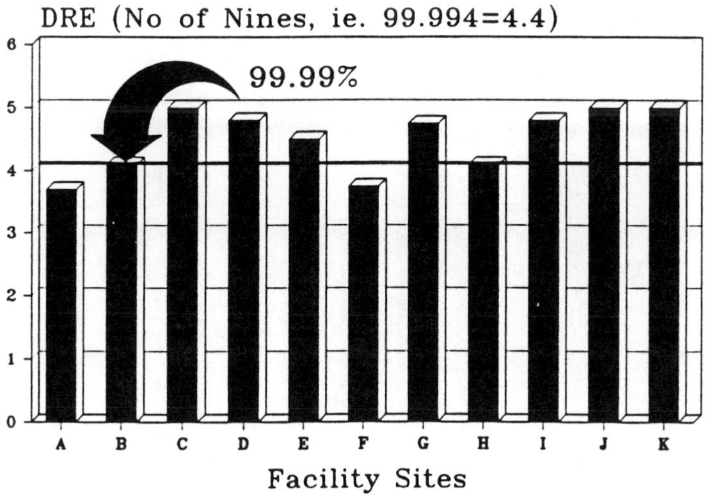

Fig. 11. Avg. DREs from 11 Boilers

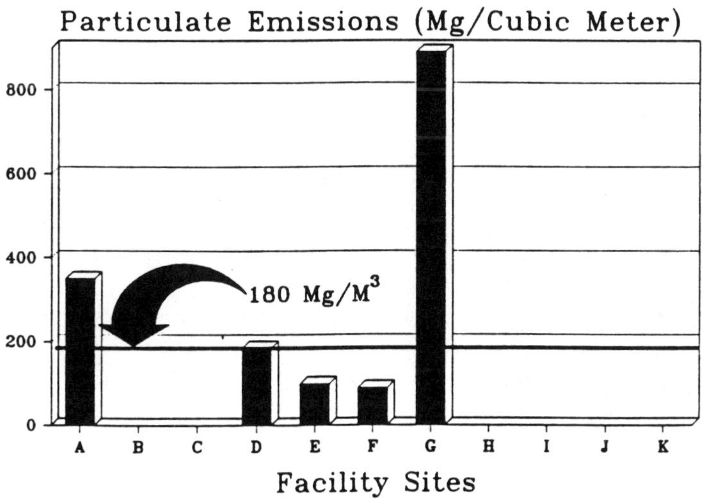

Fig. 12. Maximum Particulates from 11 Boilers

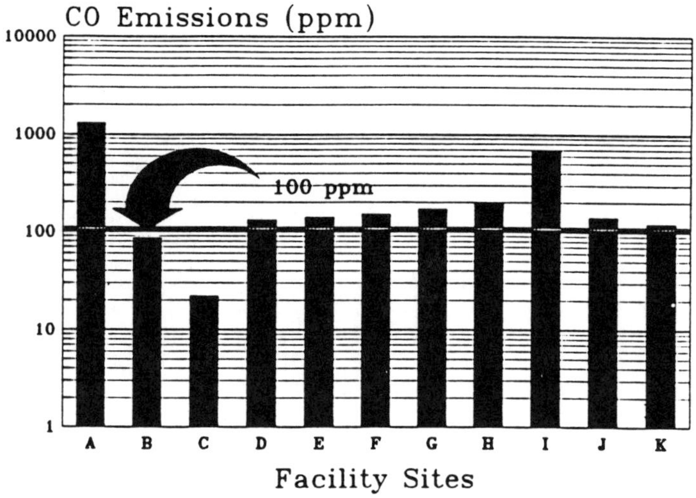

Fig. 13. Maximum CO Emissions from 11 Boilers

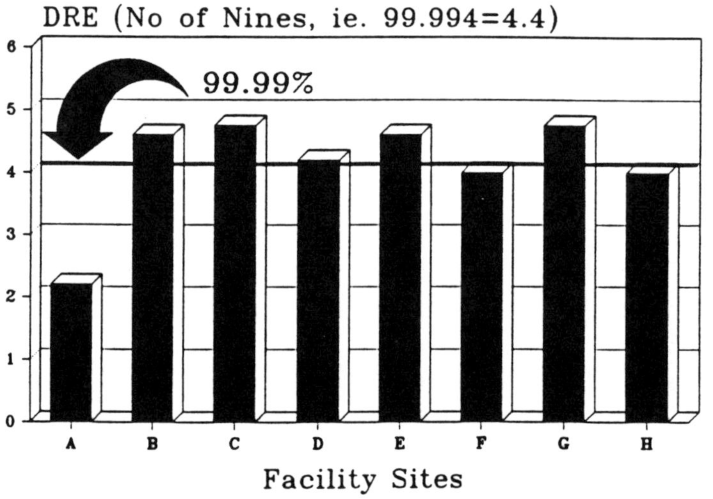

Fig. 14. Avg. DREs from 8 Industrial Furnaces

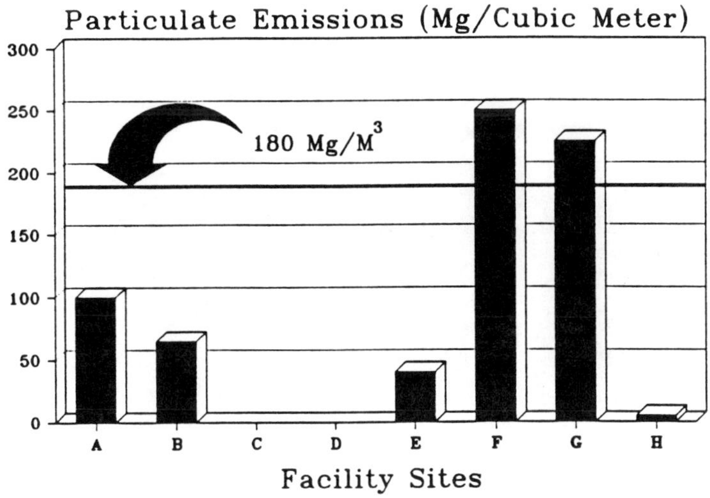

Fig. 15. Avg. Particulates from 8 Industrial Furnaces

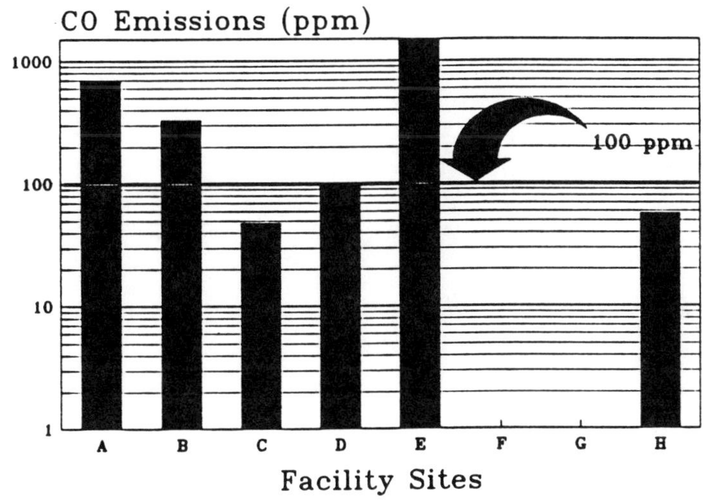

Fig. 16. Maximum CO Emissions from 8 Industrial Furnaces

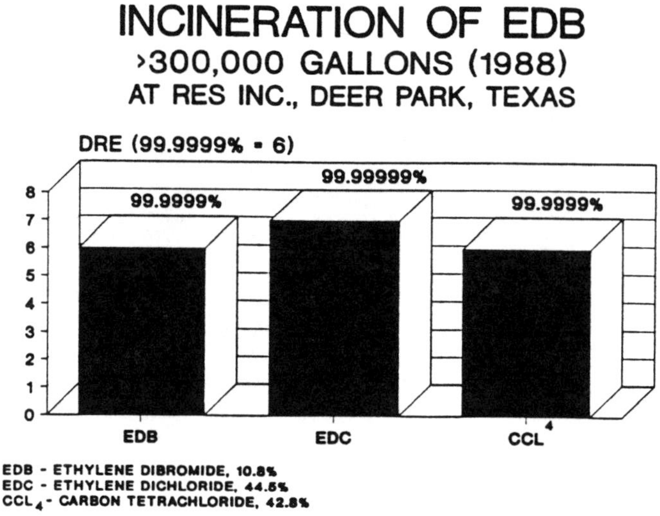

Fig. 17. Incineration of EBD

CHEMICAL OXIDATION TREATMENT OF INDUSTRIAL ORGANIC WASTE

Penny M. Wikoff and Dan F. Suciu
Environmental Research and Development, Inc.
Idaho Falls, Idaho

ABSTRACT

This paper presents chemical oxidation methods including oxidation with hydrogen peroxide, oxidation with potassium permanganate, Wet Air Oxidation, and ozonization. The benefits and limits are presented. Process demonstration and costs are given where available.

INTRODUCTION

With the increased emphasis for use of nontoxic, nonchlorinated solvents for cleaning, degreasing, or depainting operations and the increased cost for disposal of the spent solvents, an increased emphasis has been placed on recovery and/or treatment of the spent solvents. Recovery becomes more difficult because many of the solvents are not single component systems and in some cases one or more of the components may be consumed or degraded during use. Methods available to treat the spent solvents include chemical oxidation, biological oxidation and adsorption processes such as granular activated carbon or ion exchange. The optimum method is dependent on the process, process discharge requirements, volume to be treated, the concentration of the spent solvent, and whether the facility has an existing industrial waste treatment plant (for example one with an activated sludge system for treating the organics).

The focus of this paper will be chemical oxidation treatment processes. In chemical oxidation processes, the toxic organic compounds are oxidized to less toxic or non-toxic organic compounds or carbon dioxide. The chemical oxidation methods include oxidation with hydrogen peroxide in the presence of a catalyst, oxidation with potassium permanganate, incineration, Wet Air Oxidation, and ozonization. Incineration will not be discussed in this paper since it is being presented by Dr. Philip C. L. Lin in this proceeding.

OXIDATION WITH HYDROGEN PEROXIDE

Oxidation of organic waste with hydrogen peroxide (H_2O_2) is usually a slow process. However in the presence of a catalyst, the oxidation process proceeds rapidly.[1] Catalysts include ferrous iron, copper, aluminum, and chromium. Ferrous iron is a good selection since it is not listed as a heavy or toxic metal and its use and subsequent precipitation does not create further hazardous waste problems. Hydrogen peroxide in combination with a ferrous iron salt is commonly referred to as Fenton Reagent. This reagent acts as an oxidizing agent through the formation of hydroxyl free radicals:[1]

$$Fe^{+2} + H_2O_2 \rightarrow Fe^{+3} + OH^- + \cdot OH \quad (1)$$

The Fenton Reagent can act as an oxidizing agent for compounds such as alcohols, ketones, benzene, and phenols. The reaction with phenol proceeds as shown in Reaction 2.[1]

Phenol → Catechol → o-Quinone → Muconic Acid (2)

There are no reports in the literature indicating the oxidation of phenol proceeds beyond muconic acid. Optimum oxidation of phenol is achieved when 1 mole of phenol is treated with 1 mole of ferrous salt and 3 moles of hydrogen peroxide. The reaction is unaffected by pH in the range 3 to 7. However, in the presence of acetate or phosphate buffers the reaction proceeds more slowly. With a greater degree of substitution, the reaction also proceeds more slowly, particularly when the substitution is in the ortho and para position. Halogenated phenols are oxidized rapidly with the reaction rate decreasing according to the following order: Cl > Br > I.[1]

The formation of intermediate products (semiquinones) produces a dark brown color. At the completion of the reaction, the solution is a yellow or orange color due to the presence of the ferric iron. The ferric iron can be precipitated by increasing the pH either with lime or caustic. Alum can be added to aid in floc formation. There is some co-precipitation of the remaining organics. Depending on the effluent discharge requirements, further treatment may be required to reduce the chemical oxygen demand (COD) to discharge limits. The effluent, however, may be applicable for discharge to a domestic sewage treatment plant or biological industrial waste treatment plant.

The Fenton Reagent has been used to treat stripped refinery effluent and steel plant effluent. The stripped refinery effluent had an initial COD due to phenol of 280 mg/L with a total initial COD of 970 mg/L. (Note: Each mg/L of phenol (C_6H_6O) is equivalent to 6 mg/L of COD.) Nine moles of hydrogen peroxide and 1 mole of ferrous ammonium sulfate was added per 1 mole of phenol. the COD remaining due to phenol after reaction at pH 4 and 50°C for 30 minutes was 2 mg/L. The total COD remaining was 246 mg/L.[1]

CHEMICAL OXIDATION WITH POTASSIUM PERMANGANATE

Certain naturally occurring organic refractories or residual organics can be readily oxidized by potassium permanganate.[2] Functional groups are critical in determining whether or not a compound can be oxidized by potassium permanganate. The carboxyl group, carbonyl group, and hydroxyl group of alcohols resist oxidation, while the carbonyl group of aldehydes, the amino group of amines and the carbon-carbon double bond of unsaturated compounds are readily oxidized by potassium permanganate. However oxidation does not proceed beyond the acid.[2]

Aromatic compounds require greater concentration of potassium permanganate. Approximately 9.3 moles of potassium permanganate were required to oxidize one mole of phenol to malaic acid. The reaction proceeds as follows:[2]

Phenol → Hydroquinone → p-Quinone → Malaic Acid (3)

The wastewater requires further treatment to precipitate the manganese. Depending on discharge requirements, further organic treatment may be required to further reduce the COD.

WET AIR OXIDATION

Wet Air Oxidation can effectively treat aqueous waste which is too dilute for cost effective incineration but too concentrated (too toxic) for biological treatment.[3-9] Wet Air Oxidation is the aqueous phase oxidation of organic and inorganic materials at elevated temperatures (347 to 608°F) and pressures (300 to 3000 psig). There is enhanced solubility of oxygen in aqueous solutions at these temperatures and pressures. The process relies on the heat of oxidation to raise the temperature to the required operating level. The incoming waste stream is typically preheated using a counter flow heat exchanger and excess heat from the treated stream. The energy required is the difference in enthalpy of the two streams, the energy required to heat for start up, the energy required to operate the air compressor. For the reaction to be self-sustaining (no auxiliary reaction fuel) the initial COD must be at least 15,000 mg/L. (The conditions required for autogenous incineration are 1500 to 2000°F and 300,000 to 400,000 mg/L COD.)[3]

The required oxygen is provided by compressed air. The oxidation products are primarily carbon dioxide and water. Sulfur is converted to sulfate and nitrogen to ammonia. These with the halogens stay in the aqueous phase. Metals remain in the aqueous phase and can be precipitated prior to discharge. The heat from the treated waste stream can be used to generate steam.

Halogenated aromatic compounds without other non-halogenated functional groups are relatively resistant to Wet Air Oxidation. Electron donating constituents such as hydroxyl, amino, or methyl groups make aromatic rings more susceptible to destruction by Wet Air Oxidation.

Wet Air Oxidation has been demonstrated by MODAR on the pilot scale at CECO's International, Inc., Niagara Falls, New York, Hazardous Waste Treatment and Disposal Facility.[5,6] The process was demonstrated using a dilute isopropyl alcohol stream contaminated with priority pollutants including 1,1,2-trichloroethane, nitrobenzene and 2-chlorophenol and a dielectric fluid containing polychlorinated biphenyls (PCBs). The pilot unit had an organic material flow capacity of 50 gal/day. Greater than 99.99 percent destruction of the organics was demonstrated.

Wet Air Oxidation has been demonstrated at the pilot level by several other companies, including Zimpro in Rothschild, Wisconsin.[7] Zimpro treated sewage sludge and industrial waste sludges from paper and textile mills. Approximately 3400 kw-hr/day power was required to treat 10,000 lbs of sludge containing 70 grams/liter of COD to 50 percent reduction in COD.

Full-scale operations were not found in the literature. However, MODAR is working with ABB Lummus Crest to scale-up the process and expects full-scale demonstration in the near future. An estimate of the capital equipment cost for a 10,000 gal/day waste stream containing sufficient organics to be self-sustaining was obtained from ABB Lummus Crest. The estimated capital cost was $15 million.[8]

A Wet Air Oxidation demonstration unit (10,000 to 12,000 gal/day) has been in operation for 4 years in Mississauga, Ontario, Canada. The unit has been treating waste containing up to 250,000 mg/L COD, reducing the COD by 89 percent. Oils and greases are reduced by 99 percent, naphthalene by 99 percent, cresol by 94 percent, polyethylene glycol by 98 percent, and trimethylbenzene by 99 percent.[9]

OZONIZATION OF INDUSTRIAL ORGANIC WASTE

If sufficient ozone is added, essentially all organic compounds can be oxidized to carbon dioxide.[10-15] The source of the ozone can either be by electrical production with an ozone generator or through use of ultraviolet light to convert the oxygen available in the waste stream to ozone. The use of an ozone generator, although requiring more initial capital cost, is a more effective method of treatment. Generation with ultraviolet light is limited by the solubility of oxygen in the water.[10,11] The ozone generator processes either air, oxygen enriched air or pure oxygen into ozone. The ozonization process can be enhanced by the addition of hydrogen peroxide to the waste stream.

Complete and efficient ozonization is dependent on the type of waste, the degree of oxidation desired, the reactor or contactor configuration, the contact time. Contactor configurations may include bubble towers or packed columns. The presence of carbonates may inhibit ozonization at high pH.

Approximately 2 to 6 mg/L ozone are required per 1 mg/L ozone. Solutions with higher initial phenol concentration react more rapidly. The ozone demand to oxidize phenol at a pH of 12 is half that required at a pH of 7.[12]

The degree of ozonization desired is dependent on what process may follow the ozonization process. For example, if a biological treatment process is available, it may be desirable to only treat the waste stream with ozone to degrade the organics to the level where they are not toxic to the biological system or can be degraded to discharge limits by the biological system. Ozonization of high-molecular weight biorefractory, organic compounds (humic substances) produces lower-molecular weight, more biodegradable substances.[13] At the same time more oxygen is added to the waste stream for biological treatment.

Treatment of ground water contaminated with trichloroethylene (TCE) and tetrachloroethylene (PCE) has been demonstrated on a 2000 gpm unit.[14,15] The groundwater was contaminated with 200 μg/L TCE and 20 μg/L PCE. In order to achieve 80 to 90 percent destruction of the contaminants, 4 mg/L ozone was required with a ratio of hydrogen peroxide to ozone of 0.5 by weight. The capital cost for the ozone generator was $85,000. Hydrogen peroxide cost was approximately $1.00/lb.[15]

CONCLUSIONS

Complete destruction of the toxic organics to carbon dioxide by chemical oxidation is only achieved with Wet Air Oxidation and ozonization. Wet Air Oxidation is efficient for those concentrated wastes having at least 15,000 mg/L COD. The Wet Air Oxidation process has not been demonstration full-scale, although MODAR expects demonstration in the next year. Ozonization is effective and economical, especially with lower concentrations of organics. Table 1 lists a cost comparison of the more common treatment processes for organics.[15]

REFERENCES

1. Eisenhauer, Hugh R., "Oxidation of Phenolic Wastes," Journal of Water Pollution Control Federation, 40, 11, November 1968, pp. 1887-1899.

2. Spricher, R. G. and Skrinde, R. T., "Effects of Potassium Permanganate on Pure Organic Compounds," Journal of American Water Work Association, April 1965, pp. 472-484.

3. Teletzke, G. H. et al., "Components of Sludge and its Wet Air Oxidation Products," Journal of Water Pollution Control Federation, 39, 6, June 1967, pp. 994-1004.

4. Dietrich, M. J., Randall, T. L., and Canny, P. J., "Wet Air Oxidation of Hazardous Organics in Wastewater," Environmental Progress, 4, 3, August 1985, pp. 171-177.

5. Swallow, Kathleen C. et al., "The MODAR Process for the Destruction of Hazardous Organic Wastes-Field Test of a Pilot-Scale Unit," Waste Management, 9, 1989, pp. 19-26.

6. Staszak, Carl N. and Malinowski, Kenneth C., "The Pilot-scale Demonstration of the MODAR Oxidation Process for the Destruction of Hazardous Organic Waste Materials," Environmental Progress, 6, 1, February 1987, pp. 39-43.

7. Teletzke, G. H., "Wet Air Oxidation," Chemical Engineering Progress, 60, 1, January 1964, pp. 33-38.

8. Evans, Brian, ABB Lummus Crest, Houston, TX, Telephone Conversation, February 15, 1991.

9. "Wet Air Oxidation Process Operates at Low Pressures, Company Says,

"HazTech News, 5, 21, October 18, 1990, pp. 159-160.

10. Sargent, J. W. And Sanks, R. L., "Light-Energized Oxidation of Organic Wastes,", Journal of Water Pollution Control Federation, 46, 11, November 1974, pp. 2547-2554.

11. Otake, Tsutao et al., "Photo-Oxidation of Phenols with Ozone," Journal of Chemical Engineering of Japan, 12, 4, 1979, pp. 289-295.

12. Niegowski, S. J., "Destruction of Phenols by Oxidation," Industrial and Engineering Chemistry, 45, 3, 1953, pp. 632-634.

13. Jones, Bonnie M., Sakaji, Richard H., and Daughten, Christian G., "Effects of Ozonization and Ultraviolet Irradiaton on Biodegradability of Oil Shale Wastewater Organic Solutes," Water Resource, 19, 11, 1985, pp. 1241-1428.

14. Glaze, William H. and Kang, Joon-Wun, "Advanced Oxidation Processes for Treating Groundwater Contaminated with TCE and PCE Laboratory Studies," Journal of American Water Work Association, Research Technology, May 1988, pp. 57-63.

15. Aieta, E. Marco et al., "Advanced Oxidation Processes for Treating Groundwater Contaminated with TCE and PCE: Pilot-Scale Evaluations," Journal of American Water Work Association, Research and Technology, May 1988, pp. 64-72.

TABLE 1. TREATMENT COST COMPARISON[15]

Cost Type	Air Stripping	Air Stripping Gas Phase GAC	Liquid Phase GAC	Peroxide Ozone AOP
Capital Cost - $(Annualized Over 10 Years)	$ 48,400	$108,000	$192,500	$ 35,000
Operating and Maintenance Costs - $	$ 30,200	$ 78,300	$ 13,900	$ 63,900
GAC Replacement Cost - $		$105,100	$210,000	
Total Annualized Cost - $	$ 78,600	$291,400	$416,800	$ 98,900
Cost Per 1000 GAL $	(0.075)	(0.277)	(0.397)	(0.094)

MEDIATED ELECTROCHEMICAL OXIDATION OF ORGANICS

Leonard W. Gray, Robert G. Hickman, and Joseph C. Farmer
Lawrence Livermore National Laboratory
Livermore, California

INTRODUCTION

Replacement of halogenated with non-halogenated and toxic with non-toxic solvents is a worthy goal being sought within the Department of Energy (DOE) and Department of Defense (DOD) complexes. Today, however, the substitutions that would be suitable are not all identified. Consequently, there is continuing production of both hazardous waste and mixed (hazardous radioactive) waste. Furthermore, existing back-log inventories of halogenated solvents at many sites must be addressed, probably through some kind of destruction technology. Commercial technology exists for destruction of very dilute solutions of halocarbons in water. We are seeking non-incineration technologies suitable for destroying concentrated streams of various toxic organics.

OPERATING PRINCIPLES

Some metal ions, when dissolved in nitric or some other mineral acid, can have their valence state raised at the anode of an electrochemical cell. One such metal ion is Ag^+ which can be raised to Ag^{++} without significant oxidation of water at the anode. This is particularly true if the anode is made of a material that exhibits a reasonably high oxygen bubble overpotential, such as Au or Pt. Other examples of metal ions that may be raised to higher valence states and used as chemical mediators are Ce^{+++} and Fe^{+++}. Of these, silver is the most thoroughly studied.(1)

The Ag^{++} formed at the anode almost immediately forms a complex with a nitrate ion, $AgNO_3^+$ which is very strongly absorbing spectroscopically through a broad range of the visible spectrum. Formation of the complex does not significantly reduce the oxidizing power of the Ag^{++}. The complex readily attacks water to form OH free radicals and H^+, at which time the silver is reduced to its original state, Ag^+. A constant replenishment of Ag^{++} ions is accomplished by continuously circulating the electrolyte past the anode. Hydroxyl free radicals are themselves powerful oxidizers, particularly with respect to hydrocarbon molecules. In the case of ethylene glycol, the major component of antifreeze, $(CH_2OH)_2$ reacts with both $AgNO_3^+$ and OH (from water) to form CO_2, H^+, NO_3^-, and Ag^+.

On the cathode side of the electrochemical cell, NO_2 and water are produced by reduction of nitrate ion. If the NO_2 is regenerated with oxygen in a separate reactor to produce nitrate ion (which is recycled to the cell), then the overall reaction is for ethylene glycol to be converted to CO_2 and H_2O. Since the only chemical added is the O_2 used to regenerate the nitrate ion, the stoichiometry is identical to combustion, i.e. $2(CH_2OH)_2 + 5O_2 = 4CO_2 + 6H_2O$.

The differences between the process just described and combustion are obvious. One is done in an aqueous electrolyte and requires an electrochemical cell that operates near ambient conditions. Normal combustion operates near ambient pressure but at very high temperatures. Residence times in combustion are quite short. That gives some people concern about stack gas emissions, whereas the residence time in this aqueous system is quite long, and complete destruction of the starting material can be verified before anything is discharged to the environment.

LAB SCALE EXPERIMENTS

The principles described above have been checked in many small experiments. They have been used to develop several different parts of the demonstration bench-scale system that was eventually built. For example, at the anode of the electrolytic cell, it is easy to evolve molecular oxygen if the anode potential is raised too high. Scanning voltammetry using a computer controlled potentiostat allowed us to determine the maximum anode potential that could be used to produce Ag^{++} but avoid oxygen evolution, a process inefficiency. Using rotating disc electrode methods, diffusion coefficients for Ag^{+} in the electrolyte were determined so the demonstration cell could be optimized with regard to Ag^{++} production. The reaction between $AgNO_3^{+}$ and water to form OH and eventually O_2 was followed spectrophotometrically. This helped determine the way that OH attacked hydrocarbon molecules. It also provided important information that was incorporated into the numerical model of the system. Kinetics of destruction of ethylene glycol were measured and a proposed mechanism was developed.(2) It should be applicable to ketones, aldehydes, alcohols, organic acids, and the starting parent hydrocarbon such as octane. This is shown in Figure 1. Various other organic compounds also were destroyed, such as tetraphenylborate and benzene.

MODELING

Numerical models were developed for the developmental scale electrochemical cell. It is an annular design with a porous ceramic separator between the gold anode and the stainless steel cathode. Because we had measured or otherwise obtained the physical properties of the electrolyte and had measured the reaction kinetics of Ag^{++} with both water and ethylene glycol, the model was able to predict how fast a given amount of glycol could be destroyed as functions of cell dimensions, electrolyte temperature and concentrations, electrolyte flow rate, and potential. A schematic diagram is shown in Figure 2. Parametric runs were done on a small computer to calculate both the optimum cell design as well as predict the system's ability to destroy antifreeze.

DEMONSTRATION

The demonstration hardware, which included not only the electrochemical cell but other items as well, was installed in a gloved enclosure ventilated with untreated room air. The other hardware items included a demister for each side of the electrochemical cell. The cathode side allowed NO_2 to be removed without entraining liquid droplets as the gas vented to the stack. Similarly, the anode side had a comparable unit to vent the CO_2 produced and any O_2 that may have been produced. Temperature was held at about 70°C using electrical heaters immersed in the electrolyte. Each side of the cell, of course, had its own circulation pump. We chose a magnetically coupled centrifugal design, with wetted surfaces made entirely of polyvinylidene fluoride. Piping and values were of the same material or else either teflon or titanium.

The results of the demonstration were very gratifying. We had determined from the numerical model that we could destroy 0.5L ethylene glycol in 22 hours. The destruction rate would be limited by the rate at which the electrochemical cell could produce Ag^{++}. During the run, CO_2, CO, and O_2 were monitored on-line in the off-gas from the anode loop. The first two used infrared detectors and O_2 used an electrochemical method. Only CO_2 and O_2 were detected. This was later confirmed by grab samples of gas analyzed by mass spectroscopy. We also monitored $AgNO_3^{+}$ on-line in the anolyte. This was done by absorption spectroscopy using fiber optics to connect the probe to the spectrophotometer. At 20 hours, based on negligible CO_2 evolution, increased O_2 evolution, and a high concentration of $AgNO_3^{+}$ being maintained in the anolyte, the run was terminated. Using off-line total organic carbon analysis, it was determined that we had destroyed more than 99.9% of the ethylene glycol in the anolyte. The numerical model had predicted the time for destruction within 10% of the actual time, which is excellent agreement for this kind of developmental work.

SUMMARY AND FUTURE DIRECTIONS

The demonstration of the technology on this scale gives us confidence in both our ability to predict the systems performance, and also gives us some confidence in the destruction mechanism at work in the electrolyte. Other classes of organic compounds are under study, such as halogenated species. Also, destroying 0.5L in 20 hours is a scale far too small to be of practical value. A new system is being designed and constructed that has a capacity expandable to about 120 times larger than our initial demonstration system.

REFERENCES

1. D. Steele, "Electrochemical destruction of toxic organic industrial waste," Plat. Met. Rev., 34 10 (1990).

2. J. Farmer, et.al., "Initial study of the complete mediated electrochemical oxidation of ethylene glycol", Lawrence Livermore National Laboratory, Livermore, CA. Report No. UCRL-106479 (March, 1991).

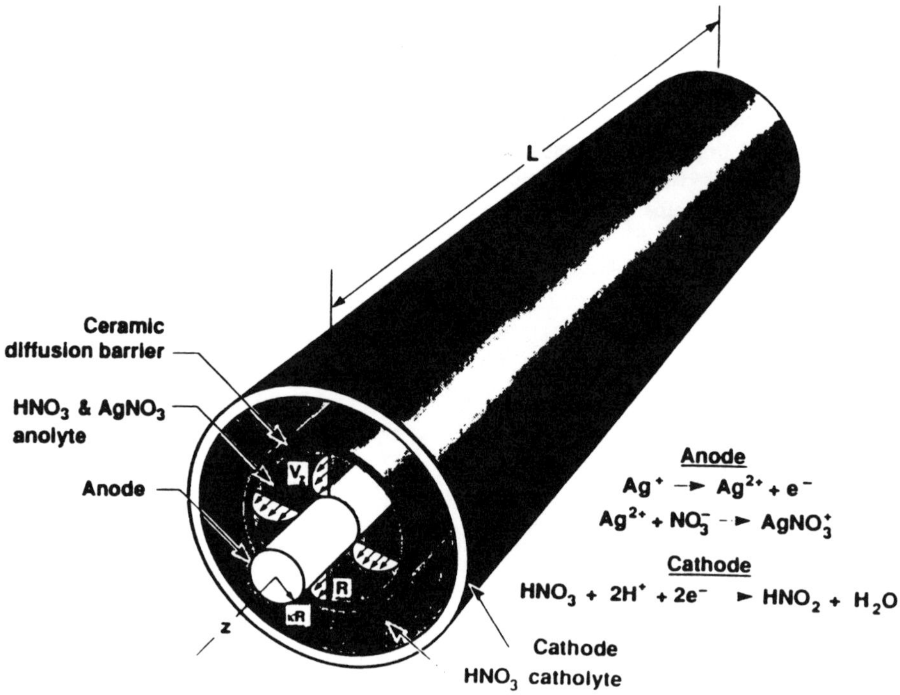

Figure 1

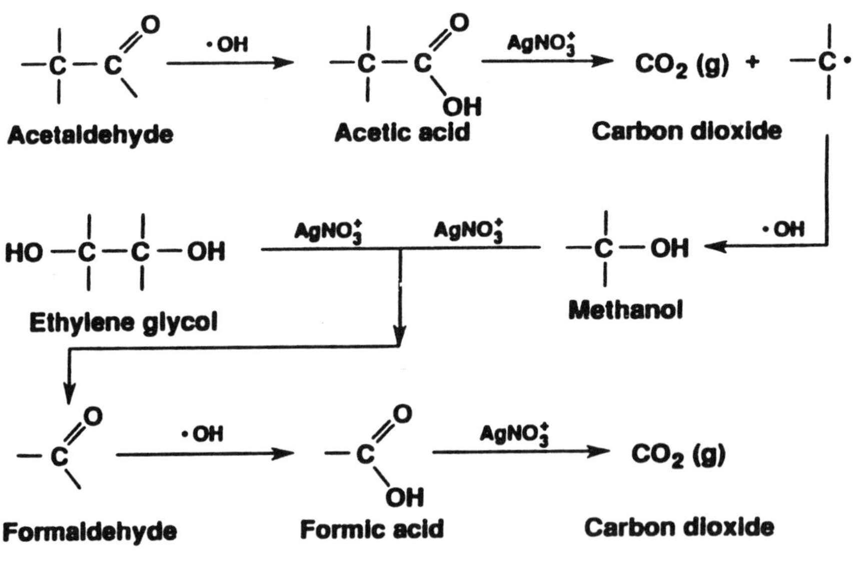

Figure 2

TOWARDS A PROTOCOL TO DETERMINE WASTE MANAGEMENT PROPERTIES OF SOLVENT SUBSTITUTES

Benerito S. Martinez Jr., Meei-Huey Li and Ricardo B. Jacquez
Department of Civil, Agricultural, and Geological Engineering
and
Walter H. Zachritz II
Southwest Technology Development Institute
New Mexico State University
Las Cruces, New Mexico

Martha I. Beach
N-CON Systems
Larchmont, New York

INTRODUCTION

Halogenated solvents such as trichlorethylene (TCE), freon, and methylene chloride, are routinely used in a wide variety of degreasing and cleaning operations (Higgens, 1989). Spent halogenated solvents released from process washwater or in concentrate can adversely impact industrial wastewater treatment plants and industrial pretreatment programs (Breton, 1988). Non-halogenated solvent substitutes with positive environmental attributes are being strongly considered for many industrial operations. However, solvent substitutes may suffer from contaminant carryover unknown impact on the industrial wastewater treatment plant such as resistance to biodegradation and physical/chemical interaction with waste components. The number of commercial products marketed as substitutes for halogenated solvents in industrial processes is rapidly increasing. Very little is known about the impact of these specific solvent formulations on potential generation of volatile organic compounds (VOC) emissions or impact on existing wastewater treatment facilities. While some methods for assessing waste management characteristics of substitute solvents have been described, a standard procedure that adequately characterizes the major environmental factors is needed. A standard method for determining waste management properties would provide broad guidelines for manufacturers and allow for more detailed disposal information in MSD sheets. At a minimum the proposed standard testing procedure should include volatilization and biodegradability as basic test procedures. The purpose of this paper is to present a preliminary protocol for the rapid determination of volatilization and biodegradability attributes of solvent substitutes.

PROTOCOL FOR VOLATILIZATION

Several methods are available to determine volatilization potential. The majority of these methods are based on ASTM standards for coatings and paints and are summarized under Environmental Protection Agency Method 24. The volatilization of a compound can have far ranging affects on the treatability and treatment control technologies. Assessment of volatilization should encompass attributes of both the solvent tested and the wastewaters receiving the solvent. Changes in pH, temperature variations or suspended solids can all impact volatilization rates of a particular compound in a given wastewater. The protocol presented here is designed to characterize the basic volatilization properties of a terpene-based solvent and evaluate the influence of pH. Additional modifications to this procedure can be used to systematically evaluate the presence of suspended and dissolved organic material or other waste parameters on volatilization.

To determine basic volatilization rates, the substitute solvent was diluted to the expected use rate concentration with distilled water. For this test, a dilution of 1:2000, based on the COD of 1,500,000 mg/L for the solvent concentrate. A 500 ml portion of this solution was placed in a 600 ml beaker and was stirred continuously for 10 hours with a 6.0 cm magnetic bar at 300 rpm. Duplicates at the least should be analyzed for each test. At intervals of 2 hrs, a 6 ml sample was taken from each beaker for COD analysis. The rate of volatilization was determined by plotting COD versus time for the test period and determining a best fit curve. The COD was measured in accordance with Standard Methods procedure 508C (APHA, 1985). The effect of pH on volatility can be rapidly evaluated by using the same procedure, but with samples adjusted over the expected pH range (pH 2 to 12 for this test) using either 2N H_2SO_4 or NaOH. Baseline volatilization data were determined using a control sample at the natural pH (9.2) of the solvent and water mixture. Using a multiple stirring plate, up to four different pH values could be evaluated at any one time.

To test this protocol, a terpene-based solvent with the major active component listed as d-limonene (Penetone Corporation, Tenafly, NJ) was used. Additional components included: ethanolamine, glycol ether, EDTA, surfactants, and water. The pH of the concentrated form was reported to be alkaline (10.4) and the Biochemical Oxygen Demand (BOD) and COD concentrations were reported as 250,000-400,000 mg/L and 1,500,000-3,700,000 mg/L, respectively (Camacho, 1989). The flash point and freezing point were reported to be 165°F (concentrate), 175°F (1:3 dilution), and 32°F, respectively.

The results, shown in Figure 1, indicated that the terpene-based, solvent mixtures decreased over time as measured by COD. This loss appeared to follow a first-order, exponential decay and after 6 to 9 hrs the COD reduction was about 46 percent. The volatilization rate determined for this data was 0.062 hr^{-1}. This solvent formulation was also tested over a pH range 2-12. The loss of COD in each mixture followed a similar, consistent pattern, but pH did not appear to affect the volatilization rate of the solvent. These results indicate that the terpene-based solvent formulation is stable over a wide pH range. Thus, the effect of pH on volatilization would be predicted to be minimal for many industrial applications. However, as the COD data indicate, the loss of unknown product components(s) was significant. Where capture of volatile organic compounds (VOCs) is a requirement, the need for appropriate control technology must be considered.

PROTOCOL FOR BIODEGRADABILITY

Recent developments in respirometric analysis have provided a new technique that can effectively determine biodegradability of volatile compounds. Jacquez et al., (1990) have provided data that describes the rapid evaluation of aqueous-based solvent substitutes. Brown et al., (1990) indicate the appropriate test culture is needed for measuring biodegradation. In developing the protocol for rapid determination of biokinetic parameters for substitute solvents, a series of batch and respirometric analysis were performed. Due to the volatile nature of the test compound, the batch studies produced inconsistent and unreproducible data. On the other hand, the respirometric technique provided reproducible and consistent data which provided for a more complete characterization of biodegradability for the solvent-substitute. A procedure using this technique will be described.

The biodegradability of the terpene-based solvent was evaluated by making respirometric measurements with microbial seeds acclimated to a variety of substrates and by evaluation of Monod kinetic coefficients. Two acclimated heterogeneous microbial seeds were prepared for the biodegradability tests. The first seed was obtained from the local wastewater treatment plant and was acclimated to peptone. This seed was maintained at a sludge age of 10 days, in a 10 L reactor mixed with air and it received a synthetic feed which contained peptone as the sole carbon source. The culture acclimated to the solvent was developed from the peptone culture and received a daily feed in which the terpene-based solvent was the sole carbon source. The two cultures received the same amount of carbon source (725 mg as COD) each day.

The biodegradation of the terpene-based solvent was then measured by using an automated respirometric technique to determine BOD and oxygen uptake rates (OUR). The instrument used in this test was an eight station COMPUT-OX respirometer (N-CON Systems Company, Inc., Larchmont, NY). For each test series, the terpene-based solvent was diluted to a concentration range 1000 to 0 ppm. A synthetic feed (Jacquez et al., 1990) was used to dilute the solution to 500 ml. In order to minimize the loss of solvent in the head space during mixing, 500 ml of solution was placed in the 690 ml respirometer bottle. Each reactor was seeded to a concentration of 10 mg/L with the test culture and the contents were stirred with a 6.0 cm magnetic stir bar.

The results of the biodegradation studies for the culture acclimated to the solvent are shown in Figure 2. For concentrations exceeding 400 ppm, the respiration curves show that oxygen uptake was greater than the maximum oxygen transfer capabilities of the instrument. This characteristic has been previously reported for this instrument (Jacquez et al., 1989). The final results indicate that initiation of respiration occurred at 3, 9, and 18 hours for the cultures acclimated to the solvent, peptone, and domestic wastewater, respectively. This increased lag period is a reflection of the specificity of the culture population developed through the acclimation process. In each case, biodegradation was apparent because the oxygen uptake was basically proportional to the solvent concentration.

Evaluation of Monod kinetic coefficients using respirometric measurements is a recent but well established procedure (Gaudy et al., 1987; Dang et al., 1989). In order to evaluate the Monod kinetic coefficients, three parameters must be measured: initial biomass concentration (X_o), growth yield (Y), and the biomass COD equivalent (O_x, mg COD/mg biomass). Y was estimated to be 0.4, from a batch growth study. O_x was determined to be 1.45 for the culture acclimated to the solvent. Based on the final results, u_{max} was determined to range between 0.045 to 0.050 hr^{-1} and K_s ranged between 45 and 70 mg/L as COD. The magnitude for these parameters is comparable to estimates determined by Dang, et al. (1989) using respirometric technique and chlorobenzene as the substrate. The mean values for u_{max}, K_s, and Y were reported to be 0.060, 3.52, and 0.42, respectively. Compared to the results obtained from the batch growth study, measuring kinetic parameters for a volatile substrate through respirometric technique is more reliable.

The respiration studies have clearly demonstrated that the solvent is readily degraded by a variety of bacterial cultures. At the concentrations tested, the terpene-based solvent may not be expected to pose a threat to either an industrial treatment plant or a Publicly Owned Treatment Works (POTW) which utilize biological processes in the treatment scheme. But from the data, the treatment plant might experience a short lived acclimation period which may extend 12-24 hours. The potential impact carryover of unknown hazardous materials (heavy metals, toxic organics, etc.) resulting from the use of the solvent substitute has not been evaluated. Currently the protocol is being refined in order to evaluate the biodegradability of these attributes of substitute solvents.

CONCLUSIONS

The use of COD for evaluating volatilization gives a direct measurement of the rate at which the solvent substitute is volatilizing. This method provides more useful information than that which can be obtained from EPA Method 24. Other waste parameters that may affect volatilization can be evaluated by using the COD method.

The protocol outlined in this paper indicated that the respirometer can be a useful tool for rapidly evaluating biodegradability of the volatile terpene-based solvent. Since the respirometer uses closed reactor vessels the test was not biased due to changes in atmospheric pressure or the loss of volatile components. Further refinements are needed to assure consistent test results. For example, test biomass either acclimated or unacclimated, must be grown in a controlled method to assure consistent determinations of lag time and kinetic parameters. Useful parameters such as growth rates, yield coefficients, or inhibition coefficients must be agreed upon and ranges of typical values determined. Additionally, methodologies addressing contaminant carryover and mobilization resulting from used solvents must be determined.

REFERENCES

1. American Public Health Association. (1985) Standard Methods for the Examination of Water and Wastewater, 16th edition., Washington, DC.

2. Breton, M., Frillici, P., Palmer, S., Spears, C., Arienti, M., Kravett, M., Shayer, A., and Suprenant, N. (1988) Treatment Technologies for Solvent Containing Wastes, Noyes Data Corporation, Park New Jersey.

3. Brown, S.C., Grady, C.P. Leslie, Tabak, H.H. (1990) "Biodegradation Kinetics of Substituted Phenolics: Demonstration of a Protocol Based on Electrolytic Respirometry." Wat. Res. 24, 853-861.

4. Comacho, R. (1989) Personal Communication, Industrial Waste Specialists, City of Austin, Texas, August.

5. Dang, J. S., Harvey D. M., Jobbagy A. and Grady C.P.L. Jr. (1989) "Evaluation of Biodegradation Kinetics with Respirometric Data." Res. J. Wat. Pollut. Control Fed. 61,1711-1721.

6. Gaudy, A.F., Jr., Rozich, A.F., Garniewske, S., Moran, N.R., and Dkambaram. (1987) "Methodology for Utilizing Respirometric Data to Assess Biodegradation Kinetic." Proceedings, 42nd Annual Industrial Waste Conference, Lewis Publishers, 573-584.

7. Higgins, J. (1989) Handbook of Hazardous Waste Minimization, CRC Press Inc., Boca Raton, Florida.

8. Jacquez, R.B., Cadena, F., Prabhakar, S., and Beach, M.I. (1989) "Gas Transfer limitations in Environmental Respirometry." Proceedings, 44th Annual Industrial Waste Conference, Lewis Publishers, 425-433.

9. Jacquez, R.B., Zachritz II, WH., Li, Meei-Huey, Beach, M.I. (1990) "Minimization of Hazardous Waste Generation: Preliminary Investigation of a Solvent Substitute for TCE." Proceedings of the 1991 National Conference on Environmental Engineering. Reno, NV.

10. U.S. EPA, CFR 40, Parts 53 to 60, Method 24, 979-981, July 1, 1990.

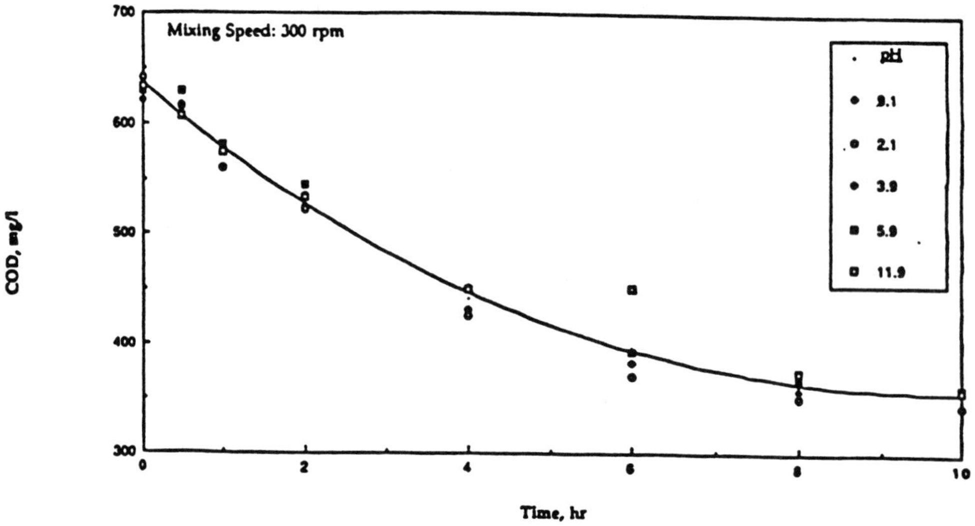

Figure 1. Volatilization of the Terpene-Based Solvent Mixture

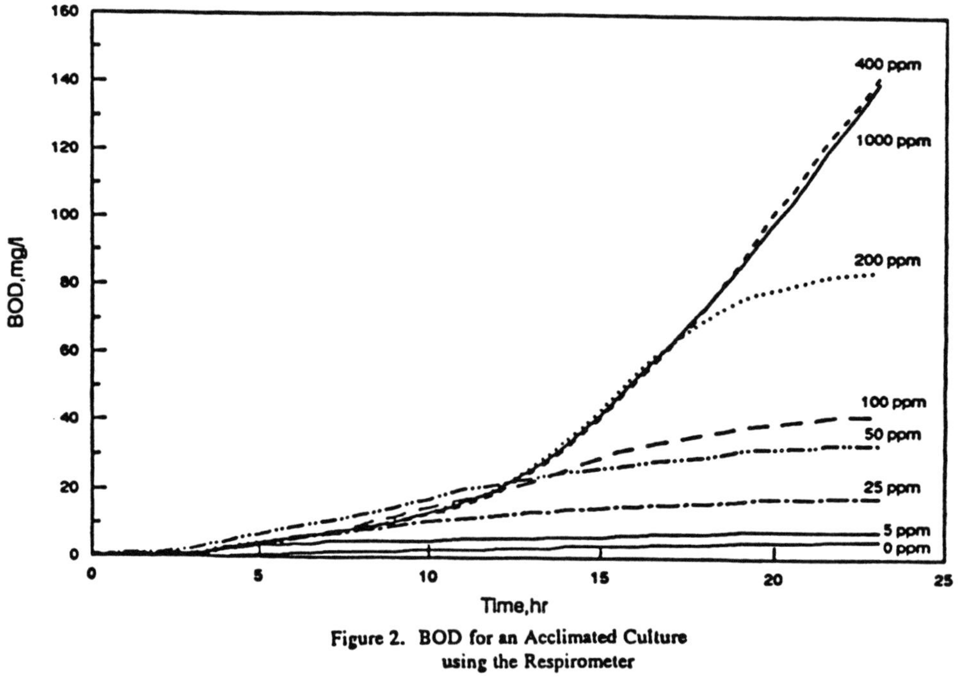

Figure 2. BOD for an Acclimated Culture using the Respirometer

Section VI

ISSUES TO CONSIDER

ALTERNATIVES TO CHLORINATED SOLVENTS: HEALTH AND ENVIRONMENTAL TRADEOFFS

Katy Wolf
Institute for Research and Technical Assistance
Los Angeles, California

INTRODUCTION

Governmental agencies have not traditionally used a systems approach to developing policies or regulations governing the use of chlorinated solvents and ozone depleting substances. As a consequence, policies have pushed users from one set of chemicals dangerous in a particular way to another set, dangerous in a different way. Policies have also transferred the problem from one medium to a less regulated medium. This paper discusses three historical case studies and focuses on two current case studies that demonstrate the lack of an integrated approach. The shift away from chlorinated solvents and ozone depleting substances is well underway and, because there is no coordinated policy, the alternatives that are adopted may eventually pose serious problems to human health and the environment.

HISTORICAL CASE STUDIES

There are numerous case studies that demonstrate the lack of integration of environmental policies. Regulations generally focus on reducing or eliminating the use of a particular chemical or set of chemicals. Users respond by adopting other substances that have not yet been targeted by regulation. They may be as dangerous as the chemical originally used, but in different ways. Frequently, different government agencies have different agendas. Even different offices within a particular government agency have different aims. This can result in inconsistent and conflicting regulations that are confusing and that may not better protect human health and the environment. The three historical case studies discussed here illustrate these points.

Ban on DBCP

In the 1970s, a chemical called dibromochloropropane or DBCP was used as a fumigant for certain crops to prevent destruction by nematodes, small worm-like soil insects. The DBCP was being produced in an Occidental Chemical plant in Lathrop, California. Several workers in the plant, through conversations, eventually realized that a large fraction of them where unable to have children. Over time, it became apparent that it was the DBCP that had caused sterility in the workers at the plant. DBCP was subsequently banned and the chemical that replaced it in many pesticide applications was ethylene dibromide or EDB, a suspect carcinogen. It, too, was eventually banned for pesticide uses. This is a situation where one chemical was banned because of a particular problem and it was replaced by another chemical that simply posed a health problem of a different kind.

VOCs and Exempt Solvents

The South Coast Air Quality Management District (AQMD) in Southern California is one of the most stringent air districts in the nation. Since the early 1970s, the District has developed a series of stringent rules designed to reduce or eliminate the use of photochemically reactive volatile organic compounds (VOCs) that contribute to smog. The rules exempt certain substances because they do not contribute to photochemical smog. These exempt chemicals include the chlorofluorocarbons (CFCs), 1,1,1-

trichloroethane (TCA) and methylene chloride (METH). Many users in the Los Angeles area have moved out of the VOCs into the CFCs, TCA and METH in response to the regulations. Indeed, the proportional use of these substances is much higher in the region than in the rest of the nation. The CFCs and TCA contribute to stratospheric ozone depletion and are scheduled to be banned over the next decade or so. METH is a suspect carcinogen and is a toxic air contaminant. Again, this is a situation where users moved from VOCs, which cause smog, to other chemicals that had different problems.

Aerosol Propellants

In the 1970s, three of the CFCs--CFC-11, CFC-12 and CFC-114--were widely used as propellants in aerosol packaging. In 1978, the U.S. unilaterally banned the use of CFCs as propellants in nonessential aerosol applications because of their contribution to stratospheric ozone depletion. Hydrocarbons like isobutane, which are flammable and photochemically reactive, replaced the CFCs as propellants in most instances. Many aerosol packers added METH to the hydrocarbons; it was an excellent co-solvent for the propellants and active ingredients. Several years later, the Consumer Product Safety Commission asked industry to label their products that contained METH with the warning that the chemical was an animal carcinogen. Rather than add this label, most aerosol formulators converted from METH to TCA. TCA will be banned because it contribute to ozone depletion. In California, the air districts are considering a ban on photochemically reactive propellants. This is a situation where there was a conversion from ozone depleters to smog producing, flammable chemicals; then a partial conversion to a suspect carcinogen; finally a partial conversion to another, less strong ozone depleter.

CURRENT CASE STUDIES

There are two very interesting case studies that are evolving today which will strongly affect nationwide chemical use patterns. The first, dealt with here in detail, involves the substitutes for ozone depleting substances. The second focuses on METH and its widespread use in paint stripping. Regulations on use of the latter chemical will become increasingly stringent over the next decade and alternatives that are dangerous in other ways will be adopted. These case studies, like the historical cases described above, demonstrate that government policies are developed without an integrated assessment of the consequences.

Ozone Depleting Substances

As mentioned earlier, TCA and CFC-113 will be banned over the next decade or so because they contribute to ozone depletion. In the light of this ban, many substitutes are being marketed in electronics, precision and general metal cleaning applications. These substitutes, fall into seven categories which are listed in Table 1, together with some of their characteristics.

The low molecular weight hydrocarbons include flammable solvents like acetone, isopropyl alcohol and methyl ethyl ketone (MEK). These substances were used widely before chlorinated and CFC solvents were introduced into the market. The latter solvents replaced them in many applications; they were perceived to be safer to workers because they were not flammable. The low molecular weight hydrocarbons are photochemically reactive and some air districts may not grant permits for their use. Their applicability as alternatives is limited because many users do not want to deal with flammable materials. EPA has issued a test rule requiring toxicity testing of isopropyl alcohol; many of the other flammable solvents have not been tested for chronic toxicity.

The high molecular weight solvents have flash points in the combustible range. They include chemicals like dibasic esters (DBE), terpenes, alkyl acetates and N-methyl pyrollidone

(NMP). These solvents are not volatile and they do not evaporate readily. They will leave a residue if they are simply oven dried. A water rinse is required to flush them completely from parts. Thus they can be used in applications where a residue does not matter and where a water rinse is feasible. The high molecular weight hydrocarbons are not exempt from photochemical smog regulations so a permit is required for their use. Because they are not volatile, however, emissions are not likely to be extremely high. D-limonene, a major ingredient of terpene formulations, has given a positive carcinogenicity test in male rats. EPA has issued a proposed test rule on NMP and few of the other combustible solvents have been scrutinized for their health and environmental effects.

The other chlorinated solvents include trichloroethylene (TCE), perchloroethylene (PERC) and METH. All three are suspect carcinogens and have been labeled toxic air contaminants under the new Clean Air Act amendments. TCE and PERC are regulated as smog contributors. These solvents have been used for many years and they perform extremely well. They are likely to be more heavily regulated in the future. William Reilly, the Administrator of EPA, has developed a list of chemicals; he is asking industry to voluntarily reduce the use of these chemicals by half over the next several years. The three chlorinated solvents are on the list.

The HCFCs are CFCs containing hydrogen; this makes them less stable in the atmosphere and they contribute less significantly to ozone depletion than do the CFCs. Those suitable for solvent use include HCFC-123, HCFC-141b and HCFC-225. The former two have extremely low boiling points; their losses will be very high and they will be extremely expensive to use. HCFC-225 is composed of two isomers, one of which is toxic. It is not clear whether a manufacturing process that selectively produces the nontoxic isomer can be developed. All three HCFCs are currently in toxicity testing and they will be fully scrutinized by the time they enter the market. Because they deplete the ozone layer to some extent, however, they are considered only interim alternatives and they will eventually be banned.

The HFCs and FCs contain no chlorine so they do not contribute to ozone depletion. They do, however, contribute to global warming, particularly the FCs which have long atmospheric lifetimes. One HFC, pentafluoropropanol, is being offered for solvent applications. It has a very low worker exposure level and is not being tested for chronic toxicity. The FCs alone are not very good cleaners and other substances must be added to them to increase their performance. They are not currently in toxicity testing and they will be very expensive to use.

Aqueous based formulations are another alternative to TCA and CFC-113. These will prove technically suitable for many applications. They have four drawbacks, however. First, water based cleaners do not dry readily and more energy will have to be used to achieve dry parts; this will exacerbate global warming. Second, these formulations virtually always contain additives and the additives are sometimes known to be dangerous or they have not been scrutinized for their health and environmental effects. Many of the additives are photochemically reactive. Third, water based cleaning will result in much more sewer loading of organics and metals and it is not clear what the consequences will be. Fourth, these cleaning processes require increased use of a scarce resource and this can be a significant problem in areas like California.

Various other processes are being proposed as alternatives to TCA and CFC-113 in various applications. Two of these--supercritical carbon dioxide and carbon dioxide snow--release carbon dioxide to the atmosphere. It is not clear whether these releases will eventually be regulated to prevent global warming.

In the years to come, as TCA and CFC-113

availability become increasingly restricted, the alternatives will be implemented widely. They generally pose known problems of a different kind or they are unscrutinized. Government agencies and offices are not taking responsibility for evaluating the alternatives and there is no evidence that they will be used properly or that their disadvantages will even be known before they are used. For example, the office of EPA responsible for banning ozone depleting substances encourages users to adopt photochemically reactive substances as substitutes. Since their agenda is to discourage the use of TCA and CFC-113, they are responsible for pushing users into substances that are unscrutinized. The AQMD in Southern California, on the other hand, discourages the use VOCs and has, for many years, encouraged the use of TCA and CFC-113. The Office of Toxic Substances at EPA has been charged with evaluating the safety and environmental characteristics of the alternatives. They have performed an initial screen of only some of the alternatives and they have received pressure from the Air office to minimize the problems with the alternatives they are examining.

The EPA has set up a special working group that was to facilitate changing Mil-Std-2000, the military standard governing the use of printed circuit board cleaning solvents. The group has tested and qualified several alternatives. In spite of the high priority assigned to changing the military specification, however, the Department of Defense (DoD) has not yet taken action.

In this case study, it is clear that different governmental agencies have different aims that may not result in better protection of human health and the environment. The result is a web of regulations which are inconsistent and conflicting and which encourage the use unscrutinized alternatives or alternatives that are known to be dangerous.

Paint Stripping

METH is an excellent paint stripper and it has been used widely for that purpose for many years. The Occupational Safety and Health Administration will lower the workplace exposure level from 500 to 25 ppm in 1991. METH is considered a suspect carcinogen and has been designated a toxic air contaminant under the Clean Air Act amendments. It will be increasingly regulated in the years to come and users will be forced to adopt alternatives.

In maintenance stripping--aircraft stripping, for instance--abrasive and fracture technologies are being investigated. Some of these are promising but they present other problems. The dry abrasive techniques require dust control which can be expensive and there remains a question of whether or not these techniques do long term substrate damage. The wet abrasive method may lead to corrosion and probably will not be adopted for whole airframes. The fracture technologies may also cause substrate damage.

In the consumer sector--household and contract stripping--several chemical stripping alternatives are being examined. They include the flammable and combustible solvents described in the last case study. EPA's Office of Toxic Substances performed an initial analysis of the alternatives and concluded that all but one alternative posed as severe or more severe health and environmental problems as METH. Many of them remain relatively unscrutinized.

CONCLUSIONS

The historical case studies discussed here demonstrate that there has not traditionally been a systems approach to regulation. Regulations have pushed users from one set of chemicals dangerous in a particular way to another set dangerous in another way. The current case studies show that this has not changed and that society will adopt unscrutinized chemicals over the next decade and will transfer the problem from the air to the sewer.

A continuing problem is that governmental agencies and offices each have a particular agenda and there is no incentive to use an integrated approach to the problems. The regulators also have only limited knowledge of the field that corresponds to their specialty. They know very little of other areas. Another problem is that regulators rarely understand the processes where chemicals are used. They grant and deny permits without knowing anything about the process itself. If this situation is allowed to continue, current policies will continue to push users into new, unscrutinized chemicals and processes and to transfer the problem to the least regulated medium.

To improve regulatory policy, there are several steps that could be taken. First there should be an admission that all chemicals and processes present problems of one kind or another and that, in fact, there is no panacea. Second, there should be an acknowledgement that there are no easy generic solutions. The alternatives will have to be chosen on a case-by-case basis with the characteristics and location of the particular operation in mind. Third, governmental policies must begin to take a systems approach to the problem. New chemicals should be tested for their health and environmental effects before they are marketed and a clear analysis of the tradeoffs should be undertaken before chemicals are heavily regulated and substitutes adopted.

TABLE 1

CHARACTERISTICS OF CHLORINATED SOLVENT ALTERNATIVES

Chemical Class	Examples	Characteristics
Low MW Hydrocarbons	Acetone: isopropanol; petroleum solvents; MEK	Flammable; photochemically reactive; evaporate readily; many not scrutinized for toxicity
High MW Hydrocarbons	Terpenes; DBE; NMP; alkyl acetates	Combustible; photochemically reactive; leave residue or require water rinse; do not dry readily; many not scrutinized for toxicity
Chlorinated Solvents	TCE; PERC; METH	No flash point; evaporate readily; health effects scrutinized; heavily regulated
HCFCs	HCFC-123; HCFC-225; HCFC-141b	No flash point; evaporate readily; low boiling point; health effects will be scrutinized; will be banned
HFCs and FCs	pentafluoropropanol	Health effects unscrutinized; contribute to global warming
Water	-	Higher energy use; increased sewer loading; additives unscrutinized; use of scarce resource

FORMATION OF SPECIFICATIONS FOR NEW PRODUCTS

Captain Daniel T. Witt
Technology and Industrial Support Directorate
San Antonio Air Logistics Center
Kelly Air Force Base, Texas

BACKGROUND

The Environmental Program Office acts as a consulting organization to the Aircraft, Engine and Support Equipment Directorates which specify the materials and processes used on San Antonio Air Logistics Center (SA-ALC) managed equipment. This includes 1/2 of the AF Engines, 60% of AF Support Equipment, All the AF Electronic Test Equipment, C-5, T-38, T-37, and OV-10 Aircraft. In conjunction with the equipment, SA-ALC currently manages over 45,000 Technical Orders. The Materials Engineering Section, which includes the TI Environmental Program, SA-ALC Corrosion Program, Nondestructive Inspection and the High Technology Metals Insertion Program Offices, recommends the materials and processes specified in all SA-ALC Technical Orders.

TECHNICAL ORDER CHANGES

The cost to change one page in a Technical Order (TO) is between 500 and 1000 dollars regardless if the change involves one word or the entire contents of the page. The reference of a specification number in a TO eliminates the arbitrary changing of manufacturers data and thus eliminates changing TO pages. To further explain, if a Commercial Item is inserted into a TO, enough information must be provided in the TO so the user can purchase the item. This includes manufacturer's name, address, phone number, and product designation. This results in several problems:

1. If the manufacturer changes any of the data, it costs the Air Force between $500 and $1000 to change the TO.

2. Sole sourcing is frowned upon, therefore, at least two products must meet the same requirements before they can be inserted in the TO.

3. When users need to purchase a non-specified product, they go out on local purchase. The installation procurement office goes out on bid based on the description of the product. He gets a LOW BID product that may not meet the requirements of the product listed in the TO.

4. The quality of the product can not be controlled. The manufacturer can change the formulation of the product for any number of reasons and not inform the organization specifying it.

By developing a specification and referencing it in the TO, the manufacturer's information can change numerous times and never affect a single TO. The manufacturer's information is changed on the Qualified Products List (QPL) the specification. Thus the cost for maintaining TOs is reduced and quality control of the product affirmed. This insures that field maintenance personnel receive a product that works.

FORMULATING A SPECIFICATION

When formulating a specification, the existing requirements for current cleaners must be incorporated along with special requirements dealing with environmental, safety, health and new technologies. Some of the existing requirements that have proven their worth are:

Hydrogen Embrittlement of High Strength Steels
Total Immersion Corrosion Tests of Aerospace Metals
Low Embrittling of Cadmium Plated Steel
Effects on Unpainted Metals
Sandwich Corrosion Tests of Aerospace Metals
Stress Crazing of Stretched Acrylics/Plastics
Effects on Painted Surfaces
Long Term Storage Stability
Effects on Aerospace Sealants and Rubbers
Hard Water Stability
Heat and Cold Stability
pH Value

As environmental and health concerns heighten, some **special requirements** are reviewed, keeping in mind new legislation and public concerns. These special requirements are:

* BIODEGRADABILITY
* TOXICITY
* DISPOSAL CHARACTERISTICS
* FLASH POINT
* VOLATILITY
* NEW TECHNOLOGIES

The largest stumbling block encountered is the lack of standard definitions for each of the **special requirements**. It becomes more difficult when writing a performance specification (versus a material specification) which does not limit the materials used in the manufacturing of the products. Added to the fact that specifications are used worldwide and regulations governing hazardous waste disposal and hazardous materials change from country to country, state to state and district to district, the task of defining common ground becomes elusive to say the least. Established guidelines concerning the **special requirements** must be standardized before a specification can satisfy the needs of every installation. But before that happens some tough questions need to be answered about each of the special requirements.

CURRENT PROJECTS

The TI Environmental Program Office, in conjunction with the Energy Management Directorate, (which manages the Air Force cleaner specifications) SA-ALC/SF, is completing research and development on specifications for cleaners passing the phase II testing conducted at Tinker AFB, hoping to quickly implement these cleaners into the AF. The following specifications have been revised or initiated for the purpose of hazardous material substitutions.

SPECIFICATION: P-D-680A Amendment 3, (added Type III)
ISSUE DATE: 13 July 90
NATIONAL STOCK NUMBER:
5 gal 6850-01-331-3349
55 gal 6850-01-331-3350

TITLE: DRY CLEANING AND DEGREASING SOLVENT

This amendment was initiated by SA-ALC/TIESM due to the numerous requests from field organizations to add a type of material with an increased flash point to over 200 degrees. The amendment added a Type III petroleum distillate solvent. The amendment was issued by SA-ALC/SF. Still in development, it is an effort to add requirements for recycled solvent into the specification.

Advantages:

•Min. 200 Degrees F flash point

•Low to no odor

•Non-chlorinated

•200 PPM Threshold Limit Value (TLV) low evaporation rate (3-4 times slower than the Type II material)

•Recyclable

Disadvantages:

•Higher cost ($8.00 per gallon vs $2.00 for Type II)
•Slower evaporation rate

SUBSTITUTION POTENTIAL: Expected to replace P-D-680 Types I and II in numerous applications. References of P-D-680 in Technical Orders at SA-ALC estimated over 300,000.

SPECIFICATION: MIL-C-87937
ISSUE DATE: 29 Oct 90
NATIONAL STOCK NUMBER: still pending

TITLE: CLEANING COMPOUND, AEROSPACE EQUIPMENT

This specification was initiated by SA-ALC/TIESM to replace solvent cleaning of aircraft, aircraft engines and support equipment with a biodegradable, water dilutable, less hazardous cleaner. The specification has two types of cleaning compounds. Type I is Terpene based (citrus based) cleaners and Type II is for general cleaners with little or no solvents.

Advantages:

•Biodegradable

•Less toxic

•Diverse uses: spot cleaning, exterior A/C washing, dip tank cleaning, carbon removal and general cleaning

•Type I recyclable

Disadvantages:

•Type I, low flash point (minimum 125 Degrees F, dilution with water raises the flash point)

•Requires water rinse except as a spot cleaner

•Higher cost ($9 a gl vs approx. $2 for current

•Cleaners but dilution is 20:1 for exterior cleaning which reduces cost to $.45 a gallon)

SUBSTITUTION POTENTIAL: Replace hazardous solvent based cleaners in exterior aircraft and engine cleaning, parts cleaning in vapor degreasing, cleaning Aerospace Ground Equipment, hazardous carbon removers and general cleaning where water rinsing is acceptable.

SPECIFICATION: MIL-C-83873
ISSUE DATE: est. 1 Mar 91
NATIONAL STOCK NUMBER: Pending

TITLE: CLEANING COMPOUND, PRECOATING SURFACE CLEANER

This specification was initiated by SA-ALC/TIESM to replace the hazardous solvents used prior to coating aircraft and support equipment. This specification takes advantage of the unique properties of enzyme based cleaners to remove contaminants.

Advantages:

•Biodegradable

•Nontoxic

•Nonhazardous

•Easily disposed of

Disadvantages:

•Cost ($65.00 a gallon, dilution ratio is 15 parts water to one part cleaner, $4.00 a gallon diluted) No handling or disposal costs.

SUBSTITUTION POTENTIAL: Expected to replace methyl ethyl ketone for exterior wipe down prior to aircraft and AGE painting.

Replacement for scuff sand and solvent wipe prior to recoating or touch up of aircraft and AGE.

SPECIFICATION: MIL-C-XXXXX
ISSUE DATE: 1 Mar 91
NATIONAL STOCK NUMBER: Pending

TITLE: ACETATE ESTER BASED SOLVENTS

This specification was initiated by SA-ALC/TIESM to replace low flash point, highly toxic solvents used to clean aircraft parts where rinsing of the parts is not feasible and for electronics cleaning. There are six types available with varying viscosities, flash points and evaporation rates.

Advantages:

- Not a Ozone Layer Depleting Substance
- Flash points range from 134 to 261 degrees F
- Low toxicity
- Low residue (27 ppm and up)

Disadvantages:

- Alcohol based

SUBSTITUTION POTENTIAL: Further testing is required to determine the full potential of the cleaning products. Expected to replace CFCs in cleaning electronics equipment and general applications on aircraft systems. Provides replacement potential for cleaning solvents used in painting equipment clean up, coating thinners, spot cleaning and solvent wiping.

APPENDIX I

Attendees

The First Annual International Workshop on Solvent Substitution

December 4-7, 1990
Phoenix, Arizona

David Albright	Environmental Protection Specialist U.S. Environmental Protection Agency OTS/CCD 401 M Street SW Washington, DC 20460 202-245-4028
Gopal Annamraju	Program Manager Wright-Patterson Air Force Base HQ-AFLC/DEVR Dayton, OH 45433 513-257-7053
Bob Anthony	Maintenance Supervisor Coors Brewing Company BC035 Golden, CO 80401 303-277-2092
Iris Artaki	AT&T Bell Laboratories P. O. Box 900 Princeton, NJ 08540 609-639-2585
Kelly K. Asada	Member of the Technical Staff Hughes Aircraft Company Radar Systems Group P. O. Box 92426 Los Angeles, CA 90009 213-334-5235
R. W. Aubert	Mgr., Environmental Resources Mgmt. General Dynamics, Air Defense Systems P. O. Box 2507 Pomona, CA 91769 714-868-3472

Daniel Axelrad	Economist, Office of Toxic Substances U.S. Environmental Protection Agency 401 M Street, SW (TS-779) Washington, DC 20460 202-382-3713
R. Page Ayres	Environmental Engineer Newport News Shipbuilding 4101 Washington Avenue Newport News, VA 23607
E. G. Baker	Battelle Pacific Northwest Laboratories Mail Stop K2-12 Richland, WA 99352 509-375-2026
Gary E. Baker	Senior Engineer SAIC 635 West 7th, Suite 403 Cincinnati, OH 45203 513-723-2611
James J. Baremore	Mgr., Manufacturing Programs Sandia National Laboratories Dept. 6610 P. O. Box 5800 Albuquerque, NM 87185 505-844-3690
Donald Bauer	Senior Manufacturing Engineer Martin Marietta Energy Systems Mail Point 150 P O. Box 5837 Orlando, FL 32855 407-356-5354
Fred Bauer	U.S. Department of Energy Technology Development Division 785 DOE Place Idaho Falls, ID 83402 208-526-0142
John Beller	Senior Project Engineer EG&G Idaho, Inc. Technology Integration P. O. Box 1625 Idaho Falls, ID 83415 208-526-1205

Mark G. Benkovich	Senior Engineer Allied-Signal Aerospace Co. P.O. Box 419159 Kansas City, MO 64141 816-997-5020
Jim Bennett	Business Manager United Technologies USBI 6000 Technology Drive Huntsville, AL 35806 205-721-5509
Richard Benson	Program Manager Los Alamos National Laboratory Mail Stop F643 Los Alamos, NM 87545 505-667-4960
Lloyd Berg	Montana State University Chemical Engineering Department Bozeman, MT 59715 406-944-2222
James Berger	Westinghouse Hanford P. O. Box 1970 Richland, WA 99352 509-376-9942
Kieran D. Bergin	Manager, Environmental Control Rockwell International Corporation P. O. Box 2515 Seal Beach, CA 90740 213-797-2412
Elliott Berkihiser	Manager, Chemical Reduction The Boeing Company P. O. Box 2707 Seattle, WA 98124 206-393-4784
James Blackburn	Associate Director, EERC University of Tennessee 423 S. Stadium Knoxville, TN 37996 615-974-1779
Jack Blakeslee	Technical Program Manager EG&G Rocky Flats, Inc. P. O. Box 464 Golden, CO 80005 303-966-4642

Brian Blakkolb	Materials Engineer TRW One Space Park Redondo Beach, CA 90278 213-813-8960
Jim Block	Creare Inc. Etna Road P. O. Box 71 Hanover, NH 03755
Elliot W. Bloom	Project Manager United Technologies Chemical Systems Division P. O. Box 49028 San Jose, CA 95161 408-365-5535
Phoebe Boelter	Weapons Complex Monitor Forums 1715 North Wells, #34 Chicago, IL 60614 312-988-7667
J. K. Bonner	Manager, Applications/Proc. Devel. Allied Signal Inc. 2001 N. Janice Avenue Melrose Park, IL 60160 708-450-3880
Robert M. Bottom	Plant Layout Engineer General Motors Corporation Allison Gas Turbine Division P. O. Box 420, Code P-21 Indianapolis, IN 46206 317-230-3618
Tom Boyce	Research Leader, Larkin Laboratory Dow Chemical Company 1691 North Swede Road Midland, MI 48640 517-636-1484
Karla M. Boyle	Hughes Aircraft Company Bldg. R7/MS 406 Los Angeles, CA 90009 213-334-4430
Raymond J. Bradley	Staff Engineer, Manufacturing Eng. Martin Marietta Manned Space Systems P. O. Box 29304 New Orleans, LA 70189 504-257-2918

James L. Breece	Vice President Safety-Kleen Corporation P. O. Box 92050 Elk Grove Village, IL 60009 312-694-2700
Owen M. Briles	Engineering Manager Sundstrand Corporation 4747 Harrison Avenue Rockford, IL 61125 815-226-2930
Walter Brodtman	Environmental Engineer U.S. Environmental Protection Agency 401 M Street RD-681 Washington, DC 20460 202-382-2615
Louis J. Brothers, Jr.	Products Manager Quaker Chemical Corporation Elm and Lee Streets Conshohocken, PA 19428 215-828-4250
Lonnie Brown	President Uniclean Products Inc. 3700 Osuna, Suite 703 Albuquerque, NM 87109 505-344-9673
Phillip G. Brown	Col., USAF Headquarters, Air Force Logistics Command Wright-Patterson Air Force Base Dayton, OH 45433 513-277-6728
Robert Brown	General Manager Uniclean Products Inc. 3700 Osuna, Suite 703 Albuquerque, NM 87109 505-344-9673
Mark E. Brynildson	Environmental Chemist Sandia National Laboratories Division 8543 P. O. Box 969 Livermore, CA 94551 415-294-3150
Marcanne Burrell	Boeing Aerospace and Electronics MS 9E-06 P. O. Box 3999 Seattle, WA 98124

Diane Bursey	Chemical Engineer Pratt & Whitney Canada 1000 Marie Victorian Longueuil, Quebec J4G1A1 Canada 514-647-3676
William Cain	Chemical Engineer, Air Logistics Ctr. Tinker Air Force Base OC-ALC/LAPEP Tinker Air Force Base, OK 73145 405-736-5986
James D. Caldwell	Industrial Engineer Hill Air Force Base 00-ALC/LAOPCR Ogden, UT 84056 801-707-2050
Joe Camahort	Senior Staff Engineer Lockheed Missiles & Space Company O/47-01 B/101 P. O. Box 3504 Sunnyvale, CA 94088 408-742-6936
Christopher J. Campbell	Executive Director Southern California Coalition Hazardous Materials Management 355 South Grand Avenue Los Angeles, CA 90071 213-683-8717
Charlie Carpenter	Engineering & Services Laboratory Tyndall Air Force Base HQ AFESC/RDVS, Building 1117 Tyndall AFB, FL 32403 904-283-6015
Robert Carrington	Manager, Waste Programs Division MSE, Inc. P. O. Box 3767 Butte, MT 59702 406-494-7100
Robert A. Carter	Staff Engineer Waste Reduction Resource Center P. O. Box 27687 Raleigh, NC 27611 800-476-8686

William L. Casper	EG&G Idaho, Inc. P. O. Box 1625, MS 3950 Idaho Falls, ID 83401 208-526-4127
Michelle C. Chairs	Sr. Mfg. Engineer General Dynamics, Convair Division P. O. Box 85377 San Diego, CA 92186 619-542-6245
Walter Chaney	Technical Sales Rep. The Rinchem Company 4115 W. Turney Avenue Phoenix, AZ 85019 602-233-2000
Sidney C. Chao	Manager, Environmental Technology Hughes Aircraft Co. P. O. Box 902 El Segundo, CA 90245 213-616-4917
Angela A. Chavez	Scientist EG&G Idaho, Inc. P. O. Box 1625, MS 1500 Idaho Falls, ID 83415 208-526-7756
Ken Clark	Senior Materials Engineer Naval Air Development Center Code 6062 Warminster, PA 18974 215-441-1508
Robert H. Clark	Senior Process Engineer Hughes Aircraft Company Missile Systems Group Bldg. 801, MS N-21 P. O. Box 11337 Tucson, AZ 85734 602-794-1788
Wayne M. Cole	Environ., Health, & Safety Programs GE Medical Systems 3114 N. Grandview Waukesha, WI 53066 414-548-2314

Tom Collipi	Vice President Advanced Sciences, Inc. 2620 San Mateo N.E., Suite D Albuquerque, NM 87110 505-883-0959
Dayle A. Conrad	Chemist Naval Aviation Depot Code 36300 Norfolk, VA 23503 804-444-8811
Anne Copeland	Chemical Engineer Tinker Air Force Base OC-ALC/EME Tinker AFB, OK 73145 405-736-5871
Weston Cox	Principal Aero-Strip, Inc. P. O. Box 166F Keller, TX 76248 817-431-2968
A. I.(Bud) Dalton	Technology Manager, CPI Air Products and Chemicals Inc. 7201 Hamilton Blvd. Allentown, PA 18195 215-482-7007
Margaret Dancey	Process Development Engineer United Technologies Chemical Systems Division P. O. Box 49028 San Jose, CA 95161 408-778-4927
Brian C. DeMonia	Environmental Scientist, ENVPD U.S. Department of Energy P. O. Box 2001 Oak Ridge, TN 37830 615-576-1674
Mary L. Delaney	Process Development Engineer United Technologies P. O. Box 49028 San Jose, CA 95161 408-778-4911

Arline Denny	Senior Environmental Engineer Rocketdyne, Rockwell Corporation 6633 Canoga Avenue, JA16 Canoga Park, CA 91304 818-773-5318
Mariann Dickman	Sales Manager Modern Chemical Box 21 Batesville, IN 47006 812-934-4463
Kenneth L. Donoff	Mechanical Engineer U.S. Air Force HQAFLC/MMES WPAFB Dayton, OH 45433 513-257-2151
Marvin Drabkin	Mgr., Waste Minimization Branch VERSAR, Inc. 6850 Versar Center Springfield, VA 22151 703-750-3000
John V. Drzewicki	Market Planner DuPont Company Barley Mill Plaza 132150 Wilmington, DE 17898 892-8151
Mark J. Duchsherer	Advanced Chemist Westinghouse Hanford Co. P. O. Box 1970 (T5-12) Richland, WA 99352 509-373-5308
Jonathan Duke	Program Manager, ERM General Dynamics Corporation Space Systems Division P. O. Box 85990 San Diego, CA 92186 619-547-4771
Frank M. Ead	Materials & Process Engineering Lockheed 86 South Cobb Drive Marietta, GA 30062 404-494-2818

Russell J. Eastenes Jr.	Project Engineer Thiokol Corporation P. O. Box 30058 Shreveport, LA 71130 318-459-5808
Patricia A. Eddy	Manager, Environ. Services General Motors Allison Gas Turbine Div. P. O. Box 420-S-44a Indianapolis, IN 46206 317-230-5456
Neal Egan	MSE, Inc. P. O. Box 3767 Butte, MT 59701 800-441-8213
Tim Ehli	Pollution Prevention Coordinator Lockheed Engineering & Sciences Co. 1050 East Flamingo Las Vegas, NV 89119 702-734-3359
Richard L. Eichholtz	Supervisory Chemical Engineer Naval Ordnance Station Strauss Avenue Indian Head, MD 20640 301-743-4365
Victor Engleman	Division Mgr., Process Technology SAIC 10240 Sorrento Valley Road, #204 San Diego, CA 92121 619-587-9071
James L. Epler	Project Manager Martin Marietta Energy Systems, Inc. P. O. Box 2003 Code 7606 Oak Ridge, TN 37831 615-435-3112
John F. Evert	Program Manager Bldg. 224-2S-25 St. Paul, MN 55144 612-733-4043
Ted Falkowski	B.F. Goodrich Aerospace 3414 South 5th Street Phoenix, AZ 85040 602/232-4000

Gregory S. Fenner	Senior Research Engineer EG&G Rocky Flats P. O. Box 464 Golden, CO 80402 303-966-4716
Suzanne Fiscus	Engineer Westinghouse Electric Corp. P. O. Box 79 West Mifflin, PA 15122 412-476-5748
Ian A. Fisher	Engineer Westinghouse Savannah River Company Savannah River Laboratory Aiken, SC 29808 803-725-8169
Paul W. Fisher	Development Staff Member Oak Ridge National Laboratory P. O. Box 2009 Oak Ridge, TN 37831 615-574-8051
A. D. FitzGerald	Director, Environmental Affairs Telecom Limited P. O. Box 458, Station A Mississauga, Ontario L5A 3A2 Canada 416-897-9000
Calvin Fong	Senior Technical Specialist Northrop Aircraft One Northrop Avenue Hawthorne, CA 90250 913-332-6661
Clyde Frank	Associate Director, OTD U.S. Department of Energy ERWM 1000 Independence Avenue Washington, DC 20585 202-586-6382
Wendy A. French	Safety Engineer Naval Aviation Depot Code 095 Marine Corps Air Station Cherry Point, NC 28570 919-466-7042

Gregory C. Frye	Sandia National Laboratories Inorganic Materials Chemical Division P. O. Box 5800 Albuquerque, NM 87185 505-844-0787
Paula M. Gallagher	Research & Develop. Engineer Phasex Corporation 360 Merrimack Street Lawrence, MA 01843 508-794-8686
Bob Garland	Environmental Engineer U.S. Air Force O3-ALL/EMR Hill Air Force Base, UT 84056 801-777-6742
Danny Gayman	Technical Dryer Corporation 24104 11th Avenue, South Des Moines, WA 98198 206-824-1261
Don M. Geering	Supervisor Los Alamos National Laboratory P. O. Box 1663, MS 0473 Los Alamos, NM 87545 505-665-1784
T. J. Gillespie	General Electric Company Neutron Devices Department MS 048 11400 S. Belcher Road Largo, FL 34649 813-541-8307
Robert Gillins	Principal Program Specialist Haz Answers, Inc. 2300 N. Yellowstone Idaho Falls, ID 83401 208-522-5526
William O. Gillum	Member, Technical Staff AT&T Bell Laboratories P. O. Box 900 Princeton, NJ 08540 609-639-2548
Kenneth V. Grams	Chemist U.S. Air Force Kelly Air Force Base San Antonio, TX 78241 512-947-3064

Barry Granoff	Supervisor Sandia National Laboratories Environ. Conscious Manufacturing Program P. O. Box 5800 Albuquerque, NM 87185 505-844-8145
Jane Gregory	General Dynamics, Convair Division P. O. Box 85377 San Diego, CA 92186 619-542-4640
William Gregory	Ebasco 143 Union Blvd., Suite 1010 Lakewood, CO 80228 303/988-2202
Gary Groenewold	Unit Manager EG&G Idaho, Inc. P. O. Box 1625-2208 Idaho Falls, ID 83415 208-526-2803
Earl C. Groshart	Senior Engineer-Chemical Technology Boeing Aerospace & Electronics Division P. O. Box 3999 MS82-32 Seattle, WA 98124 206-773-5379
Harold Gutovich	Director of Sales & Marketing IXTAL Blast Technology Corp. 627 John Street Victoria B.C. V8T 1T8 Canada
Tim Hale	Martin Marietta Energy Systems P.O. Box 2008 Bldg. K-1035, MS-7209 Oak Ridge, TN 37931 615/574-3224
John W. Hallam	Hercules, Inc. P. O. Box 210 Rocket Center, WV 26726 304-726-5476
Gerald F. Hardacre	Manager, Environmental Programs Convair, General Dynamics P. O. Box 85377 San Diego, CA 92186 619-542-7712

Leslie G. Harmon	Staff Specialist Engineering McDonnell Douglas Missile Systems Company P. O. Box 516 St. Louis, MO 63166 314-233-9337
Michael T. Haro	Health, Safety, Environ. Mgr. Allied-Signal Aerospace Company 2525 W. 190th Street Torrance, CA 90504 213-512-4798
Taryl L. Harris	Scientist EG&G Idaho, Inc. P. O. Box 1625 MD 1500 Idaho Falls, ID 83415 208-526-1382
Martin Harrison	Process Engineer Quaker Chemical Corporation Elm & Lee Streets Conshohocken, PA 19428 215-828-4250
Jean Hawkins	Chemist, Materials Engineering Lab. Naval Aviation Depot NAS Jacksonville Jacksonville, FL 32216 904-772-3444
Edward L. Helminski	Weapons Complex Monitor Forums 2014 P Street, NW Washington, DC 20036 202-296-2814
June Hennig	Chief, Technology Develop. Branch U.S. Department of Energy 750 Jadwin Richland, WA 93352 509-376-0016
Christopher Hensley	Sales Supervisor Penetone Corporation Military/Aerospace Division 74 Hudson Avenue Tenafly, NJ 07670 201-567-3000

Keith J. Herbert	Director, Industrial Catalysts Allied-Signal, Inc. P. O. Box 580970 Tulsa, OK 74158 918-266-1400
Melvin D. Herd	EG&G Idaho, Inc. MS 2208 P. O. Box 1625 Idaho Falls, ID 83415 208-526-8595
Lee D. Herriges	Environmental Specialist GE Medical Systems P. O. Box 414 (L2-08) Milwaukee, WI 53201 414-647-4152
Robert G. Hickman	Lawrence Livermore National Laboratory Isotope Separation & Materials Processing P. O. Box 808 Livermore, CA 94551 415-422-1100
Raymond C.P. Hill	Principle Engineer Westinghouse Hanford Company 1200 Jadwin (B4-53) Richland, WA 99352 509-376-7454
John S. Hoffman	Senior Technical Acct. Mgr. The DuPont Company 2312 Chamberlain Drive Plano, TX 75023 214-867-0367
Miles Holliman	Research Engineer Southern California Edison P. O. Box 800, Room 455, GO1 Rosemead, CA 91770 818-302-6222
F. Michael Hosking	Sandia National Laboratories Solder Processing Division Division 1833 P. O. Box 5800 Albuquerque, NM 87185
John B. Howard	Martin Marietta Astronautics Group P. O. Box 179 Stop 9980 Denver, CO 80201 303-977-3705

C. Fawn Hudson	Materials Engineer Allied-Signal Garrett Auxiliary Power Div. 2739 East Washington Phoenix, AZ 85034 602-220-3065
Randall B. Ivey	Chemical/Materials Engineer Robins Air Force Base Corrosion Program Office WR-ALC/CNC Robins Air Force Base, GA 31098 912-926-3284
Donald Jackson	Chemist U.S. Army Toxic and Hazardous Materials Agency Bldg. E 4460 Aberdeen Proving Ground, MD 21010 301-671-2054
Ronald Jackson	U.S. Army Toxic & Hazardous Materials Agenc Bldg. E 4460 APG-EA Pikesville, MD 21010 301/671-2054
Ricardo B. Jacquez	New Mexico State University CAGE Dept. Box 30001-Dept. 3CE Las Cruces, NM 88003 505-646-3463
D. H. Johnson	Martin Marietta Energy Systems Development Division P. O. Box 2009 Oak Ridge, TN 37831 615-574-0868
Richard E. Johnson	Member, Technical Staff Rockwell International 6432 Navajo Road Westminster, CA 92683 714-891-0319
Charles F. Joly	Scott Air Force Base Headquarters MAC/LGMWF Scott Air Force Base, IL 62225 618-256-3254

Larry N. Jones	Director of Operations University of Tennessee Waste Management Institute 428 South Stadium Knoxville, TN 37996 615-974-3379
Thomas R. Jones	Project Manager, Pollution Prevention HAZWRAP Martin Marietta Energy Systems P.O. Box 2003, MS 7606 Oak Ridge, TN 37831 615-435-3266
Janine Jorgensen	Sr. Development Engineer Dow Chemical Company 2800 Mitchell Drive Walnut Creek, CA 94590 415-944-2249
Edward F. Kay	Associate Engineer Westinghouse Savannah River Co. Savannah River Site, 707-H Aiken, SC 29808 803-557-8796
Steve K. Keipert	Senior Chemist Corp. Research Laboratories Bldg. 201-2N-19 St. Paul, MN 55144 612-736-2385
Thomas J. Kelly	Senior Technical Manager The DuPont Company 6914 Needham Drive Charlotte, NC 28270 704-365-5707
Judy Kennedy	Environmental Engineer Washington State Dept. of Ecology MS PV-11 Olympia, WA 98504 206-459-6356
Bob Kerr	Director of Industrial & Military Sales Gerolyte Systems 1657 Rollins Road Burlingame, CA 94010 415-692-9080

Ronald W. Kiehn	Senior Consultant Environmental Research & Development, Inc. 1550 Jones Street, Bldg. E Idaho Falls, ID 83401 208-522-7119
Karl Kinkade	Vice President Business Development SAIC 3100 Rollandet Avenue Idaho Falls, ID 83402 208-523-7255
David Koester	Member, Technical Staff FMC Corporation P., O. Box 580 Santa Clara, CA 95052 408-289-0766
K. M. Koester	Member, Technical Staff TRW Space & Defense Group One Space Park Redondo Beach, CA 90278 213-814-1882
Ken G. Koller	Group Manager, Technology Development EG&G Idaho, Inc. P. O. Box 1625, MS 3940 Idaho Falls, ID 83415 208-526-4847
F. E. "Gus" Kosinski	Department Superintendent Martin Marietta Energy Systems, Inc. P. O. Box 2009 Oak Ridge, TN 37831 615-574-0868
Michael Kosusko	Chemical Engineer U.S. Environmental Protection Agency Energy Engineering Research Lab. MO-61 Research Triangle Park, NC 27711 919-541-2734
Milton Krause	Director, Research and Development Sunshine Makers, Inc. 15922 Pacific Coast Highway Huntington Harbor, CA 92649 213-592-2844

Rod Kremer	Tech. Service Representative Vulcan Chemicals P. O. Box 530390 Birmingham, AL 35253 205-877-3459
Gary Kuhlman	Materials Engineer Naval Aviation Depot North Island Materials Engineering Laboratory San Diego, CA 92135 619-545-9733
Paul Kunkel	CH2M Hill P. O. Box 28440 Tempe, AZ 85285 602-966-8188
Lance H. Lankford	Environmental Engineer U.S. Air Force SM-ALC/EME McClellan AFB, CA 95652 916-643-3672
Bill Lantz	Staff Engineer Radian Corporation 5103 W. Beloit Road Milwaukee, WI 53214 414-643-3020
Richard L. Lapado	Asst. to Vice President TRW 1 Rancho Carmel San Diego, CA 92128 619-592-3569
Nona E. Larsen	Engineer, Chemical Technology Boeing Aerospace & Electronics Div. P. O. Box 3999 MS82-32 Seattle, WA 98124 206-773-1807
Abigail C. Lee	Air Pollution Engineer Puget Sound Air Pollution Control Agency 200 W. Mercer, Room 205 Seattle, WA 98119 206-296-7468
Michael W. Lewis	Aerospace Program Manager Cold Jet, Inc. 455 Wards Corner Road Loveland, OH 45140 513-831-3211

Philip Lin	Research Engineer U.S. Environmental Protection Agency Thermal Destruction Branch 26 W. Martin Luther King Drive Cincinnati, OH 45268 513-569-7931
Duane Lindner	Mgr., Materials Dept. Sandia National Laboratories Dept. 8310 Livermore, CA 94551 415-294-3306
Michael Linn	Materials Engineer Naval Aviation Depot Materials Engineering Lab Code 342, Bldg. 793 NAS Jacksonville, FL 32216 904-772-3444
Mark J. Linville	Environmental Chemist Allison Gas Turbines Division of General Motors 2001 S. Tibbs Indianapolis, IN 46206 317-230-3617
Marianne Little	EG&G Idaho, Inc. P. O. Box 1625, MS 3940 Idaho Falls, ID 83415 208-526-8163
Cindy L. Longenbaugh	Program Engineer U.S. Department of Energy P. O. Box 5400 Albuquerque, NM 87115 505-845-4557
Joseph Lucas	President Inland Technology Inc. 2612 Pae Highway E #C Tacoma, WA 98424 206-922-8932
Anthony P. Malinauskas	Program Director Martin Marietta Energy Systems, Inc. P. O. Box 2008 Oak Ridge, TN 37831 615-576-1092

Bernard Malofsky	Vice President & Chief Chemist Loctite Corporation 705 N. Mountain Road Newington, CT 06111 203-278-1280
Patrick R. Martinez	Safety Coordinator Los Alamos National Laboratory MS D475 P. O. Box 1663 Los Alamos, NM 87545 505-665-2975
Ken P. Marts	Staff Engineer Martin Marietta Astronautics Group P. O. Box 179 F4083 Denver, CO 80201 303-971-2070
Gene S. Matsushita	Supervisor, Hazardous Materials/Waste Lockheed Aeronautical Systems Co. P. O. Box 551 Burbank, CA 91502 818-847-0195
D. B. Maxie	Lead Environmental Engineer LTV Aerospace and Defense Company P. O. Box 655907 97-09 Dallas, TX 75050 214-266-5606
Maureen M. McDonald	Manager, Materials Technology EG&G Rocky Flats Plant P. O. Box 464 Golden, CO 80402 303-966-2664
Eugene F. McInerney	Business Development Analyst Hercules Incorporated Hercules Plaza 11310 SE Wilmington, DE 19894 302-594-6423
Peter Miasek	Exxon Chemical Company, Canada #1 Duncan Mill Road Don Mills Ontario M3B 122 Canada 416-968-4111

D. Bradley Miller	President 3D Incorporated 2053 Plaza Drive Benton Harbor, MI 49022 616-925-5644
Susan J. Miller	Staff Chemical Engineer Radian Corporation 3200 E. Chapel Hill Road P. O. Box 13000 Research Triangle Park, NC 27709 919-541-9100
Tom Morehouse	U.S. Air Force Headquarters USAF/LEEV Washington, DC 20332 202-767-6240
David L. Morrison	Technical Director The MITRE Corporation 7525 Colshire Drive McLean, VA 22102 703-883-7750
Tom Murphy	U.S. Environmental Protection Agency 230 South Dearborn Street Chicago, IL 60604 312-886-6874
Jack Musall	Engineer Westinghouse Savannah River Co. Bldg. 730-M Aiken, SC 27808 803-725-2774
R. Nagarajan	Contamination Control Engineer IBM Corporation 5600 Cottle Road San Jose, CA 95193
Roop Nahta	Vice President The Virkler Company 12345 Steele Creer Road Charlotte, NC 28273 704-588-8500
Gordon Nelson	Supervisor, Test Evaluation Branch MSE, Inc. P. O. Box 3767 Butte, MT 59702 406-494-7100

John A. Nepute	Supervisory Environmental Engineer Wright-Patterson Air Force Base 2750 ABW/EME Wright-Patterson AFB, OH 45433 513-257-5535
David C. Ng	Chemist TRW One Space Park Redondo Beach, CA 90278 213-813-9350
Nhan T. Nguyen	Chemical Engineer U.S. Environmental Protection Agency Mail Code TS-779 401 M Street, SW Washington, DC 20460 202-382-3695
Van Nguyen	General Engineer U.S. Department of Energy 19901 Germantown Road Germantown, MD 20585 301-353-3048
Jon Nimitz	Senior Research Scientist New Mexico Engineering Research Inst. University of New Mexico Albuquerque, NM 87131 505-768-7534
Sarah C.K. O'Connor	Environmental Engineer-Research USA-CERL P. O. Box 4005 Champaign, IL 61801 217-352-6511
K.E. O'Rourke	Environmental Engineer General Dynamics Air Defense Systems P. O. Box 50800 Ontario, CA 91761 714-945-7729
Michael F. O'Shaughnessy	Chief Chemist Sigmund Cohn Corporation 121 South Columbus Avenue Mount Vernon, NY 10553 914-664-5300

Michael C. Oborny	Sandia National Laboratories Surface Eng. and Thin Film Technology P. O. Box 5800 Albuquerque, NM 87185
Carla O. Oldham	Environmental Scientist Radian Corporation 3200 E. Chapel Hill Rd. P. O. Box 13000 Research Triangle Park, NC 27709 919-541-9100
Errol Orebaugh	Staff Chemist Westinghouse Savannah River Lab. Aiken, SC 29808
Ron Orrell	Staff Engineer Martin Marietta Astronautics Box 179 Denver, CO 80201 303-971-8606
Jim Ostrowski	Manager, Toll Processing Safety-Kleen 777 Big Timber Road Elgin, IL 60121 708-697-8460
L. Leonard Packer	Chief, Chemical Processing United Technologies Research Center 411 Silver Lane East Hartford, CT 06108
Ben C. Padgett	Chemical Engineer Lee Wan & Associates 120 S. Jefferson Circle Oak Ridge, TN 37830 615-483-9870
Keith E. Pearce	Waste Watchers Program Manager Aerojet Propulsion Division P. O. Box 13222 Sacramento, CA 95813 916-355-3941
William Pearce	Environmental Engineer Douglas Aircraft Company MC 74-41 3855 Lakewood Blvd. Long Beach, CA 90846 213-497-5167

Henry C. Peebles	Senior Member, Technical Staff Sandia National Laboratories Division 1834 P. O. Box 5800 Albuquerque, NM 87185 505-845-8021
Greg Perry	Product Manager BASF Corporation Chemical Intermediates 100 Cherry Hill Road Parsippany, NJ 07054 201-316-3878
Jerry L. Peterson	Program Manager EG&G Rocky Flats P. O. Box 464 Golden, CO 80402 303-966-5349
Keith J. Pettus	Mgr., Hazardous Waste TRW, Space and Defense Sector One Space Park 140/2302 Redondo Beach, CA 90278 213-812-1183
C. Gregory Piner	Chemist Naval Aviation Depot (Code 354) Cherry Point, NC 28533
Bryant N. Poston	Environmental Scientist Lee Wan & Associates 120 South Jefferson Circle Oak Ridge, TN 37830 615-483-9870
Carl W. Pretzel	Sr. Member, Technical Staff Sandia National Laboratories P. O. Box 969 Livermore, CA 94551 415-294-2530
Urmi Ray	AT&T Bell Labs P. O. Box 900 Princeton, NJ 08540 609-639-3054
Wilfred J. Rebello	President PAR Enterprises Inc. 12601 Clifton Hunt Lane Clifton, VA 22024 703-818-9274

Charlene Reynolds	Environmental Chemist Robins Air Force Base 101 Country Place Warner Robins, GA 31093 912-926-9277
Stanley Richter	Chief, Process Technology Sikorsky Division of UTC 6900 Main Street 5321A Stratford, CT 06601 203-386-4388
Vernon Robertson	Manager, Manufacturing Energy A.B. Chance Co. 210 N. Allen Central, MO 65240 314-682-8428
Bernard Rykaczewski	Technical Support Engineer Westinghouse Savannah River Company Bldg. 321-M Aiken, SC 29808 803-725-4703
David Sachs	Design Engineer Los Alamos Technical Associates 6363 W. 120th Street, Suite 200 Broomfield, CO 80020 303-466-0400
Robert Salerno	EG&G Mound Applied Technologies P.O. Box 3000 Miamisburg, OH 45343
Leo Salinas	Group Leader Dow Chemical Texas Oper. Bldg. A-1401 Freeport, TX 77541 409-238-4116
Robert D. Sanders	Mgr., ECRA Chem-Nuclear Geotech Box 14000 Grand Junction, CO 81502 303-248-6035
Alex Sapre	Hughes Aircraft Company Bldg. C-1, MS B145 P. O. Box 54066 Los Angeles, CA 90045 213-568-6942

Paul E. Scheihing	Program Manager U.S. Department of Energy Office of Industrial Technologies CE-221 1000 Independence Ave., SW Washington, DC 20585 202-586-7234
Cathy Scheirman	Tinker Air Force Base OC-ALC/EME Tinker Air Force Base, OK 73145 405-734-7071
Eileen K. Schmitz	Weapons Complex Monitor Forums Box 406 Lake Bluff, IL 60044 708-234-2353
Wayne Schmitz	McDonnell Douglas Corporation M/C 270 3622 P. O. Box 516 St. Louis, MO 63166 314-232-2921
Nick Schoulal	Callington Haven Ltd. Rydalmere, N.W. Australia
Jim Schreiner	Exxon Chemical Company P. O. Box 4900 Bayton, TX 77522 713-425-2115
Micki Schultz	Chemical Engineer, Environ. Mgmt. Allied Signal Aerospace Co. Garrett Fluid Systems Div. P. O. Box 22200 Tempe, AZ 85285 602-893-4619
Sean Schwartz	Project Engineer Chemical Waste Management, Inc. Waste Reduction Services 3001 Butterfield Road Oak Brook, IL 60521 708-218-1668
Robin Sellers	Manager, Organic Materials Branch Naval Avionics Center 6000 East 21st Street Indianapolis, IN 46219 317-353-3621

Mike Seybold	Materials Engineer Naval Aviation Depot North Island Materials Engineering Laboratory San Diego, CA 92135 619-545-9733
Mark D. Shepard	Program Mgr., Waste Minimization EG&G Rocky Flats P. O. Box 464 Golden, CO 80402 303-966-4014
Barry Silver	Industrial Engineer Naval Weapons Station (Code 063) Human Resources Department Code 063 Concord, CA 94520 415-246-5208
Charles E. Simpson	Chemical Reduction Boeing Corporate Safety Health & Environmental Affairs P. O. Box 3707 Seattle, WA 98124 206-393-4717
Mark Singleton	Manager, Pollution Prevention General Electric Aircraft Engines 1 Neumann Way Cincinnati, OH 45215 513-786-1871
Walter Skrabski	Engineering Specialist General Dynamics/Convair Division MZ 43-6334 P. O. Box 85357 San Diego, CA 92186 619-547-5511
J. A. Slade	Plant Engineer Atomic Energy of Canada Limited Chalk River, Ontario KOJ 1JO Canada 613-584-3311
Edward S. Smith	Senior Materials Engineer Pratt & Whitney 3181 Medinah Lake Worth, FL 33467 407-796-6431

George B. Smith	AWD Technologies, Inc. 49 Stevenson Street San Francisco, CA 94105 415-227-0822
Mark D. Smith	Allied Signal Inc. Kansas City Division Dept. 837, 2C43 P. O. Box 419159 Kansas City, MO 64141 816-997-2561
Anthony Solazzo	Business Development Manager Glitsch Package Plants 1055 Parsippany Blvd., Suite 503 Parsippany, NJ 07054 201-299-9350
Bill Spargo	GENCORP 5410 Toombs St. Fair Oaks, CA 95628 915/967-5996
Robert (Wally) Spencer	Senior Project Engineer Thiokol Corporation 1990 North Mountain North Ogden, UT 84414 801-782-1080
Monique Spgars	Chemical Engineer NEESA Code 112F3 Port Hueneme, CA 93043 805-982-3626
Johnny Springer	U.S. Environmental Protection Agency Risk Reduction Engineering Lab. 26 W. Martin Luther King Drive Cincinnati, OH 45268 513-569-7542
Robert Stiger	Manager Waste Technology Development Dept. EG&G Idaho, Inc. MC 3940 P. O. Box 1625 Idaho Falls, ID 83415 208-526-8505

David A. Strah	Senior Engineering Scientist Conoco, Inc. P. O. Box 1267 Ponca City, OK 74604 405-767-3846
Jan Strous	Training Specialist Lee Wan & Associates 120 S. Jefferson Oak Ridge, TN 37830 615-483-9870
Harold F. Sturm, Jr.	Mgr., Works Eng. & Development Savannah River Site-SAL P. O. Box 616 Aiken, SC 29802 803-725-5244
Dan Suciu	Environmental Research & Development, Inc. 1550 Jones Street, Bldg. E Idaho Falls, ID 83415 208-522-7119
Sam Suffern	Pollution Prevention Prog. Mgr. HAZWRAP Martin Marietta Energy Systems P. O. Box 2003 MS 7606 Oak Ridge, TN 37831 615-435-3239
David J. Swanberg	Boeing Aerospace and Electronics Mail Stop 82-32 P. O. Box 3999 Seattle, WA 98124 206-773-4495
Scott Sysum	Materials Engineer WPNSTA Concord Code 331 Concord, CA 94520
George H. Terrell	General Engineer Department of the Army 2806 Ringgold Court Woodbridge, VA 22192 703-274-0815
Lisa Thompson	Martin Marietta Energy Systems P. O. Box 2009 Oak Ridge, TN 37831 615-574-0868

Shelley R. Thompson	Chemical Engineer McDonnell Douglas Corporation D357/0341200j P. O. Box 516 St. Louis, MO 46316 314-233-8318
Steve Tunick	Manager Hughes Aircraft Company Materials Engineering Department Bldg. E-1 M/S F157 P. O. Box 902 El Segundo, CA 90245 213-616-6167
Isaac M. Valdez	Program Engineer U.S. Department of Energy P. O. Box 5400 Albuquerque, NM 87115 505-845-5483
John Van Name	Supervisory Environ. Engineer Naval Aviation Depot Bldg. R-51, Code 615 Norfolk, VA 23511 804-444-8398
Maria Vargas	U.S. DOE P.O. Box 928 Golden, CO 80402 303/966-5922
Gary D. Vest	Deputy Assistant Secretary United States Air Force Environment, Safety, & Occupational Health The Pentagon Room 4-C916 Washington, DC 20330 703-697-9297
John Vidic	Chemical Engineer Hill Air Force Base TIVV Bldg. 5 Hill AFB, UT 84056 801-777-3124
Alice M. Viera	Materials Engineer Naval Aviation Depot Alameda Code 0542, Bldg. 7 Alameda, CA 94517 415-263-7174

Ron Vogel	Lee Wan & Associates 120 S. Jefferson Circle Oak Ridge, TN 37830 615-483-9870
Raymond R. Wallage	President INSITU Environmental 8402 E. Redwing Road Scottsdale, AZ 85250 602-948-9209
Steven R. Walter	Senior Principle Develop. Engineer EG&G Rocky Flats Plant P. O. Box 464 Golden, CO 80402 303-966-4335
William J. Ward, Jr.	Manager, Project Engineering Hughes Aircraft Company Missile Systems Group Bldg. 801, M/S G-10 P. O. Box 11337 Tucson, AZ 85734 602-794-8243
Ed Wasson	Kelly Air Force Base Facilities Engineering SA-ALC/LABEE Kelly AFB, TX 78241 512-925-8541
Larry Watkins	Program Supervisor So. Coast Air Quality Mgmt. Dist. 9150 E. Flair Drive El Monte, CA 91731 818-572-6308
Lloyd Watson	Custom Spray Technologies, Inc. 328 N. 4000E Rigby, ID 83443 208-745-7515
Jeffrey B. Weinrach	Staff Member, Waste Minimization Los Alamos National Laboratory Mail Stop E518 Los Alamos, NM 87545 505-667-4301

Guy W. Wells	Environmental Program Mgr. U.S. Air Force Offutt Air Force Base HQ SAC/DEV Offutt AFB, NE 68113 402-294-5303
Judi A. Werner	Boeing Commercial Airplanes Mail Stop 73-40 P. O. Box 3707 Seattle, WA 98124 206-237-4778
L. L. Whinnery	Sandia National Laboratories Materials Department 8310 P. O. Box 969 Livermore, CA 94551 415-294-2167
Walter N. Whinnery	Engineer Martin Marietta Energy Systems, Inc. P. O. Box 1410 Paducah, KY 42001 502-294-1215
Paul Wichlacz	Director Idaho National Engineering Laboratory Center for Waste Technology Development P. O. Box 1625 Idaho Falls, ID 83415 208-526-1292
Penny Wikoff	Environmental Research and Development, Inc. 1550 Jones Street, Bldg. E Idaho Falls, ID 83401 208-522-7119
Leann Williams	B.F. Goodrich Aerospace 3414 South 5th Street Phoenix, AZ 85040 602/232-4000
Dan Witt	Capt., U.S. Air Force Kelly Air Force Base Environmental Management Office SA-ALC/TIESM Kelly AFB, TX 78241 512-925-8745

Lawrence F. Wojdac	Principal Engineer Westinghouse Hanford P. O. Box 1970 Richland, WA 99352 509-373-4574
Katy Wolf	Inst. for Research and Technical Assistance 1429 South Bundy Drive Los Angeles, CA 90025 213-826-4700
Joan B. Woodard	Director Sandia National Laboratories Environmental & Manufacturing R&D Programs Org. 6600 Albuquerque, NM 87185 505-844-8904
Jocelyn Woodman	Environmental Engineer U.S. Environmental Protection Agency Pollution Prevention Division 401 M Street, SW Washington, DC 20460 202-245-4165
Robert Yee	Subsystem Manager GENCORP - AEROJET P. O. Box 13222 Sacramento, CA 95813 916-355-4902
Bill Young	Senior Engineer Babcock & Wilcox Naval Nuclear Fuel Division P. O. Box 785 Lynchburg, VA 24505 804-522-6203
William Young	Senior Engineer Babcock & Wilcox Naval Nuclear Fuel Division P. O. Box 785 Lynchburg, VA 24505 804-522-6203
Walter Zachritz	Program Manager New Mexico State University SW Technology Development Institute P. O. Box 30001, Dept. 3 SOL Las Cruces, NM 88003

John Zavodjancik	Senior Manufacturing Engineer Pratt & Whitney MS 104-03 400 Main Street East Hartford, CT 06108 203-565-5030
Bill Zidar	Manufacturing Engineer Talley Defense Systems 4551 East McKellups Road Mesa, AZ 85205 602-898-2568

APPENDIX II

CONTRIBUTING AUTHORS*

K. K. Asada	Hughes Aircraft Company P. O. Box 92426 Los Angeles, CA 90009
R. S. Basu	Allied-Signal, Inc. Buffalo Research Laboratory 20 Peabody Street Buffalo, NY 14210
Martha I. Beach	N-Con Systems 2410 Boston Post Road Larchmont, NY 10538
Lisa Brown	U.S. Environmental Protection Agency Risk Reduction Engineering Laboratory 26 W. Martin Luther King Drive Cincinnati, OH 45268
Joseph C. Farmer	Lawrence Livermore National Lab. L-370 Livermore, CA 94551
P. M. Gallagher	Phasex Corporation 360 Merrimack Street Lawrence, MA 01843
Leonard W. Gray	Lawrence Livermore National Lab. L-439 Livermore, CA 94551
Ralph D. Hermansen	Hughes Aircraft Company Bldg. E-1, MS F157 2000 E. Segundo Blvd. El Segundo, CA 90245
Robert G. Hickman	Lawrence Livermore National Lab. L-439 Livermore, CA 94551
K.S. Hill	Hughes Aircraft P. O. Box 92426 Los Angeles, CA 90009
Randall B. Ivey	Air Force Corrosion Program Office WR-ALC/CNC Robins Air Force Base, GA 31098

Note: Authors in this list are those whose papers were selected under the general "Call for Papers" but whose papers were not presented at the Workshop.

Ronald Jackson	U.S. Army Toxic & Hazardous Materials Agency Aberdeen Proving Ground, MD 21010
Ricardo B. Jacquez	New Mexico State University Dept. of Civil, Agricultural, and Geological Engineering Las Cruces, NM 88001
E. M. Kenny-McDermott	Allied-Signal Aerospace Guidance Systems Division Teterboro, NJ 07608
V. J. Krukonis	Phasex Corporation 360 Merrimack Street Lawrence, MA 01843
J. M. Locklin	Douglas Aircraft Company MC 36-41 3855 Lakewood Blvd. Long Beach, CA 90846
P. B. Logsdon	Allied-Signal, Inc. Buffalo Research Laboratory 2 Peabody Street Buffalo, NY 14210
Benerito S. Martinez, Jr.	New Mexico State University Dept. of Civil, Agricultural and Geological Engineering Las Cruces, New Mexico 88001
Keturah Reinbold	U.S. Army Corps of Engineers Construction Engr. Research Lab. CECER-ENR P. O. Box 4005 Champaign, IL 61824
Robert F. Salerno	EG&G Mound Applied Technologies Organic Material/Surface Modification P. O. Box 3000 Miamisburg, OH 45343
J. T. Snyder	Martin Marietta Astronautics Group P. O. Box 179, MS 9080 Denver, CO 80201
Johnny Springer	U.S. Environmental Protection Agency Risk Reduction Engineering Lab. 26 W. Martin Luther King Drive Cincinnati, OH 45268

Ronald Stevenson	Sacramento Army Depot MC 5018 Sacramento, CA 95813
Dan Suciu	Tec/Niques International, Inc. 2053 Plaza Drive Benton Harbor, MI 49022
M. D. Wally	Hughes Aircraft Company Radar Systems Group P. O. Box 92426 Los Angeles, CA 90009
Walter N. Whinnery	Martin Marietta Energy Systems, Inc. Paducah Gaseous Diffusion Plant P. O. Box 1410 Paducah, KY 42001
Walter H. Zachritz II	New Mexico State University SW Technology Development Institute Las Cruces, NM 88001

Other Noyes Publications

HANDBOOK OF POLLUTION CONTROL PROCESSES

Edited by

Robert Noyes

This handbook presents a comprehensive and thorough overview of state-of-the-art technology for pollution control processes. It will be of interest to those engineers, consultants, educators, architects, planners, government officials, industry executives, attorneys, students and others concerned with solving environmental problems.

The pollution control processes are organized into chapters by broad **problem areas**; and appropriate technology for decontamination, destruction, isolation, etc. for each problem area is presented. Since many of these technologies are useful for more than one problem area, a specific technology may be included in more than one chapter, modified to suit the specific considerations involved.

The pollution control processes described are those that are actively in use today, as well as those innovative and emerging processes that have good future potential. An important feature of the book is that **advantages** and **disadvantages** of many processes are cited. Also, in many cases, **regulatory-driven trends** are discussed, which will impact the technology used in the future.

Where pertinent, regulations are discussed that relate to the technology under consideration. Regulations are continually evolving, frequently requiring modified or new treatment technologies. This should be borne in mind by those pursuing solutions to environmental problems.

Innovative and emerging technologies are also discussed; it is important to consider these new processes carefully, due to increasingly tighter regulatory restrictions, and possibly lower costs. For some pollutants specific treatment methods may be required; however for other pollutants, specific treatment levels must be obtained.

CONTENTS

In summary, a vast number of pollution control processes and process systems are discussed.

ISBN 0-8155-1290-2 (1991) **7" x 10"** **750 pages**

Other Noyes Publications

HANDBOOK OF TOXIC AND HAZARDOUS CHEMICALS AND CARCINOGENS

1991 — Third Edition

by Marshall Sittig
Princeton University (retired)

"... an important addition to the literature on toxic chemicals.... Marshall Sittig has made it easier for us to understand how to control the handling of toxic chemicals and the disposal of toxic wastes." —from the Foreword by *Bill Bradley, U.S. Senator, New Jersey.*

The *Third Edition* of this Handbook is a completely new, much-expanded version of this widely-accepted classic. It presents chemical, health and safety information on 1300 toxic and hazardous chemicals and carcinogens (up from 800 in the previous edition). The book will aid responsible decision making by chemical manufacturers, safety equipment producers, toxicologists, industrial safety engineers, waste disposal operators, health care professionals, and the many others who may have contact with or interest in these chemicals due to their own or third party exposure. The book is a must for industrial, medical and legal libraries.

Included in the Third Edition are substances which are:

Designated by **EPA** as **"Hazardous Substances"**, **"Hazardous Wastes"** under **RCRA**, **"Priority Toxic Pollutants"**, **"Extremely Hazardous Substances"** under **SARA** (particularly the top 100), or **"Toxic Chemicals"**;

Identified as **carcinogens** by the **U.S. National Toxicology Program** and the **IARC**;

Identified in **UN/DOT** shipping regulations;

Reviewed in **NIOSH Information Profiles** and **EPA "CHIPS"** documents;

Set forth by **OSHA;**

Whose allowable concentrations in air and other safety considerations have been considered by **NIOSH:**

Whose allowable concentrations in workplace air are adopted or proposed by **ACGIH, Deutsche Forschungsgemeinschaft,** or the **Health and Safety Executive** (U.K.);

Whose allowable concentrations in workplace air, residential air and water have been set by the **USSR-UNEP/IRPTC** project.

In addition the book covers:

Most of the chemicals in the **ILO** *Encyclopedia of Occupational Health and Safety,* the **United Nations'** *IRPTC Legal File,* or the journal *Dangerous Properties of Industrial Materials Report,* and

Many chemicals cited in "right-to-know" legislation of one or more of the 50 states in the U.S.A.

Essentially the book attempts to answer these questions about a compound (to the extent information is available): (1) What is it? (2) Where is it encountered? (3) How much can one tolerate? (4) How does one measure it? (5) What are its harmful effects? (6) How does one protect against it? (7) How does one handle it and protect against mishaps?

The chemicals are presented alphabetically and each is classified as a "carcinogen," "hazardous substance," "hazardous waste," and/or a "priority toxic pollutant"—as defined by the various federal agencies. Data and citations are furnished, to the extent currently available, on any or all of these important categories:

Chemical Description
Code Numbers
Synonyms
Potential Exposure
Incompatibilities
Permissible Exposure Limits in Air
Determination in Air
Permissible Concentration in Water
Determination in Water
Routes of Entry
Harmful Effects and Symptoms
Points of Attack
Medical Surveillance
First Aid
Personal Protective Methods
Respirator Selection
Storage
Shipping
Spill Handling
Disposal Method Suggested
References

ISBN 0-8155-1286-4 (1991) **volumes** **7" x 10"** **1685 pages**

Other Noyes Publications

SOLVENT WASTE REDUCTION

by

U.S. Environmental Protection Agency
ICF Consulting Associates, Incorporated

Pollution Technology Review No. 193

This book covers solvent waste reduction alternatives currently available to industry. Solvents are an important part of most industrial operations. Almost all plants use solvents, usually for cleaning, degreasing, painting, paint stripping, or extraction purposes. Therefore, proper management of solvent wastes is of widespread concern. There are several disposal methods available to solvent users, but in many cases the most environmentally sound and economical choice is not to dispose of the solvent at all, but to recycle it.

A brief review of the generally accepted priority list of hazardous waste management techniques indicates it is incumbent upon generators of waste, first, to institute a program to reduce the amount of waste generated. Second, after the amount of waste generated has been minimized, recycling is considered the next most appropriate alternative. Third on most lists is incineration, and fourth is chemical or biological treatment to render the waste nonhazardous. Landfills, deep-well injection, and other methods fall further down the list.

Recycling is ranked very high on the priority list of alternatives for hazardous waste management. On-site reclamation, where practical, is generally preferred to off-site reclamation to avoid the risk and liability of transportation. If on-site reclamation is not practical because of economic or quality limitations, off-site reclamation should be considered as the next alternative.

Other significant reasons for practicing waste reduction—to lower workers' exposure to vapors, to reduce solvent loss to the environment, and to comply with health and environmental laws—also have economic benefits. Since many of the benefits of solvent waste reduction relate to cost savings, more credibility is given to the assertion that voluntary waste reduction will work. Waste reduction efforts demonstrate a commitment to protecting the environment. But waste reduction is just one part of waste management. Safe, permanent waste disposal and compliance with state and federal laws still are top priorities.

The listing which follows contains **chapter titles and selected subtitles.**

ISBN 0-8155-1254-6 (1990) **7" x 10"** **158 pages**

Other Noyes Publications

HALOGENATED SOLVENT CLEANERS
Emission Control Technologies and Cost Analyses

by

Radian Corporation

Pollution Technology Review No. 187

This book contains information on the use and control of halogenated solvents in solvent cleaning applications. Described are the types of solvent cleaners manufactured, sources of solvent emissions, methods of controlling solvent emissions, and the costs associated with installation of control devices.

It also covers cleaning equipment that uses halogenated solvents rather than those which use petroleum distillates. The use of halogenated solvents to clean or otherwise condition the surface of metal parts, electronic compounds, and other nonporous substrates is well established. The five commonly used halogenated solvents (methylene chloride, trichloroethylene, perchloroethylene, trichlorotrifluoroethane, and 1,1,1-trichloroethane) possess the physical characteristics necessary to handle a variety of industrial cleaning situations. They can dissolve many common residues from manufacturing processes, have little or no flammability, and can achieve a high degree of cleanliness, even on very small or intricate parts. The popularity of halogenated solvent cleaning is evidenced by the fact that hundreds of millions of pounds of the five solvents are consumed in cleaning machines each year; however, most of the dissipated solvent ends up in the atmosphere, a highly negative environmental side effect.

Many states already regulate solvent cleaners, either for VOC control or for toxic pollutant control. However, equipment utilized to conform to present State controlled standards may be further improved by adoption of some additional control measures described in this book. A significant fraction of existing machines likely are uncontrolled. On the other hand, several equipment manufacturers currently are selling well designed solvent cleaning machines using the listed controls, and some have improved designs on the drawing board or in the prototype stage.

The information presented here is intended primarily to inform those involved in control in general, and solvent cleaner operators in particular, of available techniques to reduce halogenated solvent emissions, and of available alternative cleaning technologies that can often be used to eliminate halogenated solvent use completely.

A condensed table of contents giving **chapter titles and selected subtitles** is listed below.

ISBN 0-8155-1248-1 (1990) 6" x 9" 223 pages